TECHNISCHE UNIVERSITÄT DRESDEN
INSTITUTE OF MATHEMATICAL STOCHASTICS

DISSERTATION

LÉVY-TYPE PROCESSES UNDER UNCERTAINTY AND RELATED NONLOCAL EQUATIONS

AUTHOR: JULIAN HOLLENDER
BIRTHDATE: 27TH OCTOBER 1987
BIRTHPLACE: DRESDEN, GERMANY

DEGREE AIMED AT: DOCTOR RERUM NATURALIUM
(DR. RER. NAT.)

INSTITUTION: TECHNISCHE UNIVERSITÄT DRESDEN
FACULTY: SCHOOL OF SCIENCE

SUPERVISOR: PROF. DR. RENÉ L. SCHILLING

REFEREES: PROF. DR. RENÉ L. SCHILLING
PROF. DR. JIANG-LUN WU

SUBMISSION: 13TH JULY 2016
DEFENCE: 12TH OCTOBER 2016

»We should not write so that it is possible for our readers to understand us, but so that it is impossible for them to misunderstand us.«

> Marcus Fabius Quintilian, *Institutio Oratoria*

Contents

Abstract	**7**
Introduction	**9**
1. Approximation Theory	**19**
1.1. Friedrichs Mollification	19
1.2. Supremal Convolutions	25
1.3. Monotone Functions	44
1.4. Miscellaneous	52
2. Integro-Differential Equations	**57**
2.1. Viscosity Solutions	58
2.2. Uniqueness of Solutions	70
2.3. Hamilton–Jacobi–Bellman Equations	113
3. Sublinear Expectations	**135**
3.1. Expectation Spaces	135
3.2. Important Distributions	141
4. Stochastic Processes	**147**
4.1. Markov Processes	149
4.2. Construction of Processes	155
4.3. Stochastic Integration	196
4.4. Stochastic Differential Equations	206
A. Appendix	**219**
A.1. Convex Analysis	219
A.2. Functional Analysis	221
Outlook	**225**
Bibliography	**227**
List of Symbols	**237**
List of Statements	**244**
Index of Subjects	**245**

Abstract

In recent years, the theoretical study of nonlinear expectations became the focus of attention for applications in a variety of different fields – often with the objective to model systems under incomplete information. Especially in mathematical finance, advancements in the theory of sublinear expectations (also known as coherent risk measures) laid the theoretical foundation for many improved or novel approaches to evaluations under the presence of Knightian uncertainty. In his innovative work [102], Shige Peng studied sublinear expectations as an intrinsic generalization of classical probability spaces: One of his main contributions in this article is the introduction of stochastic processes with continuous paths and independent, stationary increments for sublinear expectations, which have a one-to-one correspondence to nonlinear local equations

$$\partial_t u(t,x) - \sup_{\alpha \in \mathcal{A}} \left(\frac{1}{2} \operatorname{tr} \left(c_\alpha D^2 u(t,x) \right) \right) = 0$$

for families $(c_\alpha)_{\alpha \in \mathcal{A}} \subset \mathbb{S}^{d \times d}_+$ of positive, symmetric matrices with arbitrary index sets $\mathcal{A}$. These so-called G-Brownian motions can be interpreted as a generalization of classical Brownian motions (and their relation to the heat equation) under volatility uncertainty, and are therefore of particular interest for applications in mathematical finance. Shige Peng's approach was later extended by Mingshang Hu and Shige Peng in [67] and by Ariel Neufeld and Marcel Nutz in [93] to introduce stochastic jump-type processes with independent, stationary increments for sublinear expectations, which correspond to nonlinear nonlocal equations

$$\partial_t u(t,x) - \sup_{\alpha \in \mathcal{A}} \left(b_\alpha D u(t,x) + \frac{1}{2} \operatorname{tr} \left(c_\alpha D^2 u(t,x) \right) \right.$$
$$\left. + \int_{\mathbb{R}^d} \left(u(t, x+z) - u(t,x) - Du(t,x) z \mathbb{1}_{|z| \leq 1} \right) F_\alpha(dz) \right) = 0$$

for families $(b_\alpha, c_\alpha, F_\alpha)_{\alpha \in \mathcal{A}} \subset \mathbb{R}^d \times \mathbb{S}^{d \times d}_+ \times \mathfrak{L}(\mathbb{R}^d)$ of Lévy triplets, and can be viewed as a generalization of classical Lévy processes under uncertainty in their Lévy triplets.

In the latter half of this thesis, we pick up these ideas to introduce a broad class of stochastic processes with jumps for sublinear expectations, that can also exhibit spatial inhomogeneities. More precisely, we introduce the concept of Markov processes for sublinear expectations and relate them to nonlinear nonlocal equations

$$\partial_t u(t,x) - G\Big(u(t, \cdot) \Big)(x) = 0$$

for some sublinear generalization $G(\cdot)$ of classical (pointwise) Markov generators.

Abstract

In addition, we give a construction for a large subclass of these Markov processes for sublinear expectations, that corresponds to equations of the special form

$$\partial_t u(t,x) - \sup_{\alpha \in \mathcal{A}} G_\alpha\big(u(t,\cdot)\big)(x) = 0$$

for families $(G_\alpha)_{\alpha \in \mathcal{A}}$ of Lévy-type generators. These processes can be interpreted as generalizations of classical Lévy-type processes under uncertainty in their semimartingale characteristics. Furthermore, we establish a general stochastic integration and differential equation theory for processes with jumps under sublinear expectations.

In the first half of this thesis, we give a purely analytical proof, showing that a large family of fully nonlinear initial value problems (including the ones from above)

$$\partial_t u(t,x) - G(u)(t,x) = 0 \quad \text{in } (0,T) \times \mathbb{R}^d$$
$$u(t,x) = \varphi(x) \quad \text{on } \{0\} \times \mathbb{R}^d$$

have at most one viscosity solution, where $G : C^2((0,T) \times \mathbb{R}^d) \to C((0,T) \times \mathbb{R}^d)$ are second-order, nonlocal degenerate elliptic operators – i.e. they satisfy

$$G(\phi)(t,x) \leq G(\psi)(t,x)$$

whenever $\phi - \psi$ has a global maximum in $(t,x) \in (0,T) \times \mathbb{R}^d$ with $(\phi - \psi)(t,x) \geq 0$. The notion of viscosity solutions, which we are using, was originally introduced by Michael Crandall and Pierre-Louis Lions in [36] for local equations. We prove a comparison principle for unbounded viscosity solutions of parabolic, second-order, nonlocal degenerate elliptic equations. In order to cover the equations needed for the stochastic part, we have to generalize the maximum principle from Espen Jakobsen and Kenneth Karlsen in [75], since our operators have significantly weaker continuity properties.

[36] CRANDALL, M., LIONS, P. Viscosity solutions of Hamilton-Jacobi equations. *Transactions of the American Mathematical Society 277*, 1 (1983), 1–42.

[67] HU, M., PENG, S. G-Lévy processes under sublinear expectations. arXiv:0911.3533v1, 2009.

[75] JAKOBSEN, E., KARLSEN, K. A "maximum principle for semicontinuous functions" applicable to integro-partial differential equations. *Nonlinear Differential Equations and Applications 13*, 2 (2006), 137–165.

[93] NEUFELD, A., NUTZ, M. Nonlinear Lévy processes and their characteristics. arXiv:1401.7253v2, 2015.

[102] PENG, S. G-expectation, G-Brownian motion and related stochastic calculus of Itô type. In *Stochastic Analysis and Applications - The Abel Symposium 2005* (2007), vol. 2, Springer, 541–567.

Introduction

Motivation

In his pioneering work [82], the economist Frank Knight postulated a distinction between two inherently different types of threats in the evaluation of unknown parameters: On the one hand, there are parameters under so-called Knightian risk (or risk for short), for which one can reasonably assign probabilities to the possible values of the unknown parameter, despite the lack of knowledge of its true value. Typical examples include the tossing of fair coins or the rolling of perfect dice. On the other hand, there are parameters under so-called Knightian uncertainty (or uncertainty for short), for which it seems not feasible to assign distinct probabilities to each observable value of the unknown parameter, often caused by the absence of relevant information or due to a lack of full understanding of the underlying dependencies. Typical examples include the pricing of contingent claims and their underpinning financial modeling of stock prices.

	Experiment 1		Experiment 2		
Color	Option 1A	Option 1B	Option 2A	Option 2B	Ratio
Red	$100	-	$100	-	33%
Black	-	$100	-	$100	66%
Yellow	-	-	$100	$100	

Figure 0.1.: Payoff matrix for Ellsberg's experiments

In order to clarify the difference between Knightian risk and uncertainty, and to motivate its relevance for the related mathematical modeling, let us take a look at the following illustrative example: Daniel Ellsberg popularized in [49] the evaluation of two experiments in the context of decision theory, which were originally proposed by John Maynard Keynes in [81]. In both experiments a ball is drawn from an urn that contains a mix of nine red, black and yellow balls. The participants are asked to choose between two different options for each experiment in advance, which determine the payoffs according to the color of the ball drawn, as described in Figure 0.1. Before the participants pick their options, however, they are provided with the information that the urn contains three red and six other balls, which are either black or yellow. As a result of this, the unknown payoff parameter is under Knightian risk for options 1A and 2B, since it only depends on the event that a red ball is drawn. In fact, it seems reasonable to assume that the chances for a red ball to be drawn are exactly one third. For options 1B and 2A, on the other hand, the unknown payoff parameter is under Knightian uncertainty, since the ratio between the black and yellow balls is unknown and it is therefore not

Introduction

feasible to assign probabilities to the different payoffs. Empirical data suggests (cf. [23, Section 3.1] for a literature review) that the majority of participants prefer the options under Knightian risk over those under Knightian uncertainty – an effect that is often attributed to the ambiguity aversion of the participants.

Decision theory, as a subfield of economics, is concerned with models that describe certain characteristics in the behavior of rational agents, such as the ambiguity aversion of participants in Ellsberg's two experiments. One idea at the heart of decision theory is the expected utility hypothesis, which asserts that rational agents prefer one option over another if they assign a higher expected utility to it. Suppose that

$$u : \mathbb{R} \longrightarrow \mathbb{R}$$

is a function (usually referred to as utility function), which assigns the utility (for rational agents) to possible payoff values. The preference of option 1A over 1B by the majority of participants in Ellsberg's first experiment translates to

$$E\left(u(\pi_{1A})\right) > E\left(u(\pi_{1B})\right)$$

under the expected utility hypothesis, where π_{1A} and π_{1B} denote (the random variables that represent) the payoff for option 1A or 1B, respectively, and $E(\cdot)$ is the expectation operator. Analogously, the preference of option 2B over 2A in Ellsberg's second experiment correspond to the inequality

$$E\left(u(\pi_{2A})\right) < E\left(u(\pi_{2B})\right)$$

under the expected utility hypothesis. A simple calculation shows that, if the expectation operator $E(\cdot)$ used for modeling was linear, these two inequalities could be rewritten as

$$E\left(u(\pi_{1A})\right) > E\left(u(\pi_{1B})\right) \iff E\left(\mathbb{1}_{\Omega_R}\right)\left(u(\$100) - u(\$0)\right) > E\left(\mathbb{1}_{\Omega_B}\right)\left(u(\$100) - u(\$0)\right)$$
$$E\left(u(\pi_{2A})\right) < E\left(u(\pi_{2B})\right) \iff E\left(\mathbb{1}_{\Omega_R}\right)\left(u(\$100) - u(\$0)\right) < E\left(\mathbb{1}_{\Omega_B}\right)\left(u(\$100) - u(\$0)\right),$$

where the random variables $\mathbb{1}_{\Omega_R}$ and $\mathbb{1}_{\Omega_B}$ are equal to one if a red or black ball is drawn, respectively, and zero otherwise. In particular, the conflicting inequalities demonstrate that in case $u(\$100) \neq u(\$0)$, i.e. the utility assigned to \$100 differs from the one assigned to \$0, it is not possible to use linear expectation operators in accordance with the expected utility hypothesis, to reproduce the ambiguity aversion observed in Ellsberg's experiments. To a certain extent, this insight does not come as a surprise, since linear expectations are usually based on one fixed probability measure, which assigns exact probabilities to every observable value of an unknown parameter and seems therefore only suitable for modeling parameters under Knightian risk.

Driven, among other motivations, by the demand to model unknown parameters under Knightian uncertainty, several nonlinear generalizations of linear expectation operators were proposed and studied over the years (cf. [43] for an overview). Especially during the last decades, the popularity of nonlinear expectation operators intensified in many fields, since recent observations suggest that the incorporation of Knightian uncertainty has a significant impact on the quality of models (cf. [116] for evaluations in finance). In some applications, such as mathematical finance or statistical physics, stochastic processes are usually at the center of attention for modeling unknown parameters. As large parts of the established theory for stochastic processes are strongly intertwined with the linearity underpinning classical probability theory, it is therefore valuable to generalize the existing theory to nonlinear expectation operators, in order to model stochastic processes under Knightian uncertainty. Beyond these applications in stochastic modeling under Knightian uncertainty, such generalizations usually also advance other fields inside of mathematics, by employing strong connections between stochastic processes and related objects, such as operator semigroups or integro-differential equations.

Main Results

The new results in this thesis fall into two categories: In the stochastic part, we construct and characterize a broad class of stochastic jump-type processes for sublinear expectations, which can be interpreted as worst-case or best-case bounds for evaluations of stochastic processes under Knightian uncertainty. Sublinear expectations, in this respect, are merely sublinear functionals that are monotone and constant-preserving, and play an important role (under the name of coherent risk measures) in decision theory, for the evaluation of unknown parameters under Knightian uncertainty. In the analytical part, we develop a comparison principle and uniqueness theory for a large family of nonlinear partial integro-differential equations, which includes the characterizing equations for the novel processes from the stochastic part.

Let us start with a summary of the main results for the stochastic part of this thesis: In his innovative article [102], Shige Peng introduced an intrinsic approach to study stochastic processes with continuous paths and independent, stationary increments for sublinear expectations (i.e. sublinear functionals that are monotone and constant-preserving), which extended the strong connection between classical Brownian motions and the heat equation. These so-called G-Brownian motions and their related central limit theorem can be interpreted as a generalization of classical Brownian motion under volatility uncertainty, and were studied and applied extensively in a variety of different contexts since their introduction (cf. [105] for a first overview). The intrinsic approach towards G-Brownian motions was later extended by Mingshang Hu and Shige Peng in [67] to introduce G-Lévy processes, i.e. stochastic processes with jumps and independent, stationary increments for sublinear expectations. Their construction was later generalized by Ariel Neufeld and Marcel Nutz in [93]. In this thesis, we generalize this approach even further to study Markov processes (with jumps) for sublinear expectations and their related sublinear Markov semigroups. First of all, we present a novel construction for a large class of sublinear Markov semigroups $(T_t)_{t \geq 0}$, i.e. a family of monotone and constant-preserving sublinear operators

$$T_t : \mathcal{H} \longrightarrow \mathcal{H}$$

on convex cones $\mathcal{H}$ of real-valued functions on $\mathbb{R}^d$ that satisfy $T_{t+s} = T_t T_s$ and $T_0 = \text{id}$,

Introduction

similar to Nisio semigroups from Makiko Nisio in [95]. The family of associated d-dimensional Markov processes $(X_t)_{t\geq 0}$ for sublinear expectations defined by

$$E^x\big(\varphi(X_t)\big) = T_t\varphi(x)$$

can be interpreted as a generalization of classical Markov processes under uncertainty in their characteristics, and contains all previously constructed stochastic processes for sublinear expectations (such as G-Brownian motions and G-Lévy processes) from [102], [67] and [93]. We prove an explicit representation of their associated sublinear generators

$$A\varphi(x) = \lim_{\delta\downarrow 0}\frac{T_\delta\varphi(x) - T_0\varphi(x)}{\delta} = \lim_{\delta\downarrow 0}\frac{E^x(\varphi(X_t)) - \varphi(x)}{\delta}$$

in terms of integro-differential operators, and show for $\varphi \in C_b(\mathbb{R}^d)$ that the functions

$$u^\varphi(t,x) = T_t\varphi(x) = E^x\big(\varphi(X_t)\big)$$

are viscosity solutions of the nonlinear nonlocal generator equation in $(0,\infty)\times \mathbb{R}^d$

$$\partial_t u^\varphi(t,x) - A(u^\varphi(t,\,\cdot\,))(x) = 0,$$

if the C_b-Feller property $T_t C_b(\mathbb{R}^d) \subset C_b(\mathbb{R}^d)$ for all $t\geq 0$ holds. Furthermore, we introduce a pathwise stochastic calculus for sublinear expectations and employ it to prove the existence of a large class of sublinear Markov semigroups that satisfy the C_b-Feller property, in the following way: Suppose that

$$(b_\alpha, c_\alpha, F_\alpha)_{\alpha\in\mathcal{A}} \subset \mathbb{R}^d \times \mathbb{S}^{d\times d}_+ \times \mathfrak{M}^+(\mathbb{R}^d\setminus\{0\})$$

is a convex and closed set, where $\mathfrak{M}^+(\mathbb{R}^d\setminus\{0\})$ is the family of non-negative Borel measures on $\mathbb{R}^d\setminus\{0\}$ equipped with the C_b-weak convergence of measures, such that

$$\sup_{\alpha\in\mathcal{A}}\left(|b_\alpha| + |c_\alpha| + \int_{\mathbb{R}^d}\big(|z|\wedge |z|^2\big)\,F_\alpha(dz)\right) < \infty$$

$$\lim_{r\to 0}\sup_{\alpha\in\mathcal{A}}\left(\int_{|z|<r}|z|^2\,F_\alpha(dz)\right) = 0$$

holds. As a special case of our general construction for sublinear Markov semigroups, we recover the construction of G-Lévy processes from [93] and obtain the existence of a stochastic process $(X_t)_{t\geq 0}$ with càdlàg paths starting at the origin $X_0 = 0$ on a measurable space $(\Omega, \mathcal{A})$ for the sublinear expectation $E(\cdot) := \sup_{\mathbb{P}\in\mathcal{P}}\mathbb{E}_\mathbb{P}(\cdot)$ with

$$\mathcal{P} := \left\{\mathbb{P} \in \mathfrak{P}^{\mathrm{ac}}_{\mathrm{sem}}(\Omega)\,\bigg|\, (b^\mathbb{P}_t, c^\mathbb{P}_t, F^\mathbb{P}_t)(\omega) \in \bigcup_{\alpha\in\mathcal{A}}(b_\alpha, c_\alpha, F_\alpha)\quad \lambda(dt)\times\mathbb{P}(d\omega)\text{-a.e.}\right\},$$

where $\mathfrak{P}^{\mathrm{ac}}_{\mathrm{sem}}(\Omega)$ is the family of probability measures such that $(X_t)_{t\geq 0}$ is a semimartingale with absolutely continuous semimartingale characteristics, and $(b^\mathbb{P}_t, c^\mathbb{P}_t, F^\mathbb{P}_t)_{t\geq 0}$ are their

Introduction

associated differential characteristics. This construction suggests to interpret $(X_t)_{t\geq 0}$ as a generalization of classical Lévy processes under uncertainty in their Lévy triplets. Based on that, we develop a pathwise stochastic calculus that allows us to construct k-dimensional stochastic processes $(Z_t^x)_{t\geq 0}$ for every bounded, globally Lipschitz

$$f : \mathbb{R}^k \longrightarrow \mathbb{R}^{k\times d}$$

and $x \in \mathbb{R}^k$, that are $\mathbb{P}$-almost surely the solutions of the stochastic differential equations

$$Z_t^x = x + \int_0^t f(Z_{s-}^x)\, dX_s$$

for every $\mathbb{P} \in \mathcal{P}$. Furthermore, we prove that $T_t \varphi(x) = E(\varphi(Z_t^x))$ introduces a sublinear Markov semigroup enjoying the C_b-Feller property. In particular, our general result for sublinear Markov semigroups implies that $u^\varphi(t,x) = E(\varphi(Z_t^x))$ for $\varphi \in C_b(\mathbb{R}^k)$ is a viscosity solution of the associated nonlocal generator equation

$$\partial_t u^\varphi(t,x) - \sup_{\alpha \in \mathcal{A}} \bigg(f(x) b_\alpha D u^\varphi(t,x) + \frac{1}{2} \operatorname{tr}\big(f(x) c_\alpha f(x)^T D^2 u^\varphi(t,x)\big) $$
$$+ \int_{\mathbb{R}^d} \big(u^\varphi(t, x + f(x)z) - u^\varphi(t,x) - D u^\varphi(t,x) f(x) h(z)\big) F_\alpha(dz) \bigg) = 0$$

in $(0, \infty) \times \mathbb{R}^k$. Moreover, we prove in the analytical part of this thesis that the viscosity solutions of these nonlocal initial value problems are unique. The construction suggests to interpret the Markov processes $(Z_t)_{t\geq 0}$ as a generalization of classical Lévy-type processes under uncertainty in their characteristics. It therefore seems reasonable to assume that they are of interest for applications in mathematical finance, since classical Lévy-type processes are frequently used for financial modeling (cf. [31] for an overview).

The main contributions in the analytical part of this thesis consist of a comparison principle and a uniqueness theory for a large class of nonlinear partial integro-differential equations with non-dominated integral terms: In their innovative work [36], Michael Crandall and Pierre-Louis Lions introduced the generalized notion of viscosity solutions for nonlinear partial differential equations, which naturally admit a comparison principle for degenerate elliptic equations and are therefore unique. A second-order nonlocal operator

$$G : C^2((0,T) \times \mathbb{R}^d) \longrightarrow C((0,T) \times \mathbb{R}^d)$$

with $T > 0$ is referred to as nonlocal degenerate elliptic if the inequality

$$G(\phi)(t,x) \leq G(\psi)(t,x)$$

holds, whenever $\phi - \psi$ has a global maximum in $(t,x) \in (0,T) \times \mathbb{R}^d$ with $(\phi - \psi)(t,x) \geq 0$. Based on that, an upper semicontinuous function $u \in \mathrm{USC}([0,T] \times \mathbb{R}^d)$ (lower semicontinuous function $u \in \mathrm{LSC}([0,T] \times \mathbb{R}^d)$) is a viscosity subsolution (supersolution) of

$$\partial_t u(t,x) - G(u)(t,x) = 0$$

in $(0,T) \times \mathbb{R}^d$ if $\partial_t \phi(t,x) - G(\phi)(t,x) \leq 0$ (≥ 0) holds for every $\phi \in C^2((0,T) \times \mathbb{R}^d)$ such that $u - \phi$ has a global maximum (minimum) in $(t,x) \in (0,T) \times \mathbb{R}^d$ with $(u - \phi)(t,x) = 0$.

Introduction

In this thesis, we generalize the result by Espen Jakobsen and Kenneth Karlsen in [75], to prove a comparison principle for unbounded viscosity solutions with polynomial growth for a large class of second-order nonlocal operators $G(\cdot)$ with non-dominated integral terms (and hence weaker continuity assumptions compared to the existing literature). The comparison principle shows that for viscosity subsolutions $u \in \mathrm{USC}([0,T] \times \mathbb{R}^d)$ and viscosity supersolutions $v \in \mathrm{LSC}([0,T] \times \mathbb{R}^d)$ the comparability at the initial time

$$u(0,x) \leq v(0,x)$$

for $x \in \mathbb{R}^d$ already implies $u(t,x) \leq v(t,x)$ for all $(t,x) \in (0,T) \times \mathbb{R}^d$. In particular, we obtain the uniqueness of viscosity solutions for the associated initial value problems

$$\partial_t u(t,x) - G(u)(t,x) = 0 \quad \text{in } (0,T) \times \mathbb{R}^d$$
$$u(t,x) = \varphi(x) \quad \text{on } \{0\} \times \mathbb{R}^d$$

for initial values $\varphi \in C(\mathbb{R}^d)$ with polynomial growth. This family includes Hamilton-Jacobi-Bellman and Isaacs equations with p-polynomial growth for any $p \geq 0$, for which

$$G(u)(t,x) = \sup_{\alpha \in \mathcal{A}} \inf_{\beta \in \mathcal{B}} \Big(f_{\alpha,\beta}(t,x) - \mathcal{L}_{\alpha,\beta}(u)(t,x) - \mathcal{I}_{\alpha,\beta}(u)(t,x) \Big)$$

with the linear differential operators $\mathcal{L}_{\alpha,\beta}$ and integral operators $\mathcal{I}_{\alpha,\beta}$ for $(\alpha,\beta) \in \mathcal{A} \times \mathcal{B}$

$$\mathcal{L}_{\alpha,\beta}(u)(t,x) = c_{\alpha,\beta}(t,x)u(t,x) + b_{\alpha,\beta}^T(t,x)Du(t,x) + \mathrm{tr}\left(\sigma_{\alpha,\beta}(t,x)\sigma_{\alpha,\beta}^T(t,x)D^2u(t,x)\right)$$

$$\mathcal{I}_{\alpha,\beta}(u)(t,x) = \int \big(u(x+j_{\alpha,\beta}(t,x,z)) - u(x) - Du(x)j_{\alpha,\beta}(t,x,z)\mathbf{1}_{|z|\leq 1}\big)\, m_{\alpha,\beta}(dz)$$

such that the following assumptions hold: The inequality $c_{\alpha,\beta}(t,x) \leq 0$ and boundedness

$$\sup_{(\alpha,\beta) \in \mathcal{A} \times \mathcal{B}} \left(|f_{\alpha,\beta}(t,x)| + |c_{\alpha,\beta}(t,x)| + |b_{\alpha,\beta}(t,x)| + |\sigma_{\alpha,\beta}(t,x)|\right) < \infty$$

$$\sup_{(\alpha,\beta) \in \mathcal{A} \times \mathcal{B}} \int \left(|z|^2 \mathbf{1}_{|z|\leq 1} + |z|^q \mathbf{1}_{|z|>1}\right) m_{\alpha,\beta}(dz) < \infty$$

hold for $(t,x) \in (0,T) \times \mathbb{R}^d$ and $(\alpha,\beta) \in \mathcal{A} \times \mathcal{B}$, where either $q = p = 0$ or $q > p > 0$ with $q \geq 2$. Furthermore, the tightness assumption

$$\lim_{r \to 0} \sup_{(\alpha,\beta) \in \mathcal{A} \times \mathcal{B}} \int_{|z| \leq r} |z|^2 \, m_{\alpha,\beta}(dz) = 0 = \lim_{R \to \infty} \sup_{(\alpha,\beta) \in \mathcal{A} \times \mathcal{B}} \int_{R < |z|} |z|^p \, m_{\alpha,\beta}(dz)$$

holds, and there exists $\omega : \mathbb{R}_+ \to \mathbb{R}_+$ with $\omega(0) = \lim_{h \to 0} \omega(h) = 0$ and $C > 0$ such that

$$\sup_{(\alpha,\beta) \in \mathcal{A} \times \mathcal{B}} \left(|\sigma_{\alpha,\beta}(t,x) - \sigma_{\alpha,\beta}(s,y)| + |b_{\alpha,\beta}(t,x) - b_{\alpha,\beta}(s,y)|\right) \leq \omega(|t-s|) + C|x-y|$$

$$\sup_{(\alpha,\beta) \in \mathcal{A} \times \mathcal{B}} \left(|f_{\alpha,\beta}(t,x) - f_{\alpha,\beta}(s,y)| + |c_{\alpha,\beta}(t,x) - c_{\alpha,\beta}(s,y)|\right) \leq \omega(|t-s|) + \omega(|x-y|)$$

$$\sup_{(\alpha,\beta) \in \mathcal{A} \times \mathcal{B}} |j_{\alpha,\beta}(t,x,z) - j_{\alpha,\beta}(s,y,z)| \leq |z|\big(\omega(|t-s|) + C|x-y|\big)$$

holds for all $t,s \in (0,T)$ and $x,y,z \in \mathbb{R}^d$. In particular, our results cover the uniqueness for all characterizing equations from the stochastic part of this thesis. Since the jump measures $m_{\alpha,\beta}(dz)$ can be mutually singular for $(\alpha,\beta) \in \mathcal{A} \times \mathcal{B}$, the associated integral operators $\mathcal{I}_{\alpha,\beta}$ do in general not satisfy a dominated convergence theorem uniformly in $(\alpha,\beta) \in \mathcal{A} \times \mathcal{B}$. Hence, this generalization leads to weaker continuity properties compared to the existing literature such as [1], [12] and [75], and requires substantially more involved approximation arguments. In fact, one important realization in the field of viscosity solutions is, that the precise statement of approximation results influences the generality of the equations covered by the resulting theory significantly. We therefore isolate the approximation results and its technicalities in order to clarify the presentation of the viscosity solution arguments in this thesis.

Structure

The content of this thesis is divided into four chapters: In Chapter 1, we collect and advance results from approximation theory, in the form we will need them for the analytical part of this thesis. The major novel contributions in this chapter are concerned with the one-sided approximation of semicontinuous functions by generalized supremal convolutions and the smoothing of non-decreasing functions. In Chapter 2, we introduce our generalized viscosity solution theory for nonlinear partial integro-differential equations. First of all, we discuss our assumptions on the family of covered equations and compare our approach to the existing literature in detail. Secondly, we study different definitions for viscosity solutions and show their equivalence under our general assumptions. Finally, we develop a generalization of the theory from Espen Jakobsen and Kenneth Karlsen in [75], in order to prove a comparison principle, and show how to apply it to a general class of Hamilton-Jacobi-Bellman and Isaacs equations. In Chapter 3, we give a brief introduction into sublinear expectations and recall important results from the existing literature. Moreover, we introduce an intrinsic notion of random variables and related concepts, which generalizes the ideas from Shige Peng in [102] and which is tailored to our needs in the stochastic part of this thesis. In Chapter 4, we develop a general theory for sublinear Markov semigroups and processes for sublinear expectations, and show how to relate them to nonlocal equations from Chapter 2. In particular, we generalize an approach due to Ariel Neufeld and Marcel Nutz in [93] to construct a general class of Markov processes for sublinear expectations, which can be interpreted as classical Markov semimartingales under uncertainty in their related semimartingale characteristics. Further, we develop a pathwise stochastic calculus for sublinear expectations, and show how to apply it, in order to introduce another construction for Markov processes for sublinear expectations, which can be seen as a generalization of classical Lévy-type processes under uncertainty in their characteristics. In particular, we relate these Lévy-type processes for sublinear expectations to nonlinear nonlocal equations from Chapter 2. In the appendix, we recall important results and their precise formulations from the fields of convex and functional analysis, which are used throughout this thesis. At the end of this thesis, we present a short outlook, which includes some potentially promising ideas for future research, that we stumbled upon during the creation of this thesis.

Introduction

One of the goals of this thesis was to present the material as self-contained as possible, in order to make it accessible to a broad audience. At the beginning of each chapter and section, we try to give a brief introduction into each field and motivate its relevance for this thesis. Moreover, we try to present our new results in a very general form without obscuring their statements, in order to improve their applicability. This approach leads to a slight redundancy as well as more technical proofs at times, but it also seems to shed more light on the role of the related assumptions and arguments.

In order to simplify the navigation through this thesis and highlight its contained contributions, let us list some of the important results that we obtain, and indicate the broad connection between them:

- Lemma 1.12 contains a generalization of a well-established one-sided approximation argument for semicontinuous functions using supremal convolutions, which allows us to obtain a comparison result for viscosity solutions with arbitrary polynomial growth. Furthermore, Lemma 1.15 provides a rate of convergence for this method, when it is applied to locally Lipschitz-continuous functions, which is useful to simplify some of the penalization arguments in Chapter 2.

- Lemma 1.25 is an extension of Jensen's lemma from Lemma A.8, tailored to our generalized assumptions in Chapter 2, and is one of the essential tools for our viscosity solution theory of fully nonlinear nonlocal equations.

- Theorem 1.26 shows how to extend a constructive approach for Tietze's extension theorem from Felix Hausdorff in [61] to semicontinuous functions, which allows us to show the equivalence of different notions of viscosity solutions from the existing literature with respect to their domains of definition.

- Lemma 2.4, Remark 2.5 and Lemma 2.6 show the equivalence of different notions of viscosity solutions from the existing literature with respect to the smoothness of related test functions and their domains of definition.

- Lemma 2.8 demonstrates how to extend the well-established stability of viscosity solutions under supremal convolutions to our generalized approach from Chapter 1. The antagonistic interplay between this stability and the quality of the resulting approximations are at the heart of our generalized viscosity solution theory.

- Theorem 2.14 is an extension of the generalization from Espen Jakobsen and Kenneth Karlsen in [75] of the classical maximum principle for differentiable functions to merely semicontinuous functions, which provides necessary conditions for the first- and second-order derivatives of a differentiable function in order to have a local maximum. Such generalizations of the classical maximum principles are usually one of the main tools for proving results in viscosity solution theory.

- Corollary 2.23 contains a comparison principle for viscosity solutions with arbitrary polynomial growth of our large class of fully nonlinear nonlocal equations. Moreover, Theorem 2.22 provides a comparison principle for more than two viscosity solutions

Introduction

at a time, which allows us to obtain additional results on the dependence of solutions on their initial values in Corollary 2.25 and Corollary 2.26.

- Lemma 2.32 and Proposition 2.33 show how to apply our novel maximum and comparison principle to fully nonlinear Hamilton-Jacobi-Bellman and Isaacs equations with non-dominated integral terms.

- Definition 3.10, Definition 3.11 and Definition 3.13 discuss and slightly extend the intrinsic notions of random variables and distributions for sublinear expectations from Shige Peng in [102], in order to provide a consistent framework for Chapter 4.

- Definition 4.4 and Definition 4.6 introduce a general notion of sublinear Markov semigroups and the related concept of Markov processes for sublinear expectations. Based on that, Proposition 4.10 shows how to relate those sublinear Markov semigroups to viscosity solutions of their associated sublinear generator equations under a general C_b-Feller property assumption.

- Remark 4.33 generalizes an approach due to Ariel Neufeld and Marcel Nutz in [93], in order to construct a large family of sublinear Markov semigroups, whose associated Markov processes for sublinear expectations can be interpreted as classical Markov semimartingales under uncertainty in their semimartingale characteristics. In particular, we show in Remark 4.38 how this generalization can be used to recover the construction of Lévy processes for sublinear expectations by Mingshang Hu and Shige Peng in [67] and by Ariel Neufeld and Marcel Nutz in [93]. Furthermore, Theorem 4.41 proves the compactness of the related uncertainty subsets under explicit assumptions on their uncertainty characteristics, and Lemma 4.42 and Remark 4.43 show how this compactness relates to the C_b-Feller property of the constructed sublinear Markov semigroups.

- Definition 4.51 demonstrates how a pathwise construction due to Marcel Nutz in [97] can be applied to obtain a consistent stochastic integration theory for sublinear expectations. Furthermore, we demonstrate in Proposition 4.53 that the related measurability assumptions can be substantially weakened for a large class of integrands, which plays an important role in the construction of Markov processes for sublinear expectations via stochastic differential equations.

- Theorem 4.57 and Remark 4.58 prove the existence of solutions of stochastic differential equations for sublinear expectations with uniform and functional Lipschitz coefficients, respectively. Based on that, Remark 4.62 shows how to construct a large family of Markov processes for sublinear expectations, which can be interpreted as a generalization of classical Lévy-type processes under uncertainty in their characteristics, and demonstrates how to relate those processes with viscosity solutions of nonlinear parabolic Hamilton-Jacobi-Bellman equations.

Acknowledgments

First and foremost, I like to express my gratitude to my supervisor, Prof. Dr. René L. Schilling, for his constant support and guidance throughout my research. I also want to thank all of my colleagues from the Institute of Mathematical Stochastics at my alma mater, the Technische Universität Dresden, who made the experience of working in academia greatly enjoyable on a daily basis over the last couple of years. I feel privileged that Prof. Dr. Toshihiro Uemura from the Department of Mathematics at the Kansai University (Japan) and Prof. Dr. Marcel Nutz from the Department of Statistics at the Columbia University (USA) gave me the opportunity to visit their inspiring and exceedingly friendly research teams for a prolonged time, for which they deserve my deepest thankfulness. The financial support from the German National Academic Foundation (Studienstiftung des deutschen Volkes), the German Research Foundation (Deutsche Forschungsgemeinschaft), the German Academic Exchange Service (DAAD), the ERASMUS programme and the Graduate Academy at the Technische Universität Dresden made those and other research visits possible and is thankfully acknowledged. Last but not least, I am very grateful and want to thank my family and close friends, for believing in me and supporting me in every possible way on this interesting journey.

1. Approximation Theory

The study of viscosity solutions for nonlinear nonlocal equations relies on elaborate approximation results. This chapter is devoted to establish all approximation results, which we will require for our viscosity solution theory in Chapter 2. It is divided into four relatively independent sections: Section 1.1 and Section 1.4 remind the reader of some well-known approximation results and prove minor variations, which are adjusted to our needs. Section 1.2 and Section 1.3 contain substantial generalizations of existing approximation results, which we have developed in order to treat a more general class of nonlinear nonlocal equations in Chapter 2.

1.1. Friedrichs Mollification

A classical problem in approximation theory is the approximation of integrable or continuous real-valued functions on Euclidean space by functions with derivatives of all orders. The intriguing idea of achieving this by convolving the function with an approximate identity was first introduced by Kurt Otto Friedrichs in [57], which explains why it is often referred to as *Friedrichs mollification*. Before we recall the exact statement of the approximation result, we introduce the standard mollifier – one of the most common approximate identities used for the Friedrichs mollification method.

Lemma 1.1 (Standard Mollifier). *The real-valued function* $\eta : \mathbb{R}^d \to \mathbb{R}$ *defined by*

$$\eta(x) := C \exp\left(\frac{1}{|x|^2 - 1}\right) \mathbb{1}_{B(0,1)}(x)$$

for $x \in \mathbb{R}^d$, *where the constant* $C > 0$ *is chosen so that* $\int_{\mathbb{R}^d} \eta(x)dx = 1$ *holds, is referred to as the* standard mollifier *in Euclidean space. Furthermore, for each* $\varepsilon > 0$, *the real-valued function* $\eta_\varepsilon : \mathbb{R}^d \to \mathbb{R}$ *given by*

$$\eta_\varepsilon(x) := \varepsilon^{-d}\eta(\varepsilon^{-1}x) = \varepsilon^{-d}C \exp\left(\frac{1}{\varepsilon^{-2}|x|^2 - 1}\right) \mathbb{1}_{B(0,\varepsilon)}(x)$$

for $x \in \mathbb{R}^d$ *is arbitrarily smooth, i.e.* $\eta_\varepsilon \in C^\infty(\mathbb{R}^d)$, *integrates to one*

$$\int_{\mathbb{R}^d} \eta_\varepsilon(x)dx = 1$$

and has compact support $\operatorname{supp}(\eta_\varepsilon) = B[0, \varepsilon]$.

Proof. ↪ [66, Lemma 1.2.3, p. 14] □

1. Approximation Theory

The Friedrichs mollification method presented below is based on the convolution of functions and its properties with respect to derivatives. For a comprehensive account on the convolution of functions and its important properties, we refer the interested reader to [66, Chapter 4, p. 87]. Since we are dealing with nonlocal equations in subsets of Euclidean space in Chapter 2, the approximation results presented in this chapter have a special emphasis on the respective domains of definition.

Theorem 1.2 (Mollifications). *Suppose that $\eta_\varepsilon \in C^\infty(\mathbb{R}^d)$ for $\varepsilon > 0$ is the standard mollifier from Lemma 1.1. Moreover, suppose that $\Omega \subset \mathbb{R}^d$ is open and that $f : \Omega \to \mathbb{R}^d$ is locally integrable, i.e. $f|_{\Omega'} \in L^1(\Omega')$ for all compact subsets $\Omega' \subset \Omega$. The mollification*

$$f^\varepsilon : \Omega_\varepsilon \longrightarrow \mathbb{R}$$

with $\Omega_\varepsilon := \{x \in \Omega : d(x, \partial\Omega) > \varepsilon\}$ for $\varepsilon > 0$ is defined by

$$f^\varepsilon(x) := (f * \eta_\varepsilon)(x) = \int_\Omega f(y)\eta_\varepsilon(x-y)dy = \int_{B(0,\varepsilon)} \eta_\varepsilon(y)f(x-y)dy$$

for $x \in \Omega_\varepsilon$ and approximates the original function $f : \Omega \to \mathbb{R}$ in the following way:

(i) *The mollification $f^\varepsilon : \Omega_\varepsilon \to \mathbb{R}$ has derivatives of all orders, i.e. $f^\varepsilon \in C^\infty(\Omega_\varepsilon)$, and a subsequence converges almost everywhere to $f : \Omega \to \mathbb{R}$ as $\varepsilon \to 0$, i.e.*

$$\lim_{n\to\infty} f^{\varepsilon_n} = f$$

almost everywhere in Ω for some $(\varepsilon_n)_{n\in\mathbb{N}} \subset (0, \infty)$ with $\lim_{n\to\infty} \varepsilon_n = 0$.

(ii) *The mollification $f^\varepsilon : \Omega_\varepsilon \to \mathbb{R}$ converges to $f : \Omega \to \mathbb{R}$ locally in L^p as $\varepsilon \to 0$, i.e.*

$$\lim_{\varepsilon\to 0} f^\varepsilon\big|_{\Omega'} = f\big|_{\Omega'}$$

in $L^p(\Omega')$ for all compact $\Omega' \subset \Omega$.

(iii) *If the original function $f : \Omega \to \mathbb{R}$ is continuous, i.e. $f \in C(\Omega)$, the mollification $f^\varepsilon : \Omega_\varepsilon \to \mathbb{R}$ converges uniformly to $f : \Omega \to \mathbb{R}$ on compact subsets of Ω, i.e.*

$$\lim_{\varepsilon\to 0} \sup_{x\in\Omega'} |f^\varepsilon(x) - f(x)| = 0$$

for all compact $\Omega' \subset \Omega$.

Proof. The first half of statement (i), statement (ii) and statement (iii), can be found in [66, Theorem 1.3.2, p. 17]. The second half of statement (i) follows from statement (ii) together with [56, Proposition 2.29, p. 61] and [56, Theorem 2.30, p. 61]. □

As a next step, we introduce the standard tool of smooth, locally finite partitions of unity, in order to show that (in case of continuous functions) it is even possible to obtain uniform instead of local uniform convergence. Since we have to mimic the main line of argument in the proof of Proposition 1.7 (in order to show the existence of smooth cutoff functions), we only provide a reference here.

1.1. Friedrichs Mollification

Proposition 1.3 (Partitions of Unity). *Suppose that Γ is a collection of open sets in $\mathbb{R}^d$ whose union is $\Omega \subset \mathbb{R}^d$. There exists a sequence $(\psi_n)_{n \in \mathbb{N}} \subset C_c^\infty(\mathbb{R}^d)$ of non-negative, smooth functions with compact support such that the following properties hold:*

(i) *Every function $\psi_n : \mathbb{R}^d \to [0,1]$ has its support in some member of Γ, i.e. for every $n \in \mathbb{N}$ there exists $U_n \in \Gamma$ such that $\operatorname{supp} \psi_n \subset U_n$.*

(ii) *The sequence is a partition of unity of Ω, i.e. $\sum_{n \in \mathbb{N}} \psi_n(x) = 1$ for all $x \in \Omega$.*

(iii) *The partition is locally finite, i.e. for every compact $K \subset \Omega$ there exists an open cover $U \supset K$ and $N \in \mathbb{N}$ such that $\psi_n(x) = 0$ for all $n \geq N$ and $x \in U$.*

Proof. $\hookrightarrow$ [115, Theorem 6.20, p. 162] $\square$

As announced earlier, a combination of Theorem 1.2 with Proposition 1.3 allows us to construct uniform approximations of continuous functions. Since some of the arguments in Chapter 2 depend on the precise construction of these approximations, we recall its details in the following proof.

Theorem 1.4 (Uniform Approximations). *Suppose that $\Omega \subset \mathbb{R}^d$ is open and that $f : \Omega \to \mathbb{R}$ is continuous. For every $\varepsilon > 0$, there exists a smooth approximation $g \in C^\infty(\Omega)$ such that $\sup_{x \in \Omega} |g(x) - f(x)| \leq \varepsilon$ holds.*

Proof. Due to Proposition 1.3, there exists a sequence $(\psi_n)_{n \in \mathbb{N}}$ of non-negative, smooth functions with compact support such that $\sum_{n \in \mathbb{N}} \psi_n \equiv 1$ holds on Ω. Since $f : \Omega \to \mathbb{R}$ is continuous, Theorem 1.2 (iii) implies the existence of $\delta_n > 0$ for every $n \in \mathbb{N}$ such that

$$\sup_{x \in K_n} |f^{\delta_n}(x) - f(x)| \leq \varepsilon,$$

where $f^{\delta_n} := f * \eta_{\delta_n}$ and $K_n := \operatorname{supp}(\psi_n) \subset \mathbb{R}^d$ is compact. Furthermore, the partition $(\psi_n)_{n \in \mathbb{N}}$ is locally finite according to Proposition 1.3 (iii), and hence

$$g := \sum_{n \in \mathbb{N}} f^{\delta_n} \psi_n : \Omega \longrightarrow \mathbb{R}$$

defines a smooth function. Finally, the construction and $\sum_{n \in \mathbb{N}} \psi_n \equiv 1$ imply that

$$|g(x) - f(x)| = \left| \sum_{n \in \mathbb{N}} (f^{\delta_n}(x) - f(x)) \psi_n(x) \right| \leq \sum_{n \in \mathbb{N}} |f^{\delta_n}(x) - f(x)| \, \psi_n(x)$$

$$\leq \varepsilon \sum_{n \in \mathbb{N}} \psi_n(x) = \varepsilon$$

holds for all $x \in \Omega$, as required. $\square$

One property of the uniform approximations from the preceding theorem is going to be of particular interest in Chapter 2: If the approximated function is smooth enough, the derivatives of the approximations also converge (locally uniformly) to the derivatives the approximated function.

1. Approximation Theory

Remark 1.5 (Derivatives of Approximations). If $f \in C(\Omega)$ for an open $\Omega \subset \mathbb{R}^d$, then Theorem 1.4 shows that for $\varepsilon > 0$ there exists $f^\varepsilon \in C^\infty(\Omega)$ of the form

$$f^\varepsilon = \sum_{n \in \mathbb{N}} (f * \eta_{\delta_n}) \psi_n$$

such that $\sup_{x \in \Omega} |f^\varepsilon(x) - f(x)| \leq \varepsilon$, where η_δ is the standard mollifier from Lemma 1.1, $(\psi_n)_{n \in \mathbb{N}}$ is a (smooth) partition of unity of $\Omega \subset \mathbb{R}^d$ as in Proposition 1.3 and $\delta_n(\varepsilon) \to 0$ as $\varepsilon \to 0$. A simple calculation shows that if $f \in C^n(\Omega)$ for some $n \in \mathbb{N}$, then

$$\lim_{\varepsilon \to 0} D^k f^\varepsilon = D^k f$$

locally uniform $\varepsilon \to 0$ in Ω for every $k \in \{1, \ldots, n\}$: For every term in the definition of

$$f^\varepsilon : \mathbb{R}^d \longrightarrow \mathbb{R}$$

this follows from Theorem 1.2, since

$$D^k(f * \eta_{\delta_n}) = (D^k f) * \eta_{\delta_n}$$

according to the differentiation property of the convolution. Hence, the right-hand side of

$$\sup_{x \in K} |D^k f^\varepsilon(x) - D^k f(x)| = \sup_{x \in K} \left| D^k \left(\sum_{n \in \mathbb{N}} \left((f * \eta_{\delta_n}) - f \right) \psi_n \right)(x) \right|$$

$$\leq \sup_{x \in K} \sum_{j=0}^{k} \sum_{n \in \mathbb{N}} |(D^j f * \eta_{\delta_n})(x) - D^j f(x)| \cdot |D^{k-j} \psi_n(x)|$$

converges to zero as $\varepsilon \to 0$ for every compact $K \subset \Omega$, since the partition $(\psi_n)_{n \in \mathbb{N}}$ is locally finite according to Proposition 1.3 (iii), and since continuous functions are bounded on compact sets.

The rest of this section is devoted to show the existence of smooth cutoff functions, which are sometime also referred to as bump functions. Cutoff functions can be viewed as the smooth counterparts to indicator functions and are an essential tool for the construction of smooth functions using a definition by cases.

Lemma 1.6 (Cutoff Functions with Compact Support). If $V \subset U \subset \mathbb{R}^d$ such that V is compact and U is open, then there exists $\chi \in C_c^\infty(\mathbb{R}^d)$ with compact support such that

$$\mathbb{1}_V(x) \leq \chi(x) \leq \mathbb{1}_U(x)$$

holds for all $x \in \mathbb{R}^d$. In particular, $0 \leq \chi(x) \leq 1$ for all $x \in \mathbb{R}^d$, $\chi(x) = 1$ for all $x \in V$, and $\chi(x) = 0$ for all $x \in U^C = \mathbb{R}^d \setminus U$.

Proof. Since V is compact, there exists $\varepsilon > 0$ such that the inflated set

$$V + B(0, 2\varepsilon) = \{x + y \in \mathbb{R}^d \mid x \in V \text{ and } y \in B(0, 2\varepsilon)\}$$

is still contained in U. Suppose that $\eta_\varepsilon : \mathbb{R}^d \to \mathbb{R}$ is the standard mollifier, and define

$$\chi(x) := (\mathbb{1}_{V+B(0,\varepsilon)} * \eta_\varepsilon)(x) = \int_{B(0,\varepsilon)} \eta_\varepsilon(y) \mathbb{1}_{V+B(0,\varepsilon)}(x - y) \, dy$$

for all $x \in \mathbb{R}^d$. Since $\mathbb{1}_{V+B(0,\varepsilon)}(x - y) = 1$ for all $x \in V$ and $y \in B(0, \varepsilon)$, we find

$$\chi(x) = \int_{B(0,\varepsilon)} \eta_\varepsilon(y) \cdot 1 \, dy = 1$$

for all $x \in V$. On the other hand, if $x \in \mathbb{R}^d \setminus (V + B(0, 2\varepsilon)) \supset \mathbb{R}^d \setminus U$, then

$$\mathbb{1}_{V+B(0,\varepsilon)}(x - y) = 0$$

for all $y \in B(0, \varepsilon)$. Therefore, we obtain for all $x \in \mathbb{R}^d \setminus (V + B(0, 2\varepsilon)) \supset \mathbb{R}^d \setminus U$ that

$$\chi(x) = \int_{B(0,\varepsilon)} \eta_\varepsilon(y) \cdot 0 \, dy = 0.$$

In particular, this shows that $\chi : \mathbb{R}^d \to \mathbb{R}$ has compact support, because we have proven $\operatorname{supp}(\chi) \subset V + B[0, 2\varepsilon]$ and $V + B[0, 2\varepsilon] \subset \mathbb{R}^d$ is compact. Finally, the construction yields

$$0 = \int_{B(0,\varepsilon)} \eta_\varepsilon(y) \cdot 0 \, dy \leq \chi(x) \leq \int_{B(0,\varepsilon)} \eta_\varepsilon(y) \cdot 1 \, dy = 1$$

for all $x \in \mathbb{R}^d$, and Theorem 1.2 implies $\chi \in C^\infty(\mathbb{R}^d)$, as required. $\square$

Since we want to cover viscosity solutions on the whole Euclidean space in Chapter 2, the compactness assumption in the last lemma is too restrictive for many applications. Fortunately, a localization procedure similar to the smooth, locally finite partition of unity from Proposition 1.3 shows that we can construct cutoff functions without this compactness assumption:

Proposition 1.7 (Cutoff Functions). *If $V \subset U \subset \Omega \subset \mathbb{R}^d$ such that Ω, U and $\Omega \setminus V$ are open, then there exists $\chi \in C^\infty(\Omega)$ such that*

$$\mathbb{1}_V(x) \leq \chi(x) \leq \mathbb{1}_U(x)$$

holds for all $x \in \mathbb{R}^d$. In particular, $0 \leq \chi(x) \leq 1$ for all $x \in \mathbb{R}^d$, $\chi(x) = 1$ for all $x \in V$, and $\chi(x) = 0$ for all $x \in U^C = \mathbb{R}^d \setminus U$.

1. Approximation Theory

Proof. The main idea is to partition the sets V and U into smaller, bounded pieces and then apply the well-known result from Lemma 1.6 locally. Afterwards we glue the resulting local cutoff functions together by mimicking the proof of existence for smooth, locally finite partitions of unity, cf. [115, Theorem 6.20, p. 162].

Suppose that $(\alpha, \beta) : \mathbb{N} \to \mathbb{N} \times \mathbb{N}$ is a bijection and define

$$V_n := \{x \in V \ : \ \alpha(n) - 1 \leq |x| \leq \alpha(n) \quad , \ \beta(n) - 1 \leq d(x, \mathbb{R}^d \setminus \Omega)^{-1} \leq \beta(n) \quad \}$$
$$U_n := \{x \in U \ : \ \alpha(n) - 2 < |x| < \alpha(n) + 1 \ , \ \beta(n) - 2 < d(x, \mathbb{R}^d \setminus \Omega)^{-1} < \beta(n) + 1\}$$

for all $n \in \mathbb{N}$. It is easy to check that $V_n \subset U_n \subset \Omega$ for all $n \in \mathbb{N}$, that

$$\bigcup_{n \in \mathbb{N}} V_n = V \subset U = \bigcup_{n \in \mathbb{N}} U_n,$$

and that V_n is compact and U_n is open for all $n \in \mathbb{N}$. Therefore, Lemma 1.6 implies the existence of $\phi_n \in C^\infty(\mathbb{R}^d)$ for every $n \in \mathbb{N}$ such that

$$\mathbb{1}_{V_n}(x) \leq \phi_n(x) \leq \mathbb{1}_{U_n}(x)$$

for all $x \in \mathbb{R}^d$ and $n \in \mathbb{N}$. The properties of $\phi_n : \mathbb{R}^d \to \mathbb{R}$ and the definition

$$\psi_n := \left(\prod_{j=1}^{n-1} (1 - \phi_j) \right) \cdot \phi_n$$

show that $\psi_n(x) = 0$ for all $x \in \mathbb{R}^d \setminus U_n$ and $n \in \mathbb{N}$. Furthermore, it is easy to verify that

$$\psi_1 + \ldots + \psi_n = 1 - \prod_{j=1}^{n} (1 - \phi_j)$$

and therefore $0 \leq \psi_1 + \ldots + \psi_n \leq 1$ for all $n \in \mathbb{N}$ using mathematical induction: The base case $n = 1$ follows from the definition of ψ_1. The inductive step $n \rightsquigarrow (n+1)$ is proven by adding the induction hypothesis and the definition of ψ_{n+1}. This yields

$$\psi_1 + \ldots + \psi_n \equiv 1$$

inside $V_1 \cup \ldots \cup V_n$, since $\phi_j \equiv 1$ inside V_j for $j \in \mathbb{N}$. Moreover, it also shows

$$\psi_1 + \ldots + \psi_n \equiv 0$$

outside $U_1 \cup \ldots \cup U_n$, since $\phi_j \equiv 0$ outside U_j for $j \in \mathbb{N}$. Finally, $\chi : \Omega \to \mathbb{R}$ defined by

$$\chi(x) := \sum_{n=1}^{\infty} \psi_n(x)$$

for $x \in \Omega$ is smooth, since the sum is actually locally finite: For every $x \in \Omega$ there exists $\varepsilon > 0$ such that the open ball $B(x, \varepsilon) \subset \Omega$ intersects only finitely many U_n with $n \in \mathbb{N}$. In particular, there exists a neighborhood in Ω for each $x \in \Omega$ such that all but finitely many ψ_n with $n \in \mathbb{N}$ are equal to zero in that neighborhood. $\square$

1.2. Supremal Convolutions

Supremal convolutions can be regarded as an analog to classical convolutions, where the integration operator is replaced by the supremum operator. They are most prominent in convex analysis, where they occur (mostly in the form of infimal convolutions) as *Legendre-Fenchel transformations*, which play a similar role as Fourier transformations in classical analysis. These transformations were first investigated by Werner Fenchel in [53] and are an essential component of every modern introduction to convex analysis, such as [111]. In the context of viscosity solutions, supremal convolutions first appeared (as Legendre-Fenchel transformations) in the famous *Hopf-Lax formula* in [85] and [65], where Peter Lax and Eberhard Hopf connected first-order Hamilton-Jacobi-Bellman equations with convex analysis via variational analysis.

In this thesis, supremal convolutions are used slightly different – as a generalization of Friedrichs mollification from Section 1.1 for viscosity solutions: We will prove in Chapter 2 that the supremal convolution of a viscosity solution for nonlinear nonlocal equations is again a viscosity solution, which exhibits better regularity properties and approximates the original solution. In this generality, this approach was studied for the first time by Jean-Michel Lasry and Pierre-Louis Lions in [84], and became an essential tool for the viscosity solution theory of nonlinear nonlocal equations. For the special case of convex local equations, however, it was studied even earlier by Jean-Jacques Moreau in [91] and is known as the *Moreau–Yosida approximation*.

In this section, we generalize and substantially extend the known results on supremal convolutions for the approximation of viscosity solutions. These new results enable us to develop a uniqueness theory for viscosity solutions with arbitrary polynomial growths of nonlinear nonlocal equations with weaker continuity assumptions (compared to the existing literature) in Chapter 2. We try to emphasize our contribution by highlighting important differences to the existing literature along the way.

Definition 1.8 (Supremal & Infimal Convolutions). Suppose that $\Omega \subset \mathbb{R}^d$,

$$f : \Omega \longrightarrow \mathbb{R}$$

and $\varphi : \mathbb{R}^d \to \mathbb{R}$. The *supremal convolution* $\triangle_\varphi^\varepsilon[f]$ for $\varepsilon > 0$ is defined as

$$\triangle_\varphi^\varepsilon[f] : \mathbb{R}^d \longrightarrow \mathbb{R} \cup \{+\infty\}$$
$$x \longmapsto \triangle_\varphi^\varepsilon[f](x) := \sup_{y \in \Omega} \left(f(y) - \frac{\varphi(x-y)}{\varepsilon} \right).$$

Dually, the *infimal convolution* $\nabla_\varphi^\varepsilon[f]$ for $\varepsilon > 0$ is defined as

$$\nabla_\varphi^\varepsilon[f] : \mathbb{R}^d \longrightarrow \mathbb{R} \cup \{-\infty\}$$
$$x \longmapsto \nabla_\varphi^\varepsilon[f](x) := \inf_{y \in \Omega} \left(f(y) + \frac{\varphi(x-y)}{\varepsilon} \right),$$

or equivalently as $\nabla_\varphi^\varepsilon[f](x) = -\triangle_\varphi^\varepsilon[-f](x)$ for $x \in \mathbb{R}^d$.

1. Approximation Theory

In most of the related literature on viscosity solution theory, such as [35] for local or [75] for nonlocal equations, the function $\varphi(\cdot)$ is fixed as the squared norm $|\cdot|^2$. In case of bounded (or even linearly growing) solutions $f(\cdot)$, it is easy to see that the growth of the squared norm is sufficient to ensure that the resulting supremal and infimal convolutions are real-valued. For unbounded solutions, however, it is possible that the growth cannot be compensated by the squared norm, which can lead to supremal and infimal convolutions only taking the values $\pm\infty$. In the influential paper [12] for nonlocal equations, this problem is circumvented by taking the supremum and infimum (in the definition of the supremal and infimal convolution) over a compact subset instead of the whole space Ω. Unfortunately, this leads to supremal and infimal convolutions with potential discontinuities, which are not suitable for our needs in Chapter 2, since we have to approximate functions locally uniformly and not only pointwise due to our weaker continuity assumptions. Because of this, we consider more general $\varphi(\cdot)$ instead of the squared norm $|\cdot|^2$ here, that can compensate the growth of unbounded solutions, while maintaining important properties of the approximations – such as their continuity (cf. Lemma 1.12) and their stability with respect to viscosity solutions (cf. Lemma 2.8).

Definition 1.9 (Quasidistances). A *quasidistance* (on $\mathbb{R}^d$) is a function

$$\varphi : \mathbb{R}^d \longrightarrow [0, \infty)$$

that is symmetric (i.e. $\varphi(x) = \varphi(-x)$ for $x \in \mathbb{R}^d$), definite (i.e. $\varphi(x) = 0$ if and only if $x = 0$) and satisfies a generalized triangle inequality, in that there exists $\rho \geq 1$ such that

$$\varphi(x+y) \leq \rho \cdot \Big(\varphi(x) + \varphi(y)\Big)$$

holds for all $x, y \in \mathbb{R}^d$. Constants $\rho \geq 1$ for which the generalized triangle inequality hold are referred to as *multipliers* of the quasidistance.

Note that our notion of quasidistances is different from the one of quasinorms in functional analysis (cf. e.g. [79, Chapter 25, p. 1099]), which generalizes the notion of norms by weakening the triangle inequality, but not the positive homogeneity. Moreover, quasidistances are not to be confused with seminorms or pseudonorms, that generalize norms by giving up definiteness. Typical examples of quasidistances include

$$x \longmapsto |x|^p$$

for an exponent $p > 0$, which satisfy a generalized triangle inequality with $\rho = 2^p$, since

$$|x+y|^p \leq \Big(|x|+|y|\Big)^p \leq \Big(2\max\{|x|,|y|\}\Big)^p = 2^p \max\{|x|^p, |y|^p\}$$
$$\leq 2^p \Big(|x|^p + |y|^p\Big)$$

for all $x, y \in \mathbb{R}^d$. The following result illustrates an easy procedure of how to construct new quasidistances from existing ones:

Lemma 1.10 (Stability of Quasidistances). *If $n \in \mathbb{N}$, $\lambda_i > 0$ and $\varphi_i : \mathbb{R}^d \to [0, \infty)$ are quasidistances with multipliers $\rho_i \geq 1$ for $i \in \{1, \ldots, n\}$, then*

$$\varphi : \mathbb{R}^d \longrightarrow [0, \infty)$$
$$x \longmapsto \varphi(x) := \max_{i \in \{1, \ldots, n\}} \lambda_i \cdot \varphi_i(x)$$

is a quasidistance with multiplier $\rho = \max_{i \in \{1, \ldots, n\}} \rho_i$.

Proof. It is easy to check that $\varphi = \max_{i \in \{1, \ldots, n\}} \lambda_i \cdot \varphi_i$ is again symmetric and definite. Furthermore, for every $i \in \{1, \ldots, n\}$ the inequality

$$\lambda_i \cdot \varphi_i(x+y) \leq \lambda_i \cdot \rho_i \cdot \Big(\varphi_i(x) + \varphi_i(y) \Big) \leq \Big(\max_{i \in \{1, \ldots, n\}} \rho_i \Big) \Big(\varphi(x) + \varphi(y) \Big)$$

holds for all $x, y \in \mathbb{R}^d$. This implies the generalized triangle inequality with multiplier

$$\rho = \max_{i \in \{1, \ldots, n\}} \rho_i$$

by taking the maximum over all $i \in \{1, \ldots, n\}$. □

Weierstrass' exteme value theorem for continuous functions plays an essential role in many approximation theory arguments. It is well-known and can be easily deduced from its proof (cf. e.g. [112, Chapter 1, Section C, p. 13]), that it can be generalized to semicontinuous functions $f : X \to \mathbb{R}$ on metric spaces (X, d), for which sets of the form $\{x \in X : f(x) \geq c\}$ or $\{x \in X : f(x) \leq c\}$ with $c \in \mathbb{R}$ are still closed. For notational reasons, let us quickly recall the definition of semicontinuous functions:

Definition 1.11 (Semicontinuous Functions). A real-valued function $f : X \to \mathbb{R}$ on a metric space (X, d) is called *upper semicontinuous* if

$$\limsup_{n \to \infty} f(x_n) \leq f(x)$$

for all sequences $(x_n)_{n \in \mathbb{N}} \subset X$ with $\lim_{n \to \infty} x_n = x$, and $\mathrm{USC}(X)$ denotes the family of all upper semicontinuous functions on (X, d). Dually, a function $g : X \to \mathbb{R}$ is called *lower semicontinuous* if $(-g)$ is upper semicontinuous, or equivalently

$$\liminf_{n \to \infty} g(x_n) \geq g(x)$$

for all sequences $(x_n)_{n \in \mathbb{N}} \subset X$ with $\lim_{n \to \infty} x_n = x$, and $\mathrm{LSC}(X)$ denotes the family of all lower semicontinuous functions on (X, d). For notational reasons, we write

$$\mathrm{SC}(X) := \mathrm{LSC}(X) \cup \mathrm{USC}(X)$$

for the family of all semicontinuous functions on (X, d).

1. Approximation Theory

As we will see in Chapter 2, one of the key concepts in viscosity solution theory is to scan non-smooth functions by smooth functions monotonically from above and below. Since the pointwise supremum and infimum of a family of continuous functions is in general only semicontinuous (and, as a matter of fact, every semincontinuous function is of this form, as we will see in Lemma 1.16), semicontinuity is a very natural assumption in the context of viscosity solution theory and, in fact, the best we can expect in the following approximation results.

Lemma 1.12 (Approximation by Supremal Convolutions). *If $\Omega \subset \mathbb{R}^d$ is open, if $\varphi \in \mathbb{R}^d \to [0, \infty)$ is a continuous quasidistance with multiplier $\rho \geq 1$ such that*

$$\{x \in \mathbb{R}^d : \varphi(x) \leq c\}$$

is compact for all $c > 0$, and if $C > 0$ such that $f \in USC(\Omega)$ satisfies

$$|f(x)| \leq C(1 + \varphi(x))$$

for all $x \in \Omega$, then the following properties hold for the supremal convolution:

(i) *Suppose that $0 < \varepsilon \leq \varepsilon' \leq (\rho C)^{-1}$, then*

$$\Delta_\varphi^\varepsilon[f](x) \leq \Delta_\varphi^{\varepsilon'}[f](x) \leq \rho C(1 + \varphi(x))$$

for all $x \in \mathbb{R}^d$. Moreover, $f(x) \leq \Delta_\varphi^\varepsilon[f](x)$ for all $x \in \Omega$ and $\varepsilon > 0$, and

$$\lim_{\varepsilon \to 0} \Delta_\varphi^\varepsilon[f](x) = f(x)$$

for all $x \in \Omega$.

(ii) *Suppose that $0 < \varepsilon \leq (2\rho C)^{-1}$ and define*

$$\delta(x) := 8\rho C(1 + \varphi(x))$$

for all $x \in \Omega$, then $\Delta_\varphi^\varepsilon[f] \in C(\mathbb{R}^d)$. Moreover, for every $x \in \Omega$ with

$$\inf_{y \in \mathbb{R}^d \setminus \Omega} \varphi(x - y) > \varepsilon \cdot \delta(x)$$

there exists $\overline{x} \in \Omega$ with $\varphi(x - \overline{x}) \leq \varepsilon \cdot \delta(x)$ such that

$$\Delta_\varphi^\varepsilon[f](x) = f(\overline{x}) - \frac{\varphi(x - \overline{x})}{\varepsilon}.$$

(iii) *If $\varphi(\cdot) = |\cdot|^2$ in a neighborhood of zero, then there exists $0 < \varepsilon_0 \leq (2\rho C)^{-1}$ for every convex and bounded $\Omega' \subset \Omega$ such that*

$$x \longmapsto \Delta_\varphi^\varepsilon[f](x) + \frac{|x|^2}{\varepsilon}$$

is convex in Ω' for all $0 < \varepsilon \leq \varepsilon_0$. Moreover, if $\varphi(x) = |x|^2$ for all $x \in \mathbb{R}^d$, then $\varepsilon_0 = (2\rho C)^{-1}$ for every convex $\Omega' \subset \Omega$.

1.2. Supremal Convolutions

Proof. The growth assumption for $f : \Omega \to \mathbb{R}$ together with $0 < \varepsilon \leq (\rho C)^{-1}$ imply

$$\Delta_\varphi^\varepsilon[f](x) = \sup_{y \in \Omega} \left(f(y) - \frac{\varphi(x-y)}{\varepsilon} \right)$$

$$\leq \sup_{y \in \Omega} \left(C\big(1 + \varphi(y)\big) - \rho C \varphi(x-y) \right)$$

$$\leq \sup_{y \in \Omega} \left(C + \rho C \big(\varphi(y-x) + \varphi(x)\big) - \rho C \varphi(y-x) \right)$$

$$\leq \rho C \big(1 + \varphi(x)\big)$$

for all $x \in \mathbb{R}^d$. Furthermore, the pointwise monotonicity $\Delta_\varphi^\varepsilon[f](x) \leq \Delta_\varphi^{\varepsilon'}[f](x)$ for $x \in \mathbb{R}^d$ and $0 < \varepsilon \leq \varepsilon' \leq (\rho C)^{-1}$ follows from taking the supremum in

$$f(y) - \frac{\varphi(x-y)}{\varepsilon} \leq f(y) - \frac{\varphi(x-y)}{\varepsilon'} \leq \Delta_\varphi^{\varepsilon'}[f](x).$$

over all $y \in \Omega$. Since $y = x \in \Omega$ is an admissible choice in the definition of $\Delta_\varphi^\varepsilon[f](x)$, and since $\varphi(0) = 0$ by assumption, we further obtain

$$f(x) \leq \Delta_\varphi^\varepsilon[f](x)$$

for all $x \in \Omega$ and $\varepsilon > 0$.

In order to complete the proof of property (i), it remains to show

$$\lim_{\varepsilon \to 0} \Delta_\varphi^\varepsilon[f](x) = f(x)$$

for all $x \in \Omega$: Assume that $0 < \varepsilon \leq (2\rho C)^{-1}$, $x \in \mathbb{R}^d$ and $y, y' \in \Omega$ such that

$$\varphi(y-x) > 4 \inf_{z \in \Omega} \varphi(z-x) + 8\varepsilon \rho C \big(1 + \varphi(x)\big) =: \delta_\varepsilon(x)$$

$$\varphi(y'-x) \leq \inf_{z \in \Omega} \varphi(z-x) + \varepsilon \rho C \big(1 + \varphi(x)\big).$$

These assumptions obviously imply $\varphi(y-x) > 4\varphi(y'-x) + 4\varepsilon\rho C(1+\varphi(x))$ and therefore

$$\begin{aligned} f(y) - \varepsilon^{-1}\varphi(y-x) &\leq C\big(1 + \varphi(y)\big) - \varepsilon^{-1}\varphi(y-x) \\ &\leq \rho C\big(1 + \varphi(x)\big) + \big(\rho C - \varepsilon^{-1}\big)\varphi(y-x) \\ &\leq \rho C\big(1 + \varphi(x)\big) - (2\varepsilon)^{-1}\varphi(y-x) \\ &< -\rho C\big(1 + \varphi(x)\big) - 2\varepsilon^{-1}\varphi(y'-x) \\ &\leq -\rho C\big(1 + \varphi(x) + \varphi(y'-x)\big) - \varepsilon^{-1}\varphi(y'-x) \\ &\leq -C\big(1 + \varphi(y')\big) - \varepsilon^{-1}\varphi(y'-x) \\ &\leq f(y') - \varepsilon^{-1}\varphi(y'-x), \end{aligned} \qquad (1.1)$$

using $C\rho \leq (2\varepsilon)^{-1} \leq \varepsilon^{-1}$, the growth assumption for $f : \Omega \to \mathbb{R}$ and the generalized triangle inequality of the quasidistance $\varphi : \mathbb{R}^d \to [0, \infty)$ with multiplier $\rho \geq 1$.

1. Approximation Theory

Since $\varepsilon\rho C(1+\varphi(x)) > 0$ for $x \in \mathbb{R}^d$, there exists some $y' \in \Omega$ for every $x \in \mathbb{R}^d$ such that
$$\varphi(y'-x) \leq \inf_{y\in\Omega} \varphi(y-x) + \varepsilon\rho C\Big(1+\varphi(x)\Big).$$

Consequently, Equation (1.1) and the definition of $\Delta_\varphi^\varepsilon[f] : \mathbb{R}^d \to \mathbb{R}$ yield

$$\begin{aligned}\Delta_\varphi^\varepsilon[f](x) &= \sup\left\{f(y) - \frac{\varphi(x-y)}{\varepsilon} \,\Big|\, y\in\Omega\right\} \\ &= \sup\left\{f(y) - \frac{\varphi(x-y)}{\varepsilon} \,\Big|\, y\in\Omega \text{ with } \varphi(y-x) \leq \delta_\varepsilon(x)\right\}\end{aligned} \quad (1.2)$$

for all $x \in \mathbb{R}^d$ and $0 < \varepsilon \leq (2\rho C)^{-1}$. Furthermore, if $x \in \Omega$ and $\varepsilon > 0$ such that

$$\inf_{y\in\mathbb{R}^d\setminus\Omega} \varphi(x-y) > \delta_\varepsilon(x) = 8\varepsilon\rho C\Big(1+\varphi(x)\Big) = \varepsilon \cdot \delta(x),$$

then $\{y \in \Omega \mid \varphi(y-x) \leq \delta_\varepsilon(x)\}$ is compact. Hence, the (semi-)continuity of $f : \Omega \to \mathbb{R}$ and $\varphi : \mathbb{R}^d \to [0,\infty)$ imply for every $x \in \Omega$ and $0 < \varepsilon \leq (2\rho C)^{-1}$ with

$$\inf_{y\in\mathbb{R}^d\setminus\Omega} \varphi(x-y) > \varepsilon\cdot\delta(x)$$

the existence of some $\overline{x}_\varepsilon \in \Omega$ with $\varphi(x-\overline{x}_\varepsilon) = \varphi(\overline{x}_\varepsilon - x) \leq \delta_\varepsilon(x) = \varepsilon\cdot\delta(x)$ such that

$$\begin{aligned}\Delta_\varphi^\varepsilon[f](x) &= \sup\left\{f(y) - \frac{\varphi(x-y)}{\varepsilon} \,\Big|\, y\in\Omega \text{ with } \varphi(y-x) \leq \delta_\varepsilon(x)\right\} \\ &= f(\overline{x}_\varepsilon) - \frac{\varphi(x-\overline{x}_\varepsilon)}{\varepsilon}\end{aligned}$$

holds. Note that the assumptions on $\varphi : \mathbb{R}^d \to [0,\infty)$ imply $\lim_{\varepsilon\to\infty} \overline{x}_\varepsilon = x$ for all $x \in \Omega$. The upper semicontinuity of $f : \Omega \to \mathbb{R}$ hence leads to

$$f(x) \leq \liminf_{\varepsilon\to 0} \Delta_\varphi^\varepsilon[f](x) \leq \limsup_{\varepsilon\to 0}\Delta_\varphi^\varepsilon[f](x) = \limsup_{\varepsilon\to 0}\left(f(\overline{x}_\varepsilon) - \frac{\varphi(x-\overline{x}_\varepsilon)}{\varepsilon}\right) \leq f(x),$$

for all $x \in \Omega$, which completes the proof of property (i) and of one half of property (ii). As a next step, note that our observations from Equation (1.2) imply that

$$\begin{aligned}\Delta_\varphi^\varepsilon[f](x) &= \sup_{y\in\Omega}\left\{f(y) - \varepsilon^{-1}\varphi(x-y) : \varphi(y-x) \leq \delta_\varepsilon(x)\right\} \\ &= \sup_{y\in\Omega}\Big\{f(y) - \varepsilon^{-1}\varphi(x'-y) \\ &\qquad + \varepsilon^{-1}\big(\varphi(x'-y) - \varphi(x-y)\big) : \varphi(y-x) \leq \delta_\varepsilon(x)\Big\} \\ &\leq \Delta_\varphi^\varepsilon[f](x') + \varepsilon^{-1}\sup_{y\in\Omega}\Big\{|\varphi(x'-y) - \varphi(x-y)| : \varphi(y-x) \leq \delta_\varepsilon(x)\Big\}\end{aligned}$$

holds for all $x, x' \in \mathbb{R}^d$ and $0 < \varepsilon \leq (2\rho C)^{-1}$, using the subadditivity of suprema.

An interchange of the roles of $x \in \mathbb{R}^d$ and $x' \in \mathbb{R}^d$ leads to

$$\left|\Delta_\varphi^\varepsilon[f](x) - \Delta_\varphi^\varepsilon[f](x')\right| \leq \varepsilon^{-1} \sup_{y \in K(x,x')} \left|\varphi(x'-y) - \varphi(x-y)\right|$$

with $K(x,x') := \{y \in \mathbb{R}^d : \varphi(y-x) \leq \delta_\varepsilon(x) \text{ or } \varphi(y-x') \leq \delta_\varepsilon(x')\}$ for all $x, x' \in \mathbb{R}^d$ and $0 < \varepsilon \leq (2\rho C)^{-1}$. Since continuous functions on compact sets are uniformly continuous,

$$\lim_{|x-x'|\to 0} \sup_{y \in K(x,x')} \left|\varphi(x'-y) - \varphi(x-y)\right| = 0$$

holds, which implies that $\Delta_\varphi^\varepsilon[f] \in C(\mathbb{R}^d)$ and completes the proof of property (ii).

At last, we will prove property (iii): Suppose that $\varphi(\cdot) = |\cdot|^2$ in a neighborhood of zero, i.e. there exists $r > 0$ such that

$$\varphi(x) = |x|^2 \tag{1.3}$$

for all $x \in \mathbb{R}^d$ with $\varphi(x) \leq r$, and that $\Omega' \subset \Omega$ is convex and bounded. Due to a standard covering argument, it suffices to prove that for every $x' \in \Omega'$ there exists $0 < \varepsilon_{x'} \leq (2\rho C)^{-1}$ and a neighborhood $N_{x'} \subset \mathbb{R}^d$ of $x' \in \Omega'$ such that

$$x \longmapsto \Delta_\varphi^\varepsilon[f](x) + \frac{|x|^2}{\varepsilon}$$

is convex in $N_{x'}$ for all $0 < \varepsilon \leq \varepsilon_{x'}$: Suppose that $x' \in \Omega'$ and define

$$N_{x'} := \{x \in \mathbb{R}^d : \varphi(x-x') < (6\rho)^{-1}r\}, \tag{1.4}$$

where $\rho \geq 1$ is the multiplier of $\varphi : \mathbb{R}^d \to [0,\infty)$ (cf. Definition 1.9). The generalized triangle inequality of $\varphi : \mathbb{R}^d \to [0,\infty)$ implies

$$\delta_\varepsilon(x) = 4 \inf_{y \in \Omega} \varphi(y-x) + 8\varepsilon\rho C\Big(1 + \varphi(x)\Big)$$
$$\leq 4\varphi(x-x') + 8\varepsilon\rho C\Big(1 + \rho\big(\varphi(x') + \varphi(x-x')\big)\Big)$$
$$\leq 4((6\rho)^{-1}r) + 8\varepsilon\rho C\Big(1 + \rho\big(\varphi(x') + ((6\rho)^{-1}r)\big)\Big)$$

for all $x \in N_{x'}$. Therefore, there exists $0 < \varepsilon_{x'} \leq (2\rho C)^{-1}$ such that

$$\delta_\varepsilon(x) \leq 5\left((6\rho)^{-1}r\right) \tag{1.5}$$

for all $x \in N_{x'}$ and $0 < \varepsilon \leq \varepsilon_{x'}$. If $x \in N_{x'}$ and $y \in \Omega$ with $\varphi(y-x) \leq \delta_\varepsilon(x)$, then

$$\varphi(y-x') \leq \rho\Big(\varphi(y-x) + \varphi(x-x')\Big) \leq \rho\Big(\delta_\varepsilon(x) + \varphi(x-x')\Big)$$
$$\leq \rho\Big(5\left((6\rho)^{-1}r\right) + ((6\rho)^{-1}r)\Big) = r \tag{1.6}$$

for all $0 < \varepsilon \leq \varepsilon_{x'}$, using the generalized triangle inequality, Equations (1.4) and (1.5).

1. Approximation Theory

In particular, Equation (1.2) and Equation (1.6) lead to

$$\Delta_\varphi^\varepsilon[f](x) = \sup\left\{ f(y) - \frac{\varphi(x-y)}{\varepsilon} \;\middle|\; y \in \Omega \text{ with } \varphi(y-x) \leq \delta_\varepsilon(x) \right\}$$

$$\leq \sup\left\{ f(y) - \frac{\varphi(x-y)}{\varepsilon} \;\middle|\; y \in \Omega \text{ with } \varphi(y-x') \leq r \right\}$$

$$\leq \sup\left\{ f(y) - \frac{\varphi(x-y)}{\varepsilon} \;\middle|\; y \in \Omega \right\}$$

$$= \Delta_\varphi^\varepsilon[f](x)$$

for all $x \in N_{x'}$ and $0 < \varepsilon \leq \varepsilon_{x'}$, which shows with Equation (1.3) that

$$\Delta_\varphi^\varepsilon[f](x) + \frac{|x|^2}{\varepsilon} = \sup\left\{ f(y) - \frac{\varphi(x-y)}{\varepsilon} \;\middle|\; y \in \Omega \text{ with } \varphi(y-x') \leq r \right\} + \frac{|x|^2}{\varepsilon}$$

$$= \sup\left\{ f(y) - \frac{|x-y|^2}{\varepsilon} \;\middle|\; y \in \Omega \text{ with } \varphi(y-x') \leq r \right\} + \frac{|x|^2}{\varepsilon}$$

$$= \sup\left\{ f(y) - \frac{|x-y|^2 - |x|^2}{\varepsilon} \;\middle|\; y \in \Omega \text{ with } \varphi(y-x') \leq r \right\}$$

for all $x \in N_{x'}$ and $0 < \varepsilon \leq \varepsilon_{x'}$. Finally, since

$$x \longmapsto f(y) - \frac{|x-y|^2 - |x|^2}{\varepsilon} = \frac{2y^T x}{\varepsilon} + \left(f(y) - \frac{|y|^2}{\varepsilon} \right)$$

is affine linear (and thus convex) for all $y \in \Omega$ with $\varphi(y-x') \leq r$, Lemma A.1 implies that

$$x \longmapsto \Delta_\varphi^\varepsilon[f](x) + \frac{|x|^2}{\varepsilon} = \sup\left\{ f(y) - \frac{|x-y|^2 - |x|^2}{\varepsilon} \;\middle|\; y \in \Omega \text{ with } \varphi(y-x') \leq r \right\}$$

is convex in $N_{x'} \subset \mathbb{R}^d$ for all $0 < \varepsilon \leq \varepsilon_{x'}$, as required. $\square$

Due to the duality $\Delta_\varphi^\varepsilon[-f] = -\nabla_\varphi^\varepsilon[f]$ between supremal and infimal convolutions, the preceding lemma immediately implies an approximation result for lower semicontinuous functions by infimal convolutions. However, in order to avoid redundancy, we restrict our attention in this section to approximation results for supremal convolutions.

Let us quickly highlight the significance of property (iii) from Lemma 1.12: It shows that the supremal convolution $\Delta_\varphi^\varepsilon[f]$ is (at least locally) semiconvex for small $\varepsilon > 0$ (cf. Definition A.5) if the quasinorm $\varphi(\cdot)$ equals the squared norm $|\cdot|^2$ in a neighborhood of zero. Alexandrov's theorem from convex analysis (cf. Theorem A.7) therefore implies that the supremal convolution $\Delta_\varphi^\varepsilon[f]$ is (locally) twice differentiable almost everywhere for small $\varepsilon > 0$. This idea motivates the importance of our construction for the viscosity solution theory of second-order equations. Furthermore, it suggests that we have to work with quasinorms, which not only dominate the growth of the approximated functions, but which also behave like the squared norm in a neighborhood of zero. For simplicity, we restrict our attention to the class of functions with polynomial growth:

1.2. Supremal Convolutions

Definition 1.13 (Polynomial Growth). A real-valued function $f : X \to \mathbb{R}$ on a normed space $(X, \|\cdot\|)$ is said to have (bounded) *p-polynomial growth* with order $p \geq 0$ if

$$|f(x)| \leq C(1 + \|x\|^p)$$

for some constant $C > 0$ and all $x \in X$. If $\mathcal{F}$ is a family of real-valued functions on X, then $\mathcal{F}_p$ denotes all functions in $\mathcal{F}$ with bounded p-polynomial growth. Moreover,

$$\|f\|_p := \sup\left\{ \frac{|f(x)|}{1 + \|x\|^p} \,\bigg|\, x \in X \right\} < \infty$$

denotes the *optimal growth constant* for $f \in \mathcal{F}_p$ and $p \geq 0$. In particular,

$$|f(x)| \leq \|f\|_p (1 + \|x\|^p)$$

for all $x \in X$ and $f \in \mathcal{F}_p$ with $p \geq 0$.

According to our preceding findings, we have to work with quasidistances $\varphi(\cdot)$ that equal $|\cdot|^2$ around zero and grow faster than $|\cdot|^p$ at infinity, in order to approximate functions with p-polynomial growth (with $p \geq 0$) by supremal convolutions, which are (locally) twice differentiable almost everywhere. A canonical choice for $\varphi(\cdot)$ therefore is

$$x \longmapsto |x|^2 \vee |x|^{(p \vee 2)} = \begin{cases} |x|^2 & \text{if } |x| \leq 1 \\ |x|^{(p \vee 2)} & \text{if } |x| > 1 \end{cases},$$

which is a quasidistances according to Lemma 1.10. For technical reasons, we will work with a smooth variant of this quasidistance, which we will construct in the following lemma. Note that it can be shown for all $p \geq 0$ (cf. Lemma 2.8) that the supremal convolution with respect to the quasinorm

$$x \longmapsto |x|^2 \vee |x|^{(p \vee 2)}$$

(or its smooth variant) of a function with p-polynomial growth has p-polynomial growth, which is important for the viscosity solution theory in Chapter 2.

Lemma 1.14 (Quasidistances with Polynomial Growth). *Suppose that $\eta : \mathbb{R} \to \mathbb{R}$ is the one-dimensional standard mollifier from Lemma 1.1. Define $\phi : \mathbb{R} \to [0,1]$ by*

$$\phi(r) := \left(\mathbf{1}_{B(0,5/2)} * \eta \right)(r) = \int_{\mathbb{R}} \mathbf{1}_{B(0,5/2)}(r-s)\eta(s)\,ds = \int_{-1}^{1} \mathbf{1}_{B(r,5/2)}(s)\eta(s)\,ds$$

for all $r \in \mathbb{R}$ and $\chi : \mathbb{R}^d \to [0,1]$ by $\chi(x) := \phi(|x|^2)$ for all $x \in \mathbb{R}^d$. If $p \geq 2$, then

$$\varphi_p(x) := \chi(x)|x|^2 + (1 - \chi(x))|x|^p$$

for all $x \in \mathbb{R}^d$ defines a quasidistance $\varphi_p \in C^\infty(\mathbb{R}^d)$ with multiplier $\rho = 2^{2p-2}$ and

$$2^{2-p}\left(|x|^2 \vee |x|^p\right) \leq \varphi_p(x) \leq \left(|x|^2 \vee |x|^p\right)$$
$$2^{3-p}\left(|x| \vee |x|^{p-1}\right) \leq |D\varphi_p(x)| \leq 10p\left(|x| \vee |x|^{p-1}\right)$$

for all $x \in \mathbb{R}^d$.

33

1. Approximation Theory

Proof. First of all, Theorem 1.2 shows that $\phi \in C^\infty(\mathbb{R})$. Moreover, it is easy to see (cf. the proof of Lemma 1.6) that $\mathbf{1}_{B[0,1]}(r) \leq \phi(r) \leq \mathbf{1}_{B(0,4)}(r)$ for all $r \in \mathbb{R}$, and hence

$$\mathbf{1}_{B[0,1]}(x) \leq \chi(x) \leq \mathbf{1}_{B(0,2)}(x) \tag{1.7}$$

for all $x \in \mathbb{R}^d$. Since $|\cdot|^2$ is smooth everywhere and $|\cdot|^p$ is smooth away from the origin, this implies $\varphi_p \in C^\infty(\mathbb{R}^d)$.

As a next step, we will prove the bounds for $\varphi_p : \mathbb{R}^d \to \mathbb{R}$: For $x \in \mathbb{R}^d$ with either $|x| \leq 1$ or $|x| \geq 2$, Equation (1.7) implies $\chi(x) = 1$ or $\chi(x) = 0$. Therefore, we obtain

$$\varphi_p(x) = \begin{cases} |x|^2 = (|x|^2 \vee |x|^p) & \text{if } |x| \leq 1 \\ |x|^p = (|x|^2 \vee |x|^p) & \text{if } |x| \geq 2, \end{cases} \tag{1.8}$$

using the definition of $\varphi_p : \mathbb{R}^d \to \mathbb{R}$ and $p \geq 2$. For $x \in \mathbb{R}^d$ with $1 < |x| < 2$, $p \geq 2$ implies

$$2^{2-p}|x|^p \leq |x|^{2-p} \cdot |x|^p = |x|^2$$

and $2^{2-p}|x|^p \leq |x|^p$. Therefore, $|x|^2 \leq (|x|^2 \vee |x|^p)$ and $|x|^p \leq (|x|^2 \vee |x|^p)$ lead to

$$\begin{aligned} 2^{2-p}\left(|x|^2 \vee |x|^p\right) = 2^{2-p}|x|^p &= \chi(x)\left(2^{2-p}|x|^p\right) + \left(1-\chi(x)\right)\left(2^{2-p}|x|^p\right) \\ &\leq \chi(x)\left(|x|^2\right) + \left(1-\chi(x)\right)\left(|x|^p\right) = \varphi_p(x) \\ &\leq \chi(x)\left(|x|^2 \vee |x|^p\right) + \left(1-\chi(x)\right)\left(|x|^2 \vee |x|^p\right) \\ &= \left(|x|^2 \vee |x|^p\right) \end{aligned}$$

for all $x \in \mathbb{R}^d$ with $1 < |x| < 2$. In combination with Equation (1.8), this shows

$$2^{2-p}\left(|x|^2 \vee |x|^p\right) \leq \varphi_p(x) \leq \left(|x|^2 \vee |x|^p\right) \tag{1.9}$$

for all $x \in \mathbb{R}^d$, using $2^{2-p} \leq 1$ for $p \geq 2$. Since

$$x \longmapsto \left(|x|^2 \vee |x|^p\right)$$

is a quasidistance with multiplier $2^p \vee 2^2 = 2^p$ for $p \geq 2$ according to Lemma 1.10, Equation (1.9) implies that $\varphi_p : \mathbb{R}^d \to \mathbb{R}$ for $p \geq 2$ is definite and satisfies a generalized triangle inequality. In fact, the generalized triangle inequality for $|\cdot|^2 \vee |\cdot|^p$ shows

$$\begin{aligned} \varphi_p(x+y) \leq \left(|x+y|^2 \vee |x+y|^p\right) &\leq 2^p\left(\left(|x|^2 \vee |x|^p\right) + \left(|y|^2 \vee |y|^p\right)\right) \\ &\leq 2^p\, 2^{p-2}\left(\varphi_p(x) + \varphi_p(y)\right) \end{aligned}$$

for all $x, y \in \mathbb{R}^d$, which implies that $\rho = 2^p\, 2^{p-2} = 2^{2p-2}$ is a multiplier of $\varphi_p : \mathbb{R}^d \to \mathbb{R}$. Finally, the symmetry of $\varphi_p : \mathbb{R}^d \to \mathbb{R}$ follows from the symmetry of $\chi : \mathbb{R}^d \to \mathbb{R}$.

1.2. Supremal Convolutions

At last, we will prove the bounds for $D\varphi_p : \mathbb{R}^d \to \mathbb{R}^d$: It is easy to check that

$$D\varphi_p(x) = \chi(x)\Big(2x\Big) + \Big(1 - \chi(x)\Big)\Big(p|x|^{p-2}x\Big) + \Big(|x|^2 - |x|^p\Big)D\chi(x) \quad (1.10)$$
$$= \Big(\chi(x)2 + \Big(1 - \chi(x)\Big)\Big(p|x|^{p-2}\Big) + 2\Big(|x|^2 - |x|^p\Big)D\phi(|x|^2)\Big)x$$

for all $x \in \mathbb{R}^d$. Since $D\phi(|x|^2) = 0$ for all $x \in \mathbb{R}^d$ with $|x| \le 1$ or $|x| \ge 2$, we obtain

$$\big|D\varphi_p(x)\big| = \big|\chi(x)2 + \big(1 - \chi(x)\big)\big(p|x|^{p-2}\big)\big| \cdot |x| = \chi(x)\big(2|x|\big) + \big(1 - \chi(x)\big)\big(p|x|^{p-1}\big)$$

for all $x \in \mathbb{R}^d$ with $|x| \le 1$ or $|x| \ge 2$. Analogously to Equation (1.9), this leads to

$$2\,\big(|x| \vee |x|^{p-1}\big) \le |D\varphi_p(x)| \le p\,\big(|x| \vee |x|^{p-1}\big) \quad (1.11)$$

for all $x \in \mathbb{R}^d$ with $|x| \le 1$ or $|x| \ge 2$. For $x \in \mathbb{R}^d$ with $1 < |x| < 2$, we will first obtain bounds for $D\phi(|x|^2)$: It is easy to check that

$$\int_{-1}^{1} \exp\left(\frac{1}{r^2 - 1}\right) dr \ge \int_{-1/\sqrt{2}}^{1/\sqrt{2}} \exp\left(\frac{1}{r^2 - 1}\right) dr \ge \int_{-1/\sqrt{2}}^{1/\sqrt{2}} \exp(-2)\, dr = \sqrt{2}e^{-2},$$

which implies $\eta(0) \le \big(\sqrt{2}e^{-2}\big)^{-1} e^{-1} = 2^{-1/2}e$. Therefore, we find

$$|D\phi(r)| = \left|\int_{-1}^{1} \mathbf{1}_{B(r,5/2)}(s) D\eta(s)\, ds\right| \le \int_{-1}^{0} D\eta(s)\, ds = \eta(0) \le 2^{-1/2}e \quad (1.12)$$

for all $r \in \mathbb{R}$, since $D\eta : \mathbb{R} \to \mathbb{R}$ is (as the derivative of an even function) an odd function and since $D\eta(r) \le 0$ for $r \ge 0$ (due to the monotonicity of $\eta : \mathbb{R} \to \mathbb{R}$). Similarly, $D\eta(r) \le 0$ for $r \ge 0$ implies $D\phi(r) \le 0$ for all $r \ge 0$. Hence, Equation (1.10) yields

$$|D\varphi_p(x)| = \Big(\chi(x)2 + \big(1 - \chi(x)\big)\big(p|x|^{p-2}\big) + 2\big(|x|^p - |x|^2\big)\big(-D\phi(|x|^2)\big)\Big)|x|$$
$$= \chi(x)\big(2|x|\big) + \big(1 - \chi(x)\big)\big(p|x|^{p-1}\big) + 2\big(|x|^2 - |x|^{4-p}\big)|D\phi(|x|^2)| \cdot |x|^{p-1}$$

for all $x \in \mathbb{R}^d$ with $1 < |x| < 2$. Due to Equation (1.12) and $2^{2-p}|x|^{p-1} \le |x|$ for $p \ge 2$ and $|x| > 1$, this leads to

$$2^{3-p}|x|^{p-1}$$
$$= \chi(x)\big(2 \cdot 2^{2-p}|x|^{p-1}\big) + \big(1 - \chi(x)\big)\big(2 \cdot 2^{2-p}|x|^{p-1}\big)$$
$$\le \chi(x)\big(2|x|\big) + \big(1 - \chi(x)\big)\big(p|x|^{p-1}\big) + 2\big(|x|^2 - |x|^{4-p}\big)|D\phi(|x|^2)| \cdot |x|^{p-1} = |D\varphi_p(x)|$$
$$\le \chi(x)\big(p|x|^{p-1}\big) + \big(1 - \chi(x)\big)\big(p|x|^{p-1}\big) + 2 \cdot 2^2 \cdot 2^{-1/2}e \cdot |x|^{p-1}$$
$$\le p(1 + 3 \cdot 3)|x|^{p-1}$$

for all $x \in \mathbb{R}^d$ with $1 < |x| < 2$. Finally, a combination with Equation (1.11) shows

$$2^{2-p} 2\,\big(|x| \vee |x|^{p-1}\big) \le |D\varphi_p(x)| \le 10p\,\big(|x| \vee |x|^{p-1}\big)$$

for all $x \in \mathbb{R}^d$, as required. $\square$

1. Approximation Theory

As a next step, we will obtain a rate of convergence for the approximation of fairly regular functions by supremal and infimal convolutions. Since approximations generated by supremal or infimal convolutions converge pointwise monotonically from above or below according to Lemma 1.12, Dini's theorem (cf. Theorem 1.27) already implies that this convergence is locally uniform for continuous functions. However, for some of the arguments in Chapter 2, it is important to know the exact rate of convergence for functions with higher regularity.

Lemma 1.15 (Rate of Convergence). *Suppose that $\Omega \subset \mathbb{R}^d$ is open, that for $p \geq 2$ $\varphi_p \in C^\infty(\mathbb{R}^d)$ is the smooth quasidistance with p-polynomial growth from Lemma 1.14, and that $f \in C(\Omega)$ with $C \geq 1$ such that*

$$|f(x)| \leq C(1 + \varphi_p(x))$$
$$|f(x) - f(y)| \leq C\big(1 + \sup_{t \in [0,1]} |D\varphi_p(tx + (1-t)y)|\big)|x - y|$$

for all $x, y \in \Omega$. For every $0 < \varepsilon \leq (2^{3p+1}10C)^{-1}$, the inequalities

$$0 \leq \Delta^\varepsilon[f](x) - f(x) \leq \varepsilon^{1/p} \cdot 2^{3(p+1)} pC^{3/2} \cdot (1 + \varphi_p(x))$$
$$-\varepsilon^{1/p} \cdot 2^{3(p+1)} pC^{3/2} \cdot (1 + \varphi_p(x)) \leq \nabla^\varepsilon[f](x) - f(x) \leq 0$$

hold for all $x \in \Omega$ with $\inf_{y \in \mathbb{R}^d \setminus \Omega} \varphi_p(x,y) > \varepsilon 2^{2p+1} C(1 + \varphi_p(x))$.

Proof. Due to the duality between the infimal and supremal convolution

$$\nabla^\varepsilon_{\varphi_p}[f] = -\Delta^\varepsilon_{\varphi_p}[-f],$$

it suffices to prove the inequality for $\Delta^\varepsilon_{\varphi_p}[f]$. To shorten the notation, define

$$\psi_p(x) := |x|^2 \vee |x|^p$$
$$\vartheta_p(x) := |x| \vee |x|^{p-1}$$
(1.13)

for $x \in \mathbb{R}^d$. According to Lemma 1.14, 2^{2p-2} is a multiplier of $\varphi_p : \mathbb{R}^d \to [0, \infty)$ and

$$2^{2-p} \psi_p(x) \leq \varphi_p(x) \leq \psi_p(x)$$
$$2^{3-p} \vartheta_p(x) \leq |D\varphi_p(x)| \leq 10p\, \vartheta_p(x)$$

for all $x \in \mathbb{R}^d$. Moreover, Lemma 1.10 implies that $\psi_p : \mathbb{R}^d \to \mathbb{R}$ and $\vartheta_p : \mathbb{R}^d \to \mathbb{R}$ are quasidistances with multiplier 2^p and 2^{p-1}, respectively.

Suppose that $0 < \varepsilon \leq (2^{2p+1}10C)^{-1}$ and that $x \in \Omega$ with

$$\inf_{y \in \mathbb{R}^d \setminus \Omega} \varphi_p(x, y) > \varepsilon 2^{2p+1} C\big(1 + \varphi_p(x)\big).$$

A simple calculation with $\rho := 2^p$ shows $0 < 2^{p-2}\varepsilon \leq (2^{2-p}2^{3p+1}10C)^{-1} \leq (2\rho C)^{-1}$ and

$$\inf_{y \in \mathbb{R}^d \setminus \Omega} \psi_p(x,y) \geq \inf_{y \in \mathbb{R}^d \setminus \Omega} \varphi_p(x,y) > \varepsilon 2^{2p+1} C\big(1 + \varphi_p(x)\big) \geq \varepsilon 8\rho C\big(1 + \psi_p(x)\big).$$

36

Since ρ is a multiplier of $\psi_p : \mathbb{R}^d \to \mathbb{R}$, Lemma 1.12 implies the existence of $\bar{x} \in \Omega$ with

$$\psi_p(x - \bar{x}) \leq \left(2^{p-2}\varepsilon\right)\left(8\rho C(1 + \psi_p(x))\right) = \varepsilon 2^{2p+1} C\left(1 + \psi_p(x)\right) \tag{1.14}$$

such that $\Delta^{2^{p-2}\varepsilon}_{\psi_p}[f](x) = f(\bar{x}) - (2^{p-2}\varepsilon)^{-1} \psi_p(x - \bar{x})$ holds. The assumption on the growth of the Lipschitz constant shows that

$$f(\bar{x}) - f(x) \leq C\left(1 + \sup_{t \in [0,1]} |D\varphi_p(tx + (1-t)\bar{x})|\right) |x - \bar{x}|$$

$$\leq 2^{p-1} 10pC\, |x - \bar{x}|\left(1 + \vartheta_p(x) + \vartheta_p(x - \bar{x})\right) \tag{1.15}$$

$$= 2^{p-1} 10pC\, |x - \bar{x}|\left(1 + \vartheta_p(x)\right) + 2^{p-1} 10pC \psi_p(x - \bar{x}),$$

using $|x - \bar{x}| \cdot \vartheta_p(x - \bar{x}) = \psi_p(x - \bar{x})$ and the generalized triangle inequality of $\vartheta_p : \mathbb{R}^d \to \mathbb{R}$ with multiplier 2^{p-1}. Moreover, the definition of the supremal convolution leads to

$$\Delta^\varepsilon_{\varphi_p}[f](x) = \sup_{y \in \Omega}\left(f(y) - \varepsilon^{-1} \varphi_p(x - y)\right)$$

$$\leq \sup_{y \in \Omega}\left(f(y) - \varepsilon^{-1} 2^{2-p} \psi_p(x - y)\right) = \Delta^{2^{p-2}\varepsilon}_{\psi_p}[f](x). \tag{1.16}$$

Therefore, the identity $\Delta^{2^{p-2}\varepsilon}_{\psi_p}[f](x) = f(\bar{x}) - (2^{p-2}\varepsilon)^{-1} \psi_p(x - \bar{x})$ from Equation (1.14), Equation (1.15) and Equation (1.16), and $\varepsilon \leq (2^{3p+1} 10C)^{-1} \leq (2^{2p-3} 10pC)^{-1}$ yield

$$0 \leq \Delta^\varepsilon_{\varphi_p}[f](x) - f(x)$$

$$\leq f(\bar{x}) - f(x) - \left(2^{p-2}\varepsilon\right)^{-1} \psi_p(x - \bar{x})$$

$$\leq 2^{p-1} 10pC\, |x - \bar{x}|\left(1 + \vartheta_p(x)\right) + \left(2^{p-1} 10pC - 2^{2-p}\varepsilon^{-1}\right) \psi_p(x - \bar{x}) \tag{1.17}$$

$$\leq 2^{p-1} 10pC\, |x - \bar{x}|\left(1 + \vartheta_p(x)\right).$$

Since $\varepsilon \leq (2^{3p+1} 10C)^{-1} \leq 1$ and $C \geq 1$, we obtain

$$|x - \bar{x}| = \psi_p(x - \bar{x})^{1/2} \wedge \psi_p(x - \bar{x})^{1/p}$$

$$\leq \left(\varepsilon 2^{2p+1} C(1 + \psi_p(x))\right)^{1/2} \wedge \left(\varepsilon 2^{2p+1} C(1 + \psi_p(x))\right)^{1/p}$$

$$\leq \varepsilon^{1/p} 2^{p+1} C^{1/2}\left(1 + \left(\psi_p(x)^{1/2} \wedge \psi_p(x)^{1/p}\right)\right) \tag{1.18}$$

$$= \varepsilon^{1/p} 2^{p+1} C^{1/2}\left(1 + |x|\right),$$

using the identity $|y| = \psi_p(y)^{1/2} \wedge \psi_p(y)^{1/p}$ for all $y \in \mathbb{R}^d$ and Equation (1.14). Moreover, the definitions in Equation (1.13) imply

$$(1 + |x|)(1 + \vartheta_p(x)) = 1 + |x| + \vartheta_p(x) + |x|\vartheta_p(x)$$

$$= \begin{cases} 1 + |x| + |x| + |x|^2 \leq 3(1 + |x|^2) & \text{if } |x| \leq 1 \\ 1 + |x| + |x|^{p-1} + |x|^p \leq 3(1 + |x|^p) & \text{if } |x| \geq 1 \end{cases} \tag{1.19}$$

$$\leq 3(1 + \psi_p(x)) \leq 2^{p-2} 3(1 + \varphi_p(x)).$$

1. Approximation Theory

Finally, a combination of Equation (1.17), Equation (1.18) and Equation (1.19) leads to

$$0 \leq \Delta^{\varepsilon}_{\varphi_p}[f] - f(x) \leq \varepsilon^{1/p} 2^{2p} 10 p C^{3/2} \big(1 + |x|\big)\big(1 + \vartheta_p(x)\big)$$
$$\leq \varepsilon^{1/p} 2^{3p-2} 30 p C^{3/2} \big(1 + \varphi_p(x)\big) \leq \varepsilon^{1/p} 2^{3(p+1)} p C^{3/2} \big(1 + \varphi_p(x)\big),$$

as required. □

As remarked earlier, semicontinuous functions occur naturally in the context of nonlinear analysis, since pointwise suprema and infima of continuous functions are in general only semicontinuous. Based on a result from René Baire in [6], Felix Hausdorff showed in [61] that every semicontinuous function is a pointwise supremum or infimum of continuous functions. In the following lemma, we combine our preceding findings to prove that every semicontinuous function is also a monotone limit of smooth functions.

Lemma 1.16 (Approximation of Semicontinuous Functions). *If $\Omega \subset \mathbb{R}^d$ is open and $f \in USC(\Omega)$, then there exists a sequence $(f_n)_{n \in \mathbb{N}} \subset C^{\infty}(\Omega)$ such that*

$$\lim_{n \to \infty} f_n(x) = f(x)$$

for all $x \in \Omega$ and $f_{n+1}(x) \leq f_n(x)$ for all $x \in \mathbb{R}^d$ and $n \in \mathbb{N}$.

Proof. First, suppose that there exist $a, b \in \mathbb{R}$ such that $a < f(x) < b$ for all $x \in \Omega$: Lemma 1.12 implies the existence of $(h_n)_{n \in \mathbb{N}} \subset C(\mathbb{R}^d)$ and $(\varepsilon_n)_{n \in \mathbb{N}} \subset (0, \infty)$ with

$$h_n(x) = \Delta^{\varepsilon_n}_{|\cdot|^2}[f](x) = \sup_{y \in \Omega} \Big(f(y) - \varepsilon_n^{-1}|x-y|^2\Big)$$

for all $x \in \mathbb{R}^d$ and $n \in \mathbb{N}$, such that $\varepsilon_n \downarrow 0$ and $h_n(x) \downarrow f(x)$ for all $x \in \Omega$ as $n \to \infty$. It is easy to see that

$$a < f(x) \leq h_n(x) = \Delta^{\varepsilon_n}_{|\cdot|^2}[f](x) = \sup_{y \in \Omega} \Big(f(y) - \varepsilon_n^{-1}|x-y|^2\Big) \leq b$$

for all $x \in \Omega$, which implies $a < h_n(x) \leq b$ for all $x \in \Omega$. We will now modify $h_n : \mathbb{R}^d \to \mathbb{R}$ in such a way, that it stays away from the upper boundary: Since

$$a - b \leq h_{n+k}(x) - h_{n+(k-1)}(x) \leq 0$$

for all $x \in \Omega$ and $n, k \in \mathbb{N}$, the following series

$$g_n(x) := h_n(x) + \sum_{k=1}^{\infty} k^{-2} \big(h_{n+k}(x) - h_{n+(k-1)}(x)\big)$$

converges uniformly in $x \in \Omega$ and therefore defines a continuous function $g_n \in \Omega \to \mathbb{R}$.

A simple calculation (with the local convention $0^{-2} := 1$) shows

$$f(x) = \lim_{k \to \infty} h_{n+k}(x) = h_n(x) + \sum_{k=1}^{\infty} \left(h_{n+k}(x) - h_{n+(k-1)}(x) \right)$$

$$\leq h_n(x) + \sum_{k=1}^{\infty} (k-1)^{-2} \left(h_{n+k}(x) - h_{n+(k-1)}(x) \right) = g_{n+1}(x)$$

$$\leq h_n(x) + \sum_{k=1}^{\infty} k^{-2} \left(h_{n+k}(x) - h_{n+(k-1)}(x) \right) = g_n(x)$$

$$\leq h_n(x)$$

for all $x \in \Omega$. In particular, $a < f(x) \leq g_{n+1}(x) \leq g_n(x) \leq h_n(x) \leq b$ for all $x \in \Omega$. Moreover, we find $g_n(x) < b$ for all $x \in \Omega$, because we either have $h_n(x) = f(x)$ (which implies $h_{n+k}(x) = h_n(x)$ for all $k \in \mathbb{N}$ and hence $g_n(x) = f(x) < b$) or $h_n(x) > f(x)$ (which implies $h_{n+k}(x) < h_n(x)$ for some $k \in \mathbb{N}$ and thus $g_n(x) < h_n(x) \leq b$). Altogether, we have found a sequence $(g_n)_{n \in \mathbb{N}} \subset C(\Omega)$ with

$$a < g_n(x) < b$$

for all $x \in \Omega$ such that $g_n(x) \downarrow f(x)$ for all $x \in \Omega$ as $n \to \infty$.

Next, suppose that $f \in \text{USC}(\Omega)$ is an unbounded upper semicontinuous function: It is easy to see that the continuous map $\phi : \mathbb{R} \to (-1, 1)$ with

$$\phi(x) := \frac{x}{1 + |x|}$$

is strictly increasing, since $\phi'(x) = (1 + |x|)^{-2} > 0$ for all $x \neq 0$ and

$$\phi(-x) < \phi(0) = 0 < \phi(x)$$

for all $x > 0$. Moreover, a simple calculation shows that

$$\phi^{-1}(x) = \frac{x}{1 - |x|}$$

for all $x \in (-1, 1)$, which implies that $\phi : \mathbb{R} \to (-1, 1)$ is a homeomorphism from $\mathbb{R}$ onto the open interval $(-1, 1)$. Since $\phi \circ f \in \text{USC}(\Omega)$ and

$$-1 < (\phi \circ f)(x) = \phi(f(x)) < 1$$

for all $x \in \Omega$, the first part of the proof implies the existence of $(h_n)_{n \in \mathbb{N}} \subset C(\Omega)$ with

$$-1 < h_n(x) < 1$$

for all $x \in \Omega$ such that $h_n(x) \downarrow (\phi \circ f)(x)$ for all $x \in \Omega$ as $n \to \infty$. This implies

$$g_n(x) := (\phi^{-1} \circ h_n)(x) \downarrow (\phi^{-1} \circ \phi \circ f)(x) = f(x)$$

for all $x \in \Omega$ as $n \to \infty$, and $(g_n)_{n \in \mathbb{N}} \subset C(\Omega)$.

1. Approximation Theory

Finally, Theorem 1.4 implies the existence of $\hat{g}_n \in C^\infty(\Omega)$ for every $n \in \mathbb{N}$ such that
$$\sup_{x \in \mathbb{R}^d} |\hat{g}_n(x) - g_n(x)| \leq 2^{-(2n+2)}.$$

Furthermore, the definition $f_n(x) := \hat{g}_n(x) + 2^{-(2n+1)}$ for $x \in \Omega$ leads to $f_n \in C^\infty(\Omega)$ and
$$g_n(x) \leq g_n(x) + 2^{-(2n+2)} = \left(g_n(x) - 2^{-(2n+2)}\right) + 2^{-(2n+1)}$$
$$\leq f_n(x)$$
$$\leq \left(g_n(x) + 2^{-(2n+2)}\right) + 2^{-(2n+1)} \leq g_n(x) + 2^{-2n}$$

for all $x \in \Omega$ and $n \in \mathbb{N}$. Therefore, $\lim_{n \to \infty} f_n(x) = f(x)$ for all $x \in \Omega$ and
$$f_{n+1}(x) \leq g_{n+1}(x) + 2^{-(2n+2)} \leq g_n(x) + 2^{-(2n+2)} \leq f_n(x)$$

for all $x \in \Omega$ and $n \in \mathbb{N}$, as required. $\square$

The rest of this section contains approximation results, which are tailored to our needs in Chapter 2 for developing a consistent viscosity solution theory: As mentioned earlier, one main concept in viscosity solution theory is to scan semicontinuous $u : \Omega \to \mathbb{R}$ (with some open $\Omega \subset \mathbb{R}^d$) by smooth functions $f : \Omega \to \mathbb{R}$ (which we will refer to as *scanning functions*) from above and below. The following two results show that, if our original function $u : \Omega \to \mathbb{R}$ has p-polynomial growth, then we can assume, without loss of generality, that our scanning functions $f : \Omega \to \mathbb{R}$ have p-polynomial growths as well.

Lemma 1.17 (Smooth Functions with Polynomial Growth). *For every growth exponent $p \geq 0$ and $\varepsilon > 0$, there exists a smooth function $\psi \in C^\infty(\mathbb{R}^d)$ such that*
$$(1+\varepsilon)^{-1}\left(1+|x|^p\right) \leq \psi(x) \leq \left(1+|x|^p\right)$$

for all $x \in \mathbb{R}^d$.

Proof. First of all, note that $x \mapsto |x|^p$ is (as the composition of smooth functions) smooth in all $x \neq 0$. According to Lemma 1.6, there exists $\chi \in C^\infty(\mathbb{R}^d)$ with
$$0 \leq \chi(x) \leq 1$$

for all $x \in \mathbb{R}^d$ such that $\chi(x) = 1$ for $|x| \leq \varepsilon^{1/p}/2$ and $\chi(x) = 0$ for $|x| \geq \varepsilon^{1/p}$. Therefore,
$$\psi(x) := \chi(x) + \left(1 - \chi(x)\right)\left(1 + |x|^p\right)$$

with $x \in \Omega$ defines a smooth function $\psi \in C^\infty(\mathbb{R}^d)$. For $x \in \mathbb{R}^d$ with $|x| \geq \varepsilon^{1/p}$, we find
$$\psi(x) = \chi(x) + \left(1 - \chi(x)\right)\left(1 + |x|^p\right) = \left(1 + |x|^p\right).$$

For $|x| < \varepsilon^{1/p}$, we get $(1+|x|^p) \leq (1+\varepsilon)$ and hence
$$(1+\varepsilon)^{-1}\left(1+|x|^p\right) \leq \chi(x)(1+\varepsilon)^{-1}\left(1+|x|^p\right) + \left(1-\chi(x)\right)\left(1+|x|^p\right)$$
$$\leq \chi(x) + \left(1-\chi(x)\right)\left(1+|x|^p\right) = \psi(x)$$
$$\leq \chi(x)\left(1+|x|^p\right) + \left(1-\chi(x)\right)\left(1+|x|^p\right) = \left(1+|x|^p\right),$$

which shows the required inequality. $\square$

1.2. Supremal Convolutions

Lemma 1.18 (Tamed Scanning Functions). *Suppose that $u \in SC_p(\Omega)$ with $p \geq 0$, that $\Omega \subset \mathbb{R}^d$ is open, and that $f \in C^k(\Omega)$ with $k \in \mathbb{N}$ such that*

$$u(x) \leq f(x)$$

for all $x \in \Omega$. For every $C \geq \|u\|_p = \sup_{x \in \Omega}(1+|x|^p)^{-1}u(x)$, there exists $g \in C_p^k(\Omega)$ with

$$u(x) \leq g(x) \leq \min\left(f(x), 4C(1+|x|^p)\right)$$

for all $x \in \Omega$, and $g \equiv f$ on $\{x \in \Omega : f(x) \leq C(1+|x|^p)\}$.

Proof. Due to the choice of $C > 0$, the estimate $u(x) \leq C(1+|x|^p)$ holds for all $x \in \Omega$. Using $\psi \in C_p^\infty(\mathbb{R}^d)$ from Lemma 1.17 with $\varepsilon = 1$, the construction

$$V := \{x \in \Omega : f(x) \leq 2C\psi(x)\}$$
$$U := \{x \in \Omega : f(x) < 4C\psi(x)\}$$

implies that $V \subset U \subset \Omega \subset \mathbb{R}^d$ holds and that U and $\Omega \setminus V$ are open due to the continuity of $f : \mathbb{R}^d \to \mathbb{R}$ and $\psi : \mathbb{R}^d \to \mathbb{R}$. Hence, Proposition 1.7 yields $\chi \in C^\infty(\Omega)$ with $\chi(x) = 1$ for $x \in V$ and $\chi(x) = 0$ for $x \in \Omega \setminus U$. Define $g \in C^k(\Omega)$ by

$$g(x) := \chi(x) \cdot f(x) + \bigl(1 - \chi(x)\bigr) \cdot 2C\psi(x)$$

for $x \in \Omega$. It is easy to see that

$$\{x \in \Omega : f(x) \leq C(1+|x|^p)\} \subset V$$

using $2^{-1}(1+|x|^p) \leq \psi(x)$ for $x \in \Omega$. In particular, $g \equiv f$ on $\{x \in \Omega : f(x) \leq C(1+|x|^p)\}$ because $\chi(x) = 1$ for $x \in V$. Moreover, since $f(x) \geq u(x)$ and

$$2C\psi(x) \geq 2C \cdot 2^{-1}(1+|x|^p) = C(1+|x|^p) \geq u(x)$$

for all $x \in \Omega$ from the construction of $\psi : \mathbb{R}^d \to \mathbb{R}$, we obtain $g(x) \geq u(x)$ for all $x \in \Omega$.

For $x \in V$ we have $\chi(x) = 1$ and therefore $g(x) = f(x)$ by construction. For $x \in \Omega \setminus V$, on the other hand, we have $2C\psi(x) < f(x)$ and thus

$$g(x) \leq \chi(x) \cdot f(x) + \bigl(1 - \chi(x)\bigr) \cdot f(x) = f(x),$$

using the definition of $g : \mathbb{R}^d \to \mathbb{R}$. Altogether, we find $g(x) \leq f(x)$ for all $x \in \Omega$.

Analogously, we have $f(x) < 4C\psi(x)$ for all $x \in U$ and hence

$$g(x) \leq \chi(x) \cdot 4C\psi(x) + \bigl(1 - \chi(x)\bigr) \cdot 2C\psi(x) \leq 4C\psi(x) \leq 4C(1+|x|^p),$$

using the definition of $g : \mathbb{R}^d \to \mathbb{R}$ and properties of $\psi : \mathbb{R}^d \to \mathbb{R}$. On the other hand, we have $\chi(x) = 0$ for $x \in \Omega \setminus U$ so that $g(x) = 2C\psi(x) \leq 4C(1+|x|^p)$ by the properties of $\psi : \mathbb{R}^d \to \mathbb{R}$. Altogether, this shows $g(x) \leq 4C(1+|x|^p)$ for all $x \in \Omega$, as required. □

1. Approximation Theory

As a last result in this section, we show the validity of an argument (for our generalized set-up) that is frequently used in the viscosity solution theory of nonlocal equations: Suppose that an irregular function $u : \Omega \to \mathbb{R}$ can be scanned in a point $\overline{x} \in \Omega$, i.e. there exists a scanning function $f : \Omega \to \mathbb{R}$ such that

$$u(x) \leq f(x)$$

for all $x \in \Omega$ and $u(\overline{x}) = f(\overline{x})$. Many important results in viscosity solution theory rely on the existence of a sequence of scanning functions $(f_n)_{n \in \mathbb{N}}$, which are dominated by $f : \Omega \to \mathbb{R}$ (and hence $f_n(\overline{x}) = u(\overline{x})$ for all $n \in \mathbb{N}$), and which converge

$$\lim_{n \to \infty} f_n(x) = u(x)$$

monotonically for all $x \in \Omega$.

Lemma 1.19 (Approximation of Positive Part). *There exists a sequence $(\psi_n)_{n \in \mathbb{N}}$ of non-decreasing functions $\psi_n \in C^\infty(\mathbb{R})$ such that*

$$0 \leq \psi_n(x) \leq \psi_{n+1}(x) \leq x^+ = \max\{0, x\}$$

for all $x \in \mathbb{R}$ and $n \in \mathbb{N}$, and $\lim_{n \to \infty} \sup_{x \in \mathbb{R}} |\psi_n(x) - x^+| = 0$ for all $x \in \mathbb{R}$.

Proof. Suppose that $\eta \in C^\infty(\mathbb{R})$ is the standard mollifier from Lemma 1.1 and define

$$h_n(t) := \int_{-1}^{nt-1} \eta(s)\, ds$$

$$\psi_n(x) := \int_0^x h_n(t)\, dt = \int_0^x \int_{-1}^{nt-1} \eta(s)\, ds\, dt$$

for $x, t \in \mathbb{R}$ and $n \in \mathbb{N}$. The successive application of the fundamental theorem of calculus implies with $\eta \in C^\infty(\mathbb{R})$ that $h_n \in C^\infty(\mathbb{R})$ and $\psi_n \in C^\infty(\mathbb{R})$ for all $n \in \mathbb{N}$. Since $\eta \geq 0$, we find $h_{n+1} \geq h_n \geq 0$ and thus $\psi_{n+1} \geq \psi_n \geq 0$ for all $n \in \mathbb{N}$. Furthermore, the normalization $\int \eta(s)\, ds = 1$ implies that $h_n \leq 1$ and thus $\psi_n(x) \leq x^+$ for all $x \in \mathbb{R}$. Using Fubini's theorem as well as the normalization and symmetry of $\eta \in C^\infty(\mathbb{R})$, we find

$$\psi_n(x) = \int_0^x \int_{-1}^{nt-1} \eta(s)\, ds\, dt = \int_{-1}^1 \int_{\frac{s+1}{n}}^x \eta(s)\, dt\, ds$$

$$= \int_{-1}^1 (x - n^{-1}(s+1))\eta(s)\, ds$$

$$= (x - n^{-1}) \int_{-1}^1 \eta(s)\, ds - n^{-1} \int_{-1}^1 s\eta(y)\, ds$$

$$= x - n^{-1}$$

for all $x \geq 2n^{-2}$ and $n \in \mathbb{N}$. Consequently, we obtain

$$\sup_{x \in \mathbb{R}} |\psi_n(x) - x^+| \leq 2n^{-2} \longrightarrow 0$$

as $n \to \infty$, as required. $\square$

Theorem 1.20 (Approximation by Scanning Functions). *Suppose that $\Omega \subset \mathbb{R}^d$ is open, that $u \in USC_p(\Omega)$ with $p \geq 0$, and that $f \in C^k(\Omega)$ with $k \in \mathbb{N}$ such that*

$$u(x) \leq f(x)$$

for all $x \in \Omega$. There exists a sequence $(f_n)_{n \in \mathbb{N}} \subset C_p^k(\Omega)$ such that

$$u(x) \leq f_{n+1}(x) \leq f_n(x) \leq f(x)$$

for all $x \in \Omega$ and $n \in \mathbb{N}$, and $\lim_{n \to \infty} f_n(x) = u(x)$ for all $x \in \Omega$.

Proof. Due to Lemma 1.18, we can assume without loss of generality that $f \in C_p^k(\Omega)$. Due to Lemma 1.16, there exists a sequence $(u_n)_{n \in \mathbb{N}} \subset C^\infty(\Omega)$ such that

$$u_n(x) \downarrow u(x)$$

for all $x \in \Omega$ as $n \to \infty$. Together with $(\psi_n)_{n \in \mathbb{N}} \subset C^\infty(\mathbb{R})$ from Lemma 1.19,

$$f_n(x) := f(x) - \psi_n\Big(f(x) - u_n(x)\Big)$$

with $x \in \mathbb{R}^d$ defines a function $f_n \in C^k(\Omega)$ for $n \in \mathbb{N}$: The smoothness of $f_n : \Omega \to \mathbb{R}$ follows from the chain rule of calculus. Since $\psi_{n+1} : \mathbb{R} \to \mathbb{R}$ is non-decreasing, since

$$0 \leq \psi_n(y) \leq \psi_{n+1}(y) \leq y^+ = \max\{0, y\}$$

for all $y \in \mathbb{R}$, and since $u(x) \leq u_{n+1}(x) \leq u_n(x)$ for all $x \in \Omega$, we obtain

$$u(x) = f(x) - \Big(f(x) - u(x)\Big)^+ \leq f(x) - \psi_{n+1}\Big(f(x) - u\ (x)\Big)$$
$$\leq f(x) - \psi_{n+1}\Big(f(x) - u_{n+1}(x)\Big) = f_{n+1}(x)$$
$$\leq f(x) - \psi_{n+1}\Big(f(x) - u_n\ (x)\Big)$$
$$\leq f(x) - \psi_n\ \Big(f(x) - u_n\ (x)\Big) = f_n(x) \leq f(x)$$

for all $x \in \Omega$. In particular, $f_n : \mathbb{R}^d \to \mathbb{R}$ has polynomial growth, since $f_n(x) \leq f(x)$ for all $x \in \Omega$. Finally, we find

$$\Big|u(x) - f_n(x)\Big| = \Big|f(x) - \Big(f(x) - u(x)\Big)^+ - f_n(x)\Big|$$
$$\leq \Big|\Big(f(x) - u(x)\Big)^+ - \Big(f(x) - u_n(x)\Big)^+\Big| + \sup_{y \in \mathbb{R}} \Big|y^+ - \psi_n(y)\Big|$$
$$\leq \Big|u(x) - u_n(x)\Big| + \sup_{y \in \mathbb{R}} \Big|y^+ - \psi_n(y)\Big|$$

for all $x \in \Omega$, using the subadditivity of the positive part. Hence,

$$\lim_{n \to \infty} f_n(x) = u(x)$$

for all $x \in \Omega$, as required. $\square$

1. Approximation Theory

1.3. Monotone Functions

The following *little o-notation* is frequently used in the standard literature on viscosity solution theory: Suppose that $f : X \to \mathbb{R}$ and $g : X \to \mathbb{R}$ are real-valued functions on a metric space (X, d) such that $g \neq 0$ in a neighborhood of $x_0 \in X$. We denote

$$f(x) \leq o(g(x)) \text{ as } x \to x_0 :\iff \limsup_{\substack{x \to x_0 \\ x \neq 0}} \frac{f(x)}{g(x)} \leq 0$$

$$f(x) \geq o(g(x)) \text{ as } x \to x_0 :\iff \liminf_{\substack{x \to x_0 \\ x \neq 0}} \frac{f(x)}{g(x)} \geq 0$$

and $f(x) = o(g(x))$ as $x \to x_0$ if $\lim_{x \to x_0, x \neq x_0} f(x)\, (g(x))^{-1} = 0$. A simple calculation shows that $f(x) \leq o(g(x))$ as $x \to x_0$ is equivalent to the existence of a non-decreasing function $\phi : [0, \infty) \to [0, \infty)$ with

$$\lim_{r \to 0+} \left(\frac{\phi(r)}{\sup_{x \in B[x_0, r]} g(x)} \right) = 0$$

such that $\sup\{f(x) : x \in B[x_0, r]\} \leq \phi(r)$ for all $r > 0$. A standard argument in viscosity solution theory for local equations (typically in the context of sub- and superjets, as discussed later in this section) is that the monotone function $\phi : [0, \infty) \to [0, \infty)$ can be chosen to have more regularity under certain assumptions on $g : X \to \mathbb{R}$. In this section, we establish approximation results for monotone functions and show how to apply them to obtain appropriate generalizations of these arguments for our viscosity solution theory in Chapter 2. Due to the duality of the little o-notation

$$f(x) \leq o(g(x)) \text{ as } x \to x_0 \iff -f(x) \geq o(-g(x)) \text{ as } x \to x_0,$$

we can restrict our attention to the upper-bound case, without loss of generality.

In the beginning, let us introduce an iterative smoothing procedure for monotone functions, the idea of which is somehow hidden in the proof of [112, Theorem 6.11, p. 205]. Afterwards, we will demonstrate how to apply this procedure in order to obtain a general approximation result for monotone functions.

Lemma 1.21 (Smoothing of Monotone Functions). *Suppose* $f : [0, \infty) \to [0, \infty)$ *with* $\lim_{x \downarrow 0} f(x) = f(0) = 0$ *is non-decreasing, then* $g : [0, \infty) \to \mathbb{R}$ *with* $g(0) := 0$ *and*

$$g(x) := \frac{1}{x} \int_x^{2x} f(y)\, dy$$

for $x > 0$ *is non-decreasing and continuous. Furthermore, if* $f \in C^k([0, \infty))$ *such that*

$$\lim_{x \downarrow 0} x^{-1} f^{(k)}(x) = 0$$

for some $k \in \mathbb{N}_0 = \mathbb{N} \cup \{0\}$, *then* $g \in C^{k+1}([0, \infty))$ *with the one-sided derivatives*

$$g(0) = g'(0) = \ldots = g^{(k+1)}(0) = 0.$$

Proof. Using the definition of $g(x)$ for $x > 0$, it is easy to see that
$$f(x) \leq g(x) \leq f(2x)$$
for all $x \geq 0$, which implies the continuity of $g : [0, \infty) \to \mathbb{R}$ at zero. Moreover,
$$g(x) = \frac{1}{x}\left(\int_0^{2x} f(y)dy - \int_0^x f(y)\,dy\right) \qquad (1.20)$$
for all $x > 0$ implies the continuity of $g : [0, \infty) \to \mathbb{R}$ in $(0, \infty)$. Since the function $f : [0, \infty) \to [0, \infty)$ is non-decreasing, the one-sided limits
$$f(x\pm) = \lim_{h \to 0\pm} f(x+h)$$
exist for all $x > 0$. Hence, Equation (1.20) and $g(x) \leq f(2x-)$ by definition imply
$$\lim_{h \to 0\pm} \frac{g(x+h) - g(x)}{h} = \frac{2f(2x\pm) - f(x\pm) - g(x)}{x} \geq \frac{f(2x\pm) - f(x\pm)}{x} \geq 0 \qquad (1.21)$$
for all $x > 0$. In other words, $g : [0, \infty) \to \mathbb{R}$ has non-negative one-sided derivatives, which shows that $g : [0, \infty) \to \mathbb{R}$ is non-decreasing: Suppose that $g : [0, \infty) \to \mathbb{R}$ is not non-decreasing, i.e. there exit $x > y > 0$ such that
$$g(x) < g(y)$$
holds. It is easy to see that at least one of the following inequalities
$$\frac{g(2^{-1}(x+y)) - g(y)}{2^{-1}(x-y)} \leq \frac{g(x) - g(y)}{x-y}$$
$$\frac{g(x) - g(2^{-1}(x+y))}{2^{-1}(x-y)} \leq \frac{g(x) - g(y)}{x-y}$$
has to be satisfied. An iteration of this bisection idea leads to a non-increasing sequence $(x_n)_{n \in \mathbb{N}}$ and a non-decreasing sequence $(y_n)_{n \in \mathbb{N}}$ with $x \geq x_n > y_n \geq y$,
$$\frac{g(x_n) - g(y_n)}{x_n - y_n} \leq \frac{g(x) - g(y)}{x-y} \qquad (1.22)$$
and $x_n - y_n = 2^{-n}(x_n - y_n)$ for all $n \in \mathbb{N}$. As a consequence, either the left or the right one-sided derivative of $g : [0, \infty) \to \mathbb{R}$ in
$$\sup_{n \in \mathbb{N}} x_n = \inf_{n \in \mathbb{N}} y_n \in (x, y) \subset (0, \infty)$$
has to be smaller than $(x-y)^{-1}(g(x) - g(y)) < 0$ by Equation (1.22), in contradiction to Equation (1.21).

1. Approximation Theory

At last, suppose that $f \in C^k([0,\infty))$ and $\lim_{x\downarrow 0} x^{-1} f^{(k)}(x) = 0$ for some $k \in \mathbb{N} \cup \{0\}$. The fundamental theorem of calculus and the representation

$$g(x) = \frac{1}{x}\left(\int_0^{2x} f(y)\,dy - \int_0^x f(y)\,dy\right)$$

for $x > 0$ implies that $g : [0,\infty) \to \mathbb{R}$ is $(k+1)$-times continuously differentiable in $(0,\infty)$. Furthermore, the quotient rule combined with a simple mathematical induction shows

$$g^{(j)}(x) = \frac{2^j f^{(j-1)}(2x) - f^{(j-1)}(x) - jg^{(j-1)}(x)}{x}$$

for all $x > 0$ and $j \in \{1, \ldots, k+1\}$. The fundamental theorem of calculus implies that for all $j \in \{0, \ldots, k-1\}$ and each $x > 0$ the identity

$$f^{(j)}(x) = \int_0^x f^{(j+1)}(y)\,dy$$

and hence that $\lim_{x\downarrow 0} x^{-(k-j)-1} f^{(j)}(x) = 0$ for all $j \in \{0, \ldots, k\}$. Finally, we find

$$\lim_{x\downarrow 0} x^{-(k-j)-1} g^{(j)}(x) = \lim_{x\downarrow 0} x^{-(k-j)-2}\left(2^j f^{(j-1)}(2x) - f^{(j-1)}(x) - jg^{(j-1)}(x)\right) = 0$$

for all $j \in \{1, \ldots, k+1\}$ using mathematical induction and $f(x) \leq g(x) \leq f(2x)$ for all $x > 0$ from the construction of $g(x)$. In particular, the one-sided derivatives

$$g^{(j)}(0) = \lim_{x\downarrow 0} \frac{g^{(j-1)}(x) - g^{(j-1)}(0)}{x} = \lim_{x\downarrow 0} \frac{g^{(j-1)}(x)}{x} = 0$$

exist for all $j \in \{1, \ldots, k+1\}$ using mathematical induction and

$$\lim_{x\downarrow 0} g^{(j)}(x) = 0 = g^{(j)}(0)$$

from $\lim_{x\downarrow 0} x^{-(k-j)-1} g^{(j)}(x) = 0$ for all $j \in \{1, \ldots, k+1\}$, as required. $\square$

Proposition 1.22 (Approximation of Monotone Functions). *Suppose that the function $f : [0,\infty) \to [0,\infty)$ with $f(0) = 0$ is non-decreasing and satisfies*

$$\lim_{x\downarrow 0} x^{-k} f(x) = 0$$

for some $k \in \mathbb{N}_0 = \mathbb{N} \cup \{0\}$. There exists a non-decreasing function $g \in C^k([0,\infty))$ with

$$f(x) \leq g(x)$$

for $x \geq 0$, $\lim_{x\downarrow 0} x^{-k} g(x) = 0$ and one-sided derivatives $g(0) = g'(0) = \ldots = g^{(k)}(0) = 0$.

1.3. Monotone Functions

Proof. Define $f_{-1} := f$ and $f_n : [0, \infty) \to [0, \infty)$ for $n \in \mathbb{N} \cup \{0\}$ recursively by

$$f_n(x) := \frac{1}{x} \int_x^{2x} f_{n-1}(y) dy \qquad (1.23)$$

for $x > 0$ and $f_n(0) := 0$. According to Lemma 1.21, every $f_n : [0, \infty) \to [0, \infty)$ with $n \in \mathbb{N} \cup \{0\}$ is non-decreasing and continuous. We will show that $f_n \in C^n([0, \infty))$ with

$$f_n(0) = f_n'(0) = \ldots = f_n^{(n)}(0) = 0$$

for all $n \in \{0, \ldots, k\}$ by mathematical induction and then define $g := f_k$, in order to prove the result:

First of all, since $f_0 : [0, \infty) \to [0, \infty)$ is continuous, the induction hypothesis holds for $n = 0$. Now, suppose that

$$f_j \in C^j([0, \infty))$$

for all $j \in \{0, \ldots, n-1\}$ with $n \leq k$. We have to show that $f_n \in C^n([0, \infty))$: The fundamental theorem of calculus and $f_{n-1} \in C^{n-1}([0, \infty))$ imply that

$$x \longmapsto f_n(x) = \frac{1}{x} \int_x^{2x} f_{n-1}(y) dy$$

is n-times continuously differentiable in $(0, \infty)$. The quotient rule for differentiation implies (similar as in the proof of Lemma 1.21) that

$$f_n^{(j)}(x) = \frac{2^j f_{n-1}^{(j-1)}(2x) - f_{n-1}^{(j-1)}(x) - j f_n^{(j-1)}(x)}{x}$$

for all $x > 0$, $n \in \mathbb{N}$ and $j \in \{1, \ldots, n\}$. In particular, there exist finite constants

$$c_{i,j}^{(n)} \in \mathbb{R}$$

for all $i, j \in \mathbb{N}$ (independent of $x > 0$) such that

$$f_n^{(n-1)}(x) = \sum_{i,j=0}^{n-1} c_{i,j}^{(n)} \frac{f_j(2^i x)}{x^{n-1}}$$

for all $x > 0$. The construction of $f_j : [0, \infty) \to [0, \infty)$ from Equation (1.23) implies that

$$f(x) \leq f_0(x) \leq \ldots \leq f_j(x) \leq f_{j-1}(2x) \leq \ldots \leq f_0(2^{j+1}x) \leq f(2^{j+2}x)$$

for all $x > 0$. Hence, $\lim_{x \downarrow 0} x^{-1} f_n^{(n-1)}(x) = 0$ holds due to $n \leq k$ and the assumption on the growth of $f(x)$ as $x \downarrow 0$. Finally, the result follows from Lemma 1.21. □

1. Approximation Theory

The approximation of monotone functions from Proposition 1.22 is, in a way, optimal: Let $f(x) := x^{k+1/2}$ for $x \geq 0$ and suppose that there exists a function

$$g \in C^{k+1}([0, \infty))$$

with $f(x) \leq g(x)$ for all $x \geq 0$ and one-sided derivatives $g(0) = g'(0) = \ldots = g^{(k)}(0) = 0$. Taylor's theorem implies for every $x > 0$ the existence of $\xi \in (0, x)$ such that

$$x^{k+1/2} = f(x) \leq g(x) = \sum_{n=0}^{k} \frac{g^{(n)}(0)}{n!} \cdot x^n + \frac{g^{(k+1)}(\xi)}{(k+1)!} \cdot x^{k+1} = \frac{g^{(k+1)}(\xi)}{(k+1)!} \cdot x^{k+1}$$

holds. In particular, we find

$$\limsup_{\xi \to 0+} g^{(k+1)}(\xi) \geq \limsup_{x \to 0+} (k+1)! \cdot x^{-1/2} = \infty,$$

in contradiction to our assumption that $g^{(k+1)} \in C([0, \infty))$ is continuous at zero.

Let us recall how the approximation of monotone functions is typically applied in viscosity solution theory: One important notion for local equations is the *second-order superjet* (cf. [35, Chapter 2] for details) of upper semicontinuous functions $u : \Omega \to \mathbb{R}$ in $x_0 \in \Omega \subset \mathbb{R}^d$, which consists of all elements $(p, X) \in \mathbb{R}^d \times \mathbb{S}^{d \times d}$ such that

$$u(x) \leq u(x_0) + \langle p, x - x_0 \rangle + \tfrac{1}{2} \langle X(x - x_0), (x - x_0) \rangle + o(|x - x_0|^2) \quad (1.24)$$

as $x \to x_0$. It is frequently used in the viscosity solution theory of local equations that Equation (1.24) implies the existence of $f \in C^2(\Omega)$ with $u \leq f$ around $x_0 \in \Omega$ and

$$(f(x_0), Df(x_0), D^2 f(x_0)) = (u(x_0), p, X)$$

under the assumption that $\Omega \subset \mathbb{R}^d$ is open – take e.g.

$$f(x) := u(x_0) + \langle p, x - x_0 \rangle + \tfrac{1}{2} \langle X(x - x_0), (x - x_0) \rangle + \varepsilon |x - x_0|^2$$

for $x \in \Omega$ and small enough $\varepsilon > 0$. In the context of nonlocal equations, however, we need functions $f \in C^2(\mathbb{R}^d)$ such that $u \leq f$ globally (instead of locally) and

$$(f(x_0), Df(x_0), D^2 f(x_0)) = (u(x_0), p, X),$$

which is more involved and therefore proved in Theorem 1.23. In Corollary 1.24, we establish a refinement of this general result, which is tailored to our needs in Chapter 2.

Theorem 1.23 (Existence of Scanning Functions). *Suppose that $\Omega \subset \mathbb{R}^d$ is open, that $x_0 \in \Omega$ and $u \in USC(\Omega)$. If $p \in \mathbb{R}^d$ and $X \in \mathbb{S}^{d \times d}$ such that*

$$u(x) \leq u(x_0) + \langle p, x - x_0 \rangle + \tfrac{1}{2} \langle X(x - x_0), (x - x_0) \rangle + o(|x - x_0|^2)$$

as $x \to x_0$, then there exists $f \in C^2(\Omega)$ with $u(x) \leq f(x)$ for all $x \in \Omega$, $f(x_0) = u(x_0)$, $Df(x_0) = p$ and $D^2 f(x_0) = X$.

1.3. Monotone Functions

Proof. Since $u : \Omega \to \mathbb{R}$ is upper semicontinuous, the definition

$$\phi(r) := \sup\left\{\left(u(x_0 + \xi) - u(x_0) - \langle p, \xi\rangle - \tfrac{1}{2}\langle X\xi, \xi\rangle\right)^+ : x_0 + \xi \in \Omega \text{ and } |\xi| \leq r\right\}$$

for $r \geq 0$ leads to a non-decreasing function

$$\phi : [0, \infty) \to [0, \infty)$$

with $\phi(0) = 0$. Moreover, our assumption on the behavior of $u(x)$ as $x \to x_0$ implies

$$\lim_{r \downarrow 0} r^{-2}\phi(r) = 0.$$

Hence, Proposition 1.22 shows the existence of $\psi \in C^2([0, \infty))$ with

$$\phi(r) \leq \psi(r)$$

for all $r \geq 0$ and one-sided derivatives $\psi(0) = \psi'(0) = \psi''(0) = 0$. Define $f \in C^2(\Omega)$ by

$$f(x) := u(x_0) + \langle p, x - x_0\rangle + \tfrac{1}{2}\langle X(x - x_0), (x - x_0)\rangle + \psi(|x - x_0|^2)$$

for $x \in \Omega$ and note that $f(x_0) = u(x_0)$, $Df(x_0) = p$ and $D^2 f(x_0) = X$ by the chain rule of calculus. Furthermore, for every $x \in \Omega$

$$u(x) \leq u(x_0) + \langle p, x - x_0\rangle + \tfrac{1}{2}\langle X(x - x_0), (x - x_0)\rangle + \phi(|x - x_0|^2) \leq f(x)$$

follows from $\phi \leq \psi$ and the definition of $\phi : [0, \infty) \to [0, \infty)$. $\square$

Corollary 1.24 (Adapted Scanning Functions). *Suppose that $\Omega \subset \mathbb{R}^d$ is open, that $x_0 \in \Omega$, that $u \in USC_p(\Omega)$ with $p \geq 0$, and that $f \in C^2(\Omega)$ such that $u(x_0) = f(x_0)$ and*

$$u(x) \leq f(x)$$

for all $x \in \Omega$. If $p \in \mathbb{R}^d$ and $X \in \mathbb{S}^{d \times d}$ such that

$$u(x) = u(x_0) + \langle p, x - x_0\rangle + \tfrac{1}{2}\langle X(x - x_0), (x - x_0)\rangle + o(|x - x_0|^2)$$

as $x \to x_0$, then there exists $g \in C_p^2(\Omega)$ with

$$u(x) \leq g(x) \leq f(x)$$

for all $x \in \Omega$, $g(x_0) = u(x_0)$, $Dg(x_0) = p$ and $D^2 g(x_0) = X$.

Proof. We can assume without loss of generality that $f \in C_p^2(\Omega)$: Otherwise, apply Lemma 1.18 to find $\overline{f} \in C_p^2(\Omega)$ with

$$u(x) \leq \overline{f}(x) \leq f(x)$$

for all $x \in \Omega$ and replace $f \in C^2(\Omega)$ by $\overline{f} \in C_p^2(\Omega)$ in the following arguments.

1. Approximation Theory

Since the difference $u - f$ has a global maximum in $x_0 \in \Omega$, we have $Df(x_0) = p$ and $D^2 f(x_0) \geq X$. If $D^2 f(x_0) = X$, then $g := f$ satisfies all postulates. If $D^2 f(x_0) > X$, then Theorem 1.23 implies the existence of $\phi \in C^2(\Omega)$ with

$$u(x) \leq \phi(x)$$

for all $x \in \Omega$ such that $\phi(x_0) = u(x_0) = f(x_0)$, $D\phi(x_0) = p = Df(x_0)$ and

$$D^2 \phi(x_0) = X < D^2 f(x_0).$$

Therefore, the sufficient criterion for local extrema implies that $\phi - f$ has a strict local maximum in $x_0 \in \Omega$. In particular, there exists $\delta > 0$ such that

$$\phi(x) \leq f(x)$$

for all $x \in \Omega$ with $|x - x_0| < \delta$. According to Proposition 1.7, there exists $\chi \in C^\infty(\Omega)$ such that $\chi(x) = 1$ for $x \in \Omega$ with $|x - x_0| \leq \delta/2$ and $\chi(x) = 0$ for $x \in \Omega$ with $|x - x_0| \geq \delta$. Finally, it is easy to see that

$$g(x) := \chi(x)\phi(x) + (1 - \chi(x))f(x)$$

for $x \in \Omega$ satisfies the required properties. $\square$

At last, we demonstrate how our preceding results can be used to extend Jensen's lemma (cf. Lemma A.8) to match the requirements of our generalized set-up in Chapter 2. Originally formulated by Robert Jensen in [77, Lemma 3.15] for the viscosity solution theory of local equations, Jensen's lemma shows that maximum points of semiconvex functions (cf. Definition A.5) can be approximated by regularity points. The following proof is based on ideas from [75, Lemma 7.4]:

Lemma 1.25 (Approximate Maximum Points). *Suppose that $\Omega \subset \mathbb{R}^d$ is open, that $\phi \in C_p^2(\Omega)$ with $p \geq 0$, and that $f : \Omega \to \mathbb{R}$ is semiconvex. If $f(\overline{x}) = \phi(\overline{x})$ and*

$$f(x) < \phi(x)$$

for all $x \in \Omega$, then there exists $(x_n, p_n, X_n)_{n \in \mathbb{N}} \subset \Omega \times \mathbb{R}^d \times \mathbb{S}^{d \times d}$ with $\lim_{n \to \infty} x_n = \overline{x}$ and

$$f(x) = f(x_n) + \langle p_n, x - x_n \rangle + \tfrac{1}{2} \langle X_n(x - x_n), (x - x_n) \rangle + o(|x - x_n|^2)$$

as $x \to x_n$. Moreover, there exists $(\phi_n)_{n \in \mathbb{N}} \subset C_p^2(\Omega)$ with $\sup_{n \in \mathbb{N}} \|\phi_n\|_p < \infty$,

$$\lim_{n \to \infty} D^k \phi_n = D^k \phi$$

locally uniformly in Ω for $k \in \{0, 1, 2\}$, $f(x_n) = \phi_n(x_n)$ for all $n \in \mathbb{N}$ and

$$f(x) \leq \phi_n(x)$$

for all $x \in \Omega$ and $n \in \mathbb{N}$.

Proof. The main idea is to use Lemma A.8 locally and extend the resulting local scanning function to Ω: First, pick $r > 0$ such that $B[\bar{x}, r] \subset \Omega$. According to Lemma A.6,

$$x \longmapsto f(x) - \phi(x)$$

is semiconvex and thus Lipschitz-continuous on $B[\bar{x}, r]$ (cf. Proposition A.2). Hence, Lemma A.8 implies the existence of $x_n \in B(\bar{x}, r)$ and $q_n \in \mathbb{R}^d$ for every $n \in \mathbb{N}$ such that $|q_n| \leq n^{-1}$, such that the function

$$x \longmapsto f(x) - \phi(x) - \langle q_n, x \rangle$$

attains its maximum over $B(\bar{x}, r)$ in $x_n \in B(\bar{x}, r)$, and such that $f : \Omega \to \mathbb{R}$ is twice differentiable in $x_n \in B(\bar{x}, r)$. Since $f(x_n) \leq \phi(x_n)$ and $f(\bar{x}) = \phi(\bar{x})$ by assumption, $x_n \in \arg\max_{x \in B(\bar{x}, r)} \left(f(x) - \phi(x) - \langle q_n, x \rangle \right)$ implies

$$-\langle q_n, x_n \rangle \geq f(x_n) - \phi(x_n) - \langle q_n, x_n \rangle \geq f(\bar{x}) - \phi(\bar{x}) - \langle q_n, \bar{x} \rangle = -\langle q_n, \bar{x} \rangle$$

for all $n \in \mathbb{N}$. In particular, we obtain

$$f(\tilde{x}) = \phi(\tilde{x})$$

for all accumulation point $\tilde{x} \in \Omega$ of $(x_n)_{n \in \mathbb{N}}$. Since $\bar{x} \in \Omega$ is a strict global maximum of $f - \phi$ over Ω with $f(\bar{x}) = \phi(\bar{x})$, this implies that $\bar{x} \in \Omega$ is the only accumulation point of $(x_n)_{n \in \mathbb{N}}$. Consequently, we find

$$\lim_{n \to \infty} x_n = \bar{x},$$

using the compactness of $B[x, r]$. Define $c_n := f(x_n) - \phi(x_n) - \langle q_n, x_n \rangle$ and $\psi_n \in C^2(\Omega)$ by

$$\psi_n(x) := \phi(x) + \langle q_n, x \rangle + c_n$$

for $x \in \Omega$ and $n \in \mathbb{N}$. Since $x \mapsto f(x) - \psi_n(x) + c_n = f(x) - \phi(x) - \langle q_n, x \rangle$ attains its maximum over $B(\bar{x}, r)$ in x_n and $f(x_n) - \psi_n(x_n) = 0$, the inequality

$$f(x) \leq \psi_n(x)$$

holds for all $x \in B(\bar{x}, r)$. According to Proposition 1.7 there exists $\chi \in C^\infty(\Omega)$ with $0 \leq \chi \leq 1$ such that $\chi(x_n) = 1$ for all $n \in \mathbb{N}$ and $\chi(x) = 0$ for all $x \in \Omega \setminus B(\bar{x}, r)$. Define

$$\phi_n(x) := \chi(x)\psi_n(x) + \big(1 - \chi(x)\big)\phi(x) = \phi(x) + \chi(x)\big(\langle q_n, x \rangle + c_n\big) \quad (1.25)$$

for $x \in \Omega$ and $n \in \mathbb{N}$. We find $\phi_n(x) = \phi(x)$ for $x \in \Omega \setminus B(\bar{x}, r)$ and thus

$$\sup_{n \in \mathbb{N}} \|\phi_n\|_{C_p} \leq \|\phi\|_{C_p} + \sup_{n \in \mathbb{N}} \Big(2r|q_n| + |f(x_n)| + |\phi(x_n)| \Big) < \infty,$$

using the continuity of $f, \phi : \Omega \to \mathbb{R}$. Further, $f(x_n) = \psi_n(x_n) = \phi_n(x_n)$ for $n \in \mathbb{N}$, and

$$f(x) \leq \phi_n(x)$$

for $x \in \Omega$ and $n \in \mathbb{N}$, since $\chi(x_n) = 0$, $f \leq \psi_n$ in $B(\bar{x}, r)$ and $f \leq \phi$ in Ω.

1. Approximation Theory

It remains to show that $\lim_{n\to\infty} D^k\phi_n = D^k\phi$ locally uniformly in Ω for $k \in \{0,1,2\}$: The definition in Equation (1.25) shows that

$$\phi_n(x) - \phi(x) = \chi(x)\left(\langle q_n, x\rangle + c_n\right)$$

for all $x \in \Omega$ and $n \in \mathbb{N}$. Since $\chi(x) = 0$ for all $x \in \Omega \setminus B(\overline{x}, r)$, we find as $n \to \infty$

$$\sup_{x\in\Omega}\left|\phi_n(x) - \phi(x)\right| \leq \sup_{x\in B[\overline{x},r]} |\langle q_n, x\rangle + c_n| \leq 2r|q_n| + |f(x_n) - \phi(x_n)| \longrightarrow 0$$

$$\sup_{x\in\Omega}\left|D\phi_n(x) - D\phi(x)\right| \leq \sup_{x\in B[\overline{x},r]} \left(|q_n| + |D\chi(x)|\cdot|\langle q_n, x\rangle + c_n|\right) \longrightarrow 0$$

$$\sup_{x\in\Omega}\left|D^2\phi_n(x) - D^2\phi(x)\right| \leq \sup_{x\in B[\overline{x},r]} \left(2\cdot|q_n|\cdot|D\chi(x)| + \|D^2\chi(x)\|\cdot|\langle q_n, x\rangle + c_n|\right) \longrightarrow 0$$

from $\lim_{n\to\infty} x_n = \overline{x}$ and from the continuity of $f, \phi : \Omega \to \mathbb{R}$ with $f(\overline{x}) = \phi(\overline{x})$. □

1.4. Miscellaneous

The last section of this chapter contains a collection of miscellaneous approximation results, that do not fit in any of the other sections, but are still vital for our viscosity solution theory of nonlocal equations in Chapter 2.

The Tietze extension theorem from general topology shows that continuous functions $f : A \to \mathbb{R}$ on closed subsets $A \subset X$ of normal topological spaces X can always be extended to continuous functions on the whole space – a result that was first proved by Heinrich Tietze in [125] for metric spaces and later generalized by Paul Urysohn in [126] to normal topological spaces. In the following constructive version of the Tietze extension theorem in Euclidean spaces, we generalize an elegant proof from Felix Hausdorff in [61] to semicontinuous functions. As we will see in Chapter 2, this generalization is essential in order to show the equivalence of different notions of viscosity solutions in terms of their respective domains of definition.

Theorem 1.26 (Tietze Extension). *If $\Omega \subset \mathbb{R}^d$ is non-void and closed, and $f : \Omega \to \mathbb{R}$ with $a, b \in \mathbb{R}$ such that $a \leq f(x) \leq b$ for all $x \in \Omega$, then $g : \mathbb{R}^d \to \mathbb{R}$ with $g|_\Omega := f$ and*

$$g(x) := \inf_{y\in\Omega}\left\{f(y) + \frac{|x-y|}{d(x,\Omega)} - 1\right\}$$

for $x \in \mathbb{R}^d \setminus \Omega$ extends $f : \Omega \to \mathbb{R}$ such that $a \leq g(x) \leq b$ for all $x \in \mathbb{R}^d$ and such that $g|_{\mathbb{R}^d\setminus\Omega}$ is continuous. Moreover, if $f : \Omega \to \mathbb{R}$ is upper or lower semicontinuous, then

$$g : \mathbb{R}^d \longrightarrow \mathbb{R}$$

is also upper or lower semicontinuous, respectively. In particular, if $f : \Omega \to \mathbb{R}$ is continuous, then $g : \mathbb{R}^d \to \mathbb{R}$ is a continuous extension of $f : \Omega \to \mathbb{R}$ to the whole space.

Proof. Since $d(x,\Omega) \leq |x-y|$ for all $y \in \Omega$ by definition, we find

$$a \leq \inf_{y \in \Omega} f(y) \leq \inf_{y \in \Omega} \left\{ f(y) + \frac{|x-y|}{d(x,\Omega)} - 1 \right\} = g(x)$$

for all $x \in \mathbb{R}^d \setminus \Omega$. On the other hand, the definition of $d(x,\Omega)$ also implies

$$g(x) = \inf_{y \in \Omega} \left\{ f(y) + \frac{|x-y|}{d(x,\Omega)} - 1 \right\} \leq b - 1 + \frac{1}{d(x,\Omega)} \inf_{y \in \Omega} |x-y| = b$$

for every $x \in \mathbb{R}^d \setminus \Omega$. Note (for later use) that for all $x \in \mathbb{R}^d \setminus \Omega$ and $y \in \Omega$, the inequality

$$f(y) + \frac{|x-y|}{d(x,\Omega)} - 1 \leq g(x) + 1$$

implies $|x-y| \leq (g(x) - f(y) + 2) \cdot d(x,\Omega) \leq c \cdot d(x,\Omega)$ with $c := b - a + 2 > 0$.

At first, assume that $x_0 \in \mathbb{R}^d \setminus \Omega$ and $x \in \mathbb{R}^d \setminus \Omega$ with $0 < |x_0 - x| < 1$. According to the definition of $g(x)$, there exists $y \in \Omega$ such that

$$f(y) + \frac{|x-y|}{d(x,\Omega)} - 1 \leq g(x) + |x_0 - x|$$

and hence $|x-y| \leq c \cdot d(x,\Omega)$ hold. The definition of $g(x_0)$ therefore implies

$$g(x_0) \leq f(y) + \frac{|x_0-y|}{d(x_0,\Omega)} - 1$$
$$\leq g(x) + |x_0 - x| + \frac{|x_0-y|}{d(x_0,\Omega)} - \frac{|x-y|}{d(x,\Omega)}$$
$$\leq g(x) + |x_0 - x| + \frac{|x_0-x|}{d(x_0,\Omega)} + |x-y| \cdot \left(\frac{1}{d(x_0,\Omega)} - \frac{1}{d(x,\Omega)} \right)$$
$$\leq g(x) + |x_0 - x| \cdot \left(1 + \frac{1+c}{d(x_0,\Omega)} \right)$$

using $|d(x_0,\Omega) - d(x,\Omega)| \leq |x_0 - x|$ at the last step. Interchanging the roles of the points $x_0 \in \mathbb{R}^d \setminus \Omega$ and $x \in \mathbb{R}^d \setminus \Omega$ leads to the second inequality in

$$g(x_0) - |x_0 - x| \cdot \left(1 + \frac{1+c}{d(x_0,\Omega)} \right) \leq g(x) \leq g(x_0) + |x_0 - x| \cdot \left(1 + \frac{1+c}{d(x,\Omega)} \right)$$

and thus $g(x) \to g(x_0)$ as x tends to x_0, since the map $x \mapsto d(x,\Omega)$ is continuous. In particular, the extension $g : \mathbb{R}^d \to \mathbb{R}$ is always continuous in $\mathbb{R}^d \setminus \Omega$.

Next, suppose that $x_0 \in \Omega$ and $x \in \mathbb{R}^d \setminus \Omega$ with $0 < |x_0 - x| < 1$. According to the definition of $g(x)$, there exists $y \in \Omega$ such that

$$f(y) \leq f(y) + \frac{|x-y|}{d(x,\Omega)} - 1 \leq g(x) + |x_0 - x|$$

and therefore $|x-y| \leq c \cdot d(x,\Omega)$. Now, pick $z \in \Omega$ with $|x-z| \leq (1 + |x_0 - x|) \cdot d(x,\Omega)$.

53

1. Approximation Theory

The definition of $g(x)$ implies
$$g(x) \leq f(z) + \frac{|x-z|}{d(x,\Omega)} - 1 \leq f(z) + |x_0 - x|$$
and hence $f(y) - |x_0 - x| \leq g(x) \leq f(z) + |x_0 - x|$. It is easy to see that
$$|x_0 - y| \leq |x_0 - x| + c \cdot d(x,\Omega) \leq (1+c) \cdot |x_0 - x|$$
$$|x_0 - z| \leq |x_0 - x| + (1 + |x_0 - x|) \cdot d(x,\Omega) \leq (2 + |x_0 - x|) \cdot |x_0 - x|$$
hold and thus $x \to x_0$ with $x \in \mathbb{R}^d \setminus \Omega$ implies $y \to x_0$ as well as $z \to x_0$. This leads to
$$\liminf_{y \to x_0} f(y) \leq \liminf_{x \to x_0} g(x) \leq \limsup_{x \to x_0} g(x) \leq \limsup_{z \to x_0} f(z)$$
and since $g(x_0) = f(x_0)$ for $x_0 \in \Omega$. Thus, $g : \mathbb{R}^d \to \mathbb{R}$ is upper or lower semicontinuous in $x_0 \in \Omega$ if $f : \Omega \to \mathbb{R}$ is upper or lower semicontinuous in $x_0 \in \Omega$, respectively. □

One important feature of Theorem 1.26 is, that the construction of extensions coincides for upper and lower semicontinuous functions. Furthermore, it is easy to check that the proof also works for general metric spaces instead of Euclidean space, and that it can be generalized to unbounded functions, by using a homeomorphism between the real line and a finite interval (such as the arctangent function).

Section 1.2 shows how to monotonically approximate semicontinuous functions pointwise using supremal convolutions. Since we want to establish a viscosity solution theory for nonlocal equations with weaker regularity assumptions (compared to the existing literature) in Chapter 2, it is important to know, if this convergence can be improved under any additional assumptions. Let us therefore recall the well-known result from Ulisse Dini in [45], which shows that pointwise convergence implies local uniform convergence in the special case of monotonic sequences with continuous limits:

Theorem 1.27 (Dini). *Assume that $f : K \to \mathbb{R}$ is continuous with $K \subset \mathbb{R}^d$ compact. If $(f_n)_{n \in \mathbb{N}}$ is a sequence of continuous functions on K such that $f_n(x) \geq f_{n+1}(x)$ for $n \in \mathbb{N}$ and $\lim_{n \to \infty} f_n(x) = f(x)$ for each $x \in K$, then $\lim_{n \to \infty} \sup_{x \in K} |f_n(x) - f(x)| = 0$.*

Proof. ↪ [113, Theorem 7.13, p. 150] □

For many approximation arguments related to measure or probability theory, it is important that the approximates are chosen from a countable (or at least separable) family of functions. The Weierstrass approximation theorem, which was originally proved by Karl Weierstrass in [129], and its generalization to compact Hausdorff spaces (usually referred to as the Stone-Weierstrass theorem) by Marshall Stone in [121] are often applied in this context. In the rest of this section, we recall Weierstrass' well-known result and show how to obtain a special corollary from it, which we require for a convergence-determining class argument in Chapter 4.

Theorem 1.28 (Weierstrass). *For every continuous $f : \mathbb{R}^d \to \mathbb{R}$, there exists a sequence $(f_n)_{n \in \mathbb{N}}$ of polynomials with rational coefficients $f_n : \mathbb{R}^d \to \mathbb{R}$ such that $f_n \to f$ locally uniform, i.e. $\lim_{n \to \infty} \sup_{x \in K} |f_n(x) - f(x)| = 0$ for all compact $K \subset \mathbb{R}^d$.*

Proof. ↪ [47, Chapter 4, Section 6, Theorem 16, p. 272] □

Corollary 1.29 (Countable Families of Approximations). *Suppose that the map* $\rho : (0, \infty) \to (0, \infty)$ *is bounded and continuous with* $\rho(0) := \lim_{r \downarrow 0} \rho(r) = 0$, *then there exists a countable family*
$$\mathcal{F} \subset \{f \in C_b(\mathbb{R}^d) \mid \operatorname{supp}(f) \text{ compact and } \exists \delta > 0 \, \forall x \in B_\delta(0) : f(x) = 0\}$$
such that for every function $f \in C_b(\mathbb{R}^d)$ *with* $\sup_{x \in \mathbb{R}^d \setminus \{0\}} \rho(|x|)^{-1} |f(x)| < \infty$, *there exists a sequence* $(f_n)_{n \in \mathbb{N}} \subset \mathcal{F}$ *such that*
$$\limsup_{n \to \infty} \sup_{x \in \mathbb{R}^d \setminus \{0\}} \rho(|x|)^{-1} |f_n(x)| \leq \sup_{x \in \mathbb{R}^d \setminus \{0\}} \rho(|x|)^{-1} |f(x)|$$
as well as $\lim_{n \to \infty} \sup_{x \in K \setminus \{0\}} \rho(|x|)^{-1} |f_n(x) - f(x)| = 0$ *for every* $K \subset \mathbb{R}^d$ *compact.*

Proof. First of all, note that with the definition
$$\mathcal{F} := \{f \in C_b(\mathbb{R}^d) \mid \exists g \in \mathcal{G} \, \forall x \in \mathbb{R}^d : f(x) = \rho(|x|) g(x)\},$$
it suffices to show that there exists a countable family
$$\mathcal{G} \subset \{g \in C_b(\mathbb{R}^d) \mid \operatorname{supp}(g) \text{ compact and } \exists \delta > 0 \, \forall x \in B_\delta(0) : g(x) = 0\}$$
such that for every $g \in C_b(\mathbb{R}^d)$ with $g(0) = 0$, there exists a sequence $(g_n)_{n \in \mathbb{N}} \subset \mathcal{G}$ with $\limsup_{n \to \infty} \|g_n\|_\infty \leq \|g\|_\infty$ and $g_n \to g$ locally uniform as $n \to \infty$.

Secondly, according to Proposition 1.7, for $R \in (0, 1)$ there exists $\chi_R \in C_b^\infty(\mathbb{R}^d)$ with
$$\chi_R(x) = \begin{cases} 1 & \text{if } x \in \overline{B_{R^{-1}}(0)} \setminus B_R(0) \\ 0 & \text{if } x \in \overline{B_{R/2}(0)} \cup (\mathbb{R}^d \setminus B_{2R^{-1}}(0)) \end{cases}$$
and $0 \leq \chi_R \leq 1$. With Theorem 1.28, it is easy to check that the following family
$$\mathcal{G} := \{g : \mathbb{R}^d \to \mathbb{R} \mid \exists R \in \mathbb{Q} \cap (0, 1) \, \exists p \in \mathcal{P} \, \forall x \in \mathbb{R}^d : g(x) = \chi_R(x) p(x)\}$$
satisfies all of the required properties, where $\mathcal{P}$ is the (countable) family of polynomials with rational coefficients: For a fixed $g \in C_b(\mathbb{R}^d)$ with $g(0) = 0$, define $g_n := \chi_{n^{-1}} p_n \in \mathcal{G}$ for $n \in \mathbb{N}$, where the polynomial $p_n \in \mathcal{P}$ such that
$$\sup \{|p_n(x) - g(x)| \, : \, x \in \overline{B_{2n}(0)}\} \leq n^{-1}$$
exists due to Theorem 1.28. For $K \subset \mathbb{R}^d$ exists $N \in \mathbb{N}$ such that $K \subset B_N(0)$ and hence
$$\sup_{x \in K} |g(x) - g_n(x)| \leq \sup_{x \in B_N(0)} |g(x) - \chi_{n^{-1}}(x) p_n(x)|$$
$$\leq \sup_{x \in B_N(0)} \{|g(x) - \chi_{n^{-1}}(x) g(x)| + |\chi_{n^{-1}}(x) g(x) - \chi_{n^{-1}}(x) p_n(x)|\}$$
$$\leq 2 \cdot \sup \{|g(x)| : x \in \overline{B_{n^{-1}}(0)}\} + n^{-1}$$
for $n \geq N$, which shows that $g_n \to g$ locally uniform as $n \to \infty$. Moreover,
$$\sup_{x \in \mathbb{R}^d} |g_n(x)| = \sup_{x \in \overline{B_{2n}(0)}} |\chi_{n^{-1}}(x) p_n(x)| \leq \sup_{x \in \overline{B_{2n}(0)}} |p_n(x)| \leq n^{-1} + \sup_{x \in \mathbb{R}^d} |g(x)|$$
for every $n \in \mathbb{N}$, which implies that $\limsup_{n \to \infty} \sup_{x \in \mathbb{R}^d} |g_n(x)| \leq \sup_{x \in \mathbb{R}^d} |g(x)|$. □

2. Integro-Differential Equations

Partial integro-differential equations (or PIDEs for short) are equations that involve both integrals and derivatives of functions (defined on multi-dimensional domains) and are used in many different applications in science and engineering. They naturally extend partial differential equations (or PDEs for short) by introducing additional integral terms, that can be used to model the nonlocal behavior of systems. For an overview of modern applications of partial integro-differential equations inside and outside of mathematics and the related theory, we refer the interested reader to [99].

In this chapter, we develop the uniqueness theory for a large class of partial integro-differential equations with non-dominated nonlocal parts, that play an important role for sublinear Markov semigroups and related Markov processes for sublinear expectations, as shown in Chapter 4. As a start, Section 2.1 is devoted to state the general form of the considered nonlinear nonlocal equations

$$F(x, u(x), Du(x), D^2u(x), u(\cdot)) = 0 \tag{E1}$$

including all standing assumptions, and to introduce the notion of viscosity solutions

$$u : \mathbb{R}^d \longrightarrow \mathbb{R}$$

together with important alternative characterizations. Moreover, we try to highlight the main differences of our original results to the existing literature. In Section 2.2, we derive a nonlocal maximum principle for equations of the general form (E1) and employ it afterwards, in oder to prove that a large class of parabolic equations

$$\partial_t u(t,x) + G(t, x, u(t,x), Du(t,x), D^2u(t,x), u(t,\cdot)) = 0 \tag{E2}$$

has unique (unbounded) viscosity solutions

$$u : [0, T] \times \mathbb{R}^d \longrightarrow \mathbb{R}$$

for given initial values $u(0, \cdot) : \mathbb{R}^d \to \mathbb{R}$ and $T > 0$. At last, in Section 2.3, we introduce Hamilton-Jacobi-Bellman equations, i.e. nonlinear nonlocal equations of the form

$$\partial_t u(t,x) + \inf_{\alpha \in \mathcal{A}} G_\alpha(t, x, u(t,x), Du(t,x), D^2u(t,x), u(t,\cdot)) = 0 \tag{E3}$$

for a family $(G_\alpha)_{\alpha \in \mathcal{A}}$ of linear, nonlocal degenerate elliptic operators, and show how the results from Section 2.1 and Section 2.2 can be applied to them.

2. Integro-Differential Equations

2.1. Viscosity Solutions

An important problem in the theory of partial integro-differential equations is the appropriate notion of solutions: On the one hand, it is well-known, that there exist no classical solutions (i.e. functions that exhibit all of the required derivatives) for many interesting equations, which motivates the study of generalized solutions. On the other hand, if the generalized notion of solutions is too weak, then there might exist more than one generalized solution. An exemplary discussion of this dichotomy between the existence and uniqueness of generalized solutions for first-order, local Hamilton-Jacobi-Bellman equations can be found in [52, Section 3.3.3, p. 129].

In this chapter, we restrict our attention to the notion of viscosity solutions, which turns out to be the appropriate notion for second-order, nonlinear nonlocal equations in the context of Markov processes for sublinear expectations, as we will see in Chapter 4. The term viscosity solution was formally introduced by Michael Crandall and Pierre-Louis Lions in [36] for first-order, local Hamilton-Jacobi-Bellman equations, although the concept was already developed earlier by Lawrence Evans in [51]. Its name is derived from the first existence proofs, which were based on the method of vanishing viscosity, and is still prevalent in generalizations today for historical reasons – even though the method of vanishing viscosity is in general not applicable anymore. Despite the fact that the theory was almost immediately extended from first-order to second-order local equations by Pierre-Louis Lions in [87], most of the modern results on second-order local equations are based on a method from Robert Jensen in [77] – the so-called generalized maximum principle, which is presented in Section 2.2.

The first attempts to generalize the viscosity solution theory from local equations to (first-order) integro-differential equations were published by Halil Mete Soner in [120] (for bounded measures) and Awatif Sayah in [118] (for unbounded measures). These results were later extended to cover more general second-order nonlocal equations – notably by Olivier Alvarez and Agnès Tourin in [1], by Espen Jakobsen and Kenneth Karlsen in [75], and by Guy Barles and Cyril Imbert in [12]. Beyond these mostly analytical approaches, the viscosity solution theory for second-order nonlocal equations was also significantly extended using probabilistic methods – among others by Guy Barles, Rainer Buckdahn and Etienne Pardoux in [9] (related to backwards stochastic differential equations), by Huyên Pham in [108] (related to stochastic control with jumps), and by Mingshang Hu and Shige Peng in [67] (related to nonlinear expectations). The classes of covered equations (in the preceding articles on viscosity solution theory) typically differ with regards to the exact form and interaction between the local and nonlocal part, the behavior of the solutions at the boundary (or infinity respectively), and the singularity of the measures of the nonlocal part (and the related compensation).

During the last two decades, viscosity solutions turned out to be one the most frequently used notions for generalized solutions of nonlinear (elliptic and parabolic) partial integro-differential equations. (In fact, the comprehensive overview article [35] on viscosity solution theory for local equations by Michael Crandall, Hitoshi Ishii and Pierre-Louis Lions was one of the three most cited articles on AMS MathSciNet for every year from 2000 until 2010, cf. the statistics on [100].) However, despite its popularity, there exist

2.1. Viscosity Solutions

several different definitions of viscosity solutions for nonlocal equations in the literature. In this section, we therefore develop alternative characterizations for our definition of viscosity solutions (in Lemma 2.4, Lemma 2.6 and Lemma 2.7), which are useful for applications of our main results (contained in Section 2.2) and to compare those results to the existing literature. Moreover, we introduce our general assumptions for Section 2.2 in Remark 2.3 and highlight the main differences to the relevant literature.

Before we state our definition of viscosity solutions, let us quickly motivate its concept: Suppose that $u \in C^2(\mathbb{R}^d)$ is a classical solution in an open set $\Omega \subset \mathbb{R}^d$ of
$$F(x, u(x), Du(x), D^2u(x), u(\cdot)) = 0,$$
which satisfies a nonlocal degenerate ellipticity condition, i.e.
$$F(x, r, p, X, \phi) \geq F(x, r, p, Y, \psi)$$
whenever $(x, r, p) \in \Omega \times \mathbb{R} \times \mathbb{R}^d$, $X, Y \in \mathbb{S}^{d \times d}$ and $\phi, \psi \in C^2(\mathbb{R}^d)$ such that $X \leq Y$ and $\phi - \psi$ has a global maximum in $x \in \Omega$. Note that the partial order $X \leq Y$ for $X, Y \in \mathbb{S}^{d \times d}$ is defined in the quadratic form sense, i.e. $Y - X$ is positive semi-definite. If $\phi \in C^2(\mathbb{R}^d)$ such that $u - \phi$ has a global maximum (or global minimum) in $x \in \Omega$ with $u(x) = \phi(x)$, then $D(u - \phi)(x) = 0$ and $D^2(u - \phi)(x) \leq 0$ according to the classical maximum principle (also known as derivative test), and therefore
$$F(x, u(x), D\phi(x), D^2\phi(x), \phi) \leq 0 \ (\geq 0)$$
using the degenerate ellipticity condition. In fact, it is easy to check that a smooth function $u \in C^2(\mathbb{R}^d)$ is a classical solution in an open set $\Omega \subset \mathbb{R}^d$ if and only if
$$F(x, u(x), D\phi(x), D^2\phi(x), \phi) \leq 0 \ (\geq 0)$$
holds for all $\phi \in C^2(\mathbb{R}^d)$ such that $u - \phi$ has a global maximum (or global minimum, respectively) in $x \in \Omega$. Such functions $\phi \in C^2(\mathbb{R}^d)$ are usually referred to as *scanning functions* (or *test functions*). The main idea is to use the preceding characterization of classical solutions (by scanning them from above and below by smooth functions), which does not include any derivatives of $u : \mathbb{R}^d \to \mathbb{R}$, to define a generalized notion of solutions for degenerate elliptic equations.

Definition 2.1 (Viscosity Solutions). Assume that $p \geq 0$, $\Omega \subset \mathbb{R}^d$ is open and that the operator $F : \mathbb{R}^d \times \mathbb{R} \times \mathbb{R}^d \times \mathbb{S}^{d \times d} \times C_p^2(\mathbb{R}^d) \to \mathbb{R}$ is given. An upper semicontinuous function $u \in \text{USC}_p(\mathbb{R}^d)$ is a *viscosity subsolution* in Ω of the nonlocal equation
$$F(x, u(x), Du(x), D^2u(x), u(\cdot)) = 0, \tag{E1}$$
if for all $\phi \in C_p^2(\mathbb{R}^d) = C^2(\mathbb{R}^d) \cap C_p(\mathbb{R}^d)$ such that $u - \phi$ has a global maximum in $x \in \Omega$,
$$F(x, u(x), D\phi(x), D^2\phi(x), \phi(\cdot)) \leq 0.$$
A *viscosity supersolution* in Ω of (E1) is a lower semicontinuous function $v \in \text{LSC}_p(\mathbb{R}^d)$ such that for all $\phi \in C_p^2(\mathbb{R}^d)$, for which $v - \phi$ has a global minimum in $x \in \Omega$,
$$F(x, v(x), D\phi(x), D^2\phi(x), \phi(\cdot)) \geq 0.$$
A *viscosity solution* in Ω of (E1) is both a viscosity sub- and supersolution in Ω of (E1).

2. Integro-Differential Equations

There are a few subtleties to our definition of viscosity solutions: First of all, we restrict our attention to the natural choice of viscosity solutions with p-polynomial growth at infinity (as introduced in Definition 1.13), since we typically have to impose some integrability condition in order for the nonlocal part to exist. Secondly, since we only consider degenerate elliptic equations which are translation invariant (as defined in Remark 2.3), the definition will not change if we only consider scanning functions

$$\phi : \mathbb{R}^d \to \mathbb{R}$$

with $\phi(x) = u(x)$ at the global extremum $x \in \Omega$ of $u - \phi$ (matching the scanning motivation from before). Thirdly, the equation separates the local from the nonlocal part (i.e. the last parameter) in order to simplify certain compactness arguments in Section 2.2 and to obtain weaker assumptions in Section 2.3.

Remark 2.2 (Notations for Viscosity Solutions). Due to the duality of viscosity sub- and supersolutions, there exist situations in which it is convenient to formulate statements for both viscosity sub- and supersolutions at the same time. In this chapter, this is done by putting the corresponding changes for the dual statement in parentheses. In addition to this, we adopt the usual convention from the existing literature and call a function a viscosity solution of

$$F(x, u(x), Du(x), D^2u(x), u(\cdot)) \leq 0 \ (\geq 0)$$

if it is a viscosity subsolution (viscosity supersolution) of Equation (E1).

As a next step, we state the precise assumptions for the equations, which we will cover in Section 2.2. Instead of considering concrete examples of nonlocal equations directly (as it is done in many influential articles on nonlocal equations such as [1] or [12]), we follow the approach of [75] to state the assumptions subject to a more general operator and check the applicability for concrete examples on a case-by-case basis.

Remark 2.3 (Assumptions on F). Suppose that $p \geq 0$ and that $\Omega \subset \mathbb{R}^d$ is open. Throughout this chapter, we will assume that for each given operator

$$F : \Omega \times \mathbb{R} \times \mathbb{R}^d \times \mathbb{S}^{d \times d} \times C_p^2(\Omega) \longrightarrow \mathbb{R}$$

and every $0 < \kappa < 1$, there exist operators (the so-called *generalized operators*)

$$F^\kappa : \Omega \times \mathbb{R} \times \mathbb{R}^d \times \mathbb{S}^{d \times d} \times \mathrm{SC}_p(\Omega) \times C^2(\Omega) \longrightarrow \mathbb{R},$$

on which we will impose the following assumptions:

(A1) (Consistency) The operators F^κ are generalizations of F in the sense that

$$F^\kappa(x, r, q, X, \phi, \phi) = F(x, r, q, X, \phi)$$

holds for all $\phi \in C_p^2(\Omega)$.

2.1. Viscosity Solutions

(A2) (Degenerate Ellipticity) The operators F^κ are *nonlocal degenerate elliptic*, i.e.
$$F^\kappa(x, r, q, X, u, \phi) \geq F^\kappa(x, r, q, Y, v, \psi)$$
holds if $u - v$ and $\phi - \psi$ have global maxima in $x \in \Omega$, and if $X \leq Y$.

(A3) (Translation Invariance) The operators F^κ are *translation invariant*, i.e.
$$F^\kappa(x, r, q, X, u + c_1, \phi + c_2) = F^\kappa(x, r, q, u, \phi)$$
holds for all constants $c_1, c_2 \in \mathbb{R}$.

(A4) (Continuity) The operators F^κ meet certain continuity properties. Namely,
$$\lim_{n \to \infty} F^\kappa(x_n, r_n, q_n, X_n, u_n, \phi_n) = F^\kappa(x, r, q, X, u, \phi)$$
holds if $\lim_{n \to \infty}(x_n, r_n, q_n, X_n) = (x, r, q, X)$, if $\lim_{n \to \infty} D^k\phi_n = D^k\phi$ locally uniformly for all $k \in \{0, 1, 2\}$, and if $\lim_{n \to \infty} u_n = u$ locally uniformly with $u \in C_p(\Omega)$ and $\sup_{n \in \mathbb{N}} \|u_n\|_p < \infty$.

(A5) (Monotonicity) The operators F^κ are non-decreasing in the second parameter, i.e.
$$F^\kappa(x, r, q, X, u, \phi) \leq F^\kappa(x, s, q, X, u, \phi)$$
holds for all $r \leq s$.

Note that the preceding statements are supposed to hold for every $x \in \Omega$, $(x_n)_{n \in \mathbb{N}} \subset \Omega$, $r, s \in \mathbb{R}$, $q \in \mathbb{R}^d$, $X, Y \in \mathbb{S}^{d \times d}$, $u, v \in SC_p(\Omega)$, $(u_n)_{n \in \mathbb{N}} \subset SC_p(\Omega)$, $\phi, \psi \in C^2(\Omega)$ and $(\phi_n)_n \subset C^2(\Omega)$, if not stated otherwise.

In order to compare our assumptions with the existing literature, let us take a look at an example: An archetypical operator in the context of sublinear Markov semigroups is

$$F^\kappa(x, r, p, X, u, \phi) = -\sup_{\alpha \in \mathcal{A}} \Bigg(\int_{|z| \leq \kappa} \left(\phi(x + z) - \phi(x) - D\phi(x)z \, \mathbb{1}_{|z| \leq 1}\right) m_\alpha(dz)$$
$$+ \int_{|z| > \kappa} \left(u(x + z) - u(x) - D\phi(x)z \, \mathbb{1}_{|z| \leq 1}\right) m_\alpha(dz) \Bigg)$$

for a family of (non-negative) Borel measures $(m_\alpha)_{\alpha \in \mathcal{A}} \subset \mathfrak{M}^+(\mathbb{R}^d \setminus \{0\})$ such that
$$\sup_{\alpha \in \mathcal{A}} \int \left(1 \wedge |z|^2\right) m_\alpha(dz) < \infty.$$

This operator corresponds to a (pure-jump) Lévy process for sublinear expectations (as we will see in Chapter 4), which can be interpreted as a classical (pure-jump) Lévy process under uncertainty in its associated jump measures. One reasonable choice of uncertain jump measures would be $(m_\alpha)_{\alpha \in \mathcal{A}} = (\delta_h)_{|h| \in (1,2)}$, which leads to

$$F^\kappa(x, r, p, X, u, \phi) = -\sup_{|h| \in (1,2)} (u(x + h) - u(x)) = \inf_{|h| \in (1,2)} (u(x) - u(x + h)),$$

and corresponds to a classical Poisson process under uncertainty in its jump-height.

It is easy to check that this operator satisfies all assumptions from Remark 2.3. In particular, for fixed $x \in \mathbb{R}^d$ the map

$$C(\mathbb{R}^d) \ni u \longmapsto F^\kappa(x,r,p,X,u,\phi) = \inf_{|h| \in (1,2)} (u(x) - u(x+h))$$

is continuous with respect to local uniform convergence, but not with respect to (majorized) pointwise convergence. However, in all relevant articles known to the author (such as [1], [108], [75] or [12]) uniqueness results are restricted to the case, where the related operators are continuous with respect to (majorized) pointwise convergence. This usually stems from the fact that those articles assume that the Borel measures

$$(m_\alpha)_{\alpha \in \mathcal{A}} \subset \mathfrak{M}^+(\mathbb{R}^d \setminus \{0\})$$

are dominated by a single Borel measure, which allows them to employ the dominated convergence theorem – an assumption that is typically satisfied for applications related to stochastic control or (backwards) stochastic differential equations. The only exceptions known to the author are the results in [67] and the subsequent work [93], which deal with Lévy processes for sublinear expectations. However, these articles only consider spatially homogeneous equations on the whole space, together with stronger assumptions on the singularity of the related measures and on the behavior of the solutions at infinity. Since the continuity assumption (A4) from Remark 2.3 is at the heart of many proofs related to uniqueness results, we generalize the existing theory from [108] and [75] in this chapter to cover all equations under our broader assumptions. Beyond the generalized continuity assumption, our results also extend the existing theory by covering solutions with arbitrary polynomial growth at infinity and the comparison of multiple viscosity solutions at the same time (e.g. to obtain convexity results). Throughout this chapter, we will provide additional, more detailed comparisons of our results to the existing literature, while discussing the relevant parts.

As mentioned before, there exist several slightly different notions of viscosity solutions in the literature that are (under suitable conditions on the operators F and F^κ) equivalent to the one given in Definition 2.1. We start this discussion by showing that it suffices to scan viscosity solutions with a much smaller class of functions.

Lemma 2.4 (Choice of Scanning Functions). *Suppose that $p \geq 0$, that $\Omega \subset \mathbb{R}^d$ is open and that the consistency (A1), translation invariance (A3) and continuity assumption (A4) from Remark 2.3 hold. A function $u \in SC_p(\mathbb{R}^d)$ is a viscosity subsolution (viscosity supersolution) of (E1) in Ω if and only if*

$$F(x, u(x), D\phi(x), D^2\phi(x), \phi(\cdot)) \leq 0 \ (\geq 0) \tag{2.1}$$

is satisfied for each $\phi \in C_p^\infty(\mathbb{R}^d)$ such that $u - \phi$ has a strict global maximum (strict global minimum) in $x \in \Omega$ with $u(x) = \phi(x)$.

Proof. The only-if part is trivial because it states that inequality (2.1) holds for a smaller class of scanning functions. The opposite direction can be proved as follows:

Suppose that $\phi \in C_p^2(\mathbb{R}^d)$ such that $u - \phi$ has a global maximum in $\bar{x} \in \Omega$. A standard convolution argument (cf. Theorem 1.2) shows that there exists $\delta_n \in (0, 1)$ such that

$$\sup_{|x-\bar{x}|\leq 2n} |\phi_n(x) - \phi(x)| \leq \tfrac{1}{2} n^{-8}$$

holds with $\phi_n := \phi * \eta_{\delta_n} \in C_p^\infty(\mathbb{R}^d)$ for $n \in \mathbb{N}$, where $\eta_\delta \in C_b^\infty(\mathbb{R}^d)$ for $\delta > 0$ is the standard mollifier as in Lemma 1.1. Moreover, according to Proposition 1.7 there exists $\chi \in C^\infty(\mathbb{R}^d)$ with $\mathbf{1}_{B[0,1]}(x) \leq \chi(x) \leq \mathbf{1}_{B(0,2)}(x)$ for all $x \in \mathbb{R}^d$. Define

$$\psi_n(x) := \phi_n(x) + \left(2\hat{C}\left(1 + |x - \bar{x}|^p\right) + (\phi_n(\bar{x}) - u(\bar{x}))^+\right)(1 - \chi_n(x - \bar{x}))$$
$$+ n^{-4} |x - \bar{x}|^4 \chi_{2n}(x - \bar{x})$$

for $x \in \mathbb{R}^d$ and $n \in \mathbb{N}$ with $\chi_n(x) := \chi(n^{-1} x)$. The constant $\hat{C} > 0$ is chosen such that

$$|\phi_n(x)| = \left|\int_{|y|\leq \delta_n} \eta_{\delta_n}(y)\phi(x-y)\,dy\right| \leq \sup_{|y|\leq 1}|\phi(x-y)| \leq \hat{C}(1 + |x - \bar{x}|^p)$$

as well as $|\phi(x)| \vee |u(x)| \leq \hat{C}(1 + |x - \bar{x}|^p)$ holds for all $x \in \mathbb{R}^d$ and every $n \in \mathbb{N}$.

We can show that $u - \psi_n$ has a global maximum in some $\bar{x}_n \in \mathbb{R}^d$ with $|\bar{x}_n - \bar{x}| \leq n^{-1}$: For $x \in \mathbb{R}^d$ with $|x - \bar{x}| \geq 2n$, we have $\chi_n(x - \bar{x}) = 0$ and hence

$$u(x) - \psi_n(x) \leq u(x) - \phi_n(x) - \left(2\hat{C}(1 + |x - \bar{x}|^p) + (\phi_n(\bar{x}) - u(\bar{x}))^+\right)$$
$$\leq 2\hat{C}(1 + |x - \bar{x}|^p) - \left(2\hat{C}(1 + |x - \bar{x}|^p) + (\phi_n(\bar{x}) - u(\bar{x}))\right)$$
$$= u(\bar{x}) - \phi_n(\bar{x}) = u(\bar{x}) - \psi_n(\bar{x}),$$

using the definition of the constant $\hat{C} > 0$. For $x \in \mathbb{R}^d$ with $n^{-1} \leq |x - \bar{x}| \leq 2n$, on the other hand, we have $\chi_{2n}(x - \bar{x}) = 1$ and therefore

$$u(x) - \psi_n(x) \leq u(x) - \phi_n(x) \qquad - n^{-4}|x - \bar{x}|^4$$
$$\leq u(x) - \phi(x) + \tfrac{1}{2} n^{-8} - n^{-4}|x - \bar{x}|^4$$
$$\leq u(\bar{x}) - \phi(\bar{x}) + \tfrac{1}{2} n^{-8} - n^{-4}|x - \bar{x}|^4$$
$$\leq u(\bar{x}) - \phi_n(\bar{x}) + n^{-8} - n^{-4}|x - \bar{x}|^4$$
$$\leq u(\bar{x}) - \psi_n(\bar{x}),$$

using $\bar{x} \in \arg\max (u - \phi)$, $\sup_{|x-\bar{x}|\leq 2n} |\phi_n(x) - \phi(x)| \leq \tfrac{1}{2} n^{-8}$ and $n^{-8} - n^{-4}|x - \bar{x}|^4 \leq 0$. Thus, Weierstrass' theorem together with the (semi-)continuity of $u(\cdot)$ and $\psi_n(\cdot)$ implies

$$\sup\{u(x) - \psi_n(x) : x \in \mathbb{R}^d\} = \sup\{u(x) - \psi_n(x) : |x - \bar{x}| \leq n^{-1}\}$$
$$= u(\bar{x}_n) - \psi_n(\bar{x}_n)$$

for some $\bar{x}_n \in \mathbb{R}^d$ with $|\bar{x}_n - \bar{x}| \leq n^{-1}$.

2. Integro-Differential Equations

Finally, if we define the functions $\vartheta_n \in C_p^\infty(\mathbb{R}^d)$ by

$$\vartheta_n(x) := \psi_n(x) + n^{-1}|x - \bar{x}_n|^4 \chi(x - \bar{x}) + n^{-1}(1 - \chi(x - \bar{x})) + u(\bar{x}_n) - \psi_n(\bar{x}_n)$$

for $x \in \mathbb{R}^d$ and $n \in \mathbb{N}$, then the differences $u - \vartheta_n$ have strict global maxima in $\bar{x}_n \in \mathbb{R}^d$ with $|\bar{x}_n - \bar{x}| \leq n^{-1}$ and $\hat{\psi}_n(\bar{x}_n) = u(\bar{x}_n)$. In particular, our assumption implies

$$F(\bar{x}_n, u(\bar{x}_n), D\vartheta_n(\bar{x}_n), D^2\vartheta_n(\bar{x}_n), \vartheta_n(\cdot)) \leq 0$$

for large $n \in \mathbb{N}$, since $\Omega \subset \mathbb{R}^d$ is open and therefore $\bar{x}_n \in \Omega$ for large $n \in \mathbb{N}$. It remains to prove that we can employ the continuity assumption (A4): Remark 1.5 shows that

$$\lim_{n \to \infty} D^k \phi_n = D^k \phi$$

locally uniformly for $k \in \{0, 1, 2\}$. Our construction of ϑ_n from ϕ_n for $n \in \mathbb{N}$ thus implies

$$\lim_{n \to \infty} D^k \vartheta_n = D^k \phi$$

locally uniformly for $k \in \{0, 1, 2\}$. Moreover, an easy calculation leads to

$$|\vartheta_n(x)| \leq |\psi_n(x)| + n^{-1}(3^4 + 1) + 2\hat{C}(1 + |\bar{x}_n - \bar{x}|^p)$$
$$\leq \left(|\phi_n(x)| + 2\hat{C}(1 + |x - \bar{x}|^p) + 2\hat{C}(1 + |\bar{x} - \bar{x}|^p) + n^{-4}(4n)^4\right) + 82 + 4\hat{C}$$
$$\leq 3\hat{C}(1 + |x - \bar{x}|^p) + 6\hat{C} + 338$$

for all $x \in \mathbb{R}^d$, which shows that $\sup_{n \in \mathbb{N}} \|\vartheta_n\|_p < \infty$ holds. Furthermore, note that

$$u(\bar{x}_n) = u(\bar{x}_n) - \vartheta_n(\bar{x}_n) + \vartheta_n(\bar{x}_n) \geq u(\bar{x}) - \vartheta_n(\bar{x}) + \vartheta_n(\bar{x}_n)$$

from $\bar{x}_n \in \arg\max(u - \vartheta_n)$ for $n \in \mathbb{N}$. Hence, the upper semicontinuity of $u(\cdot)$ and $\lim_{n \to \infty} \vartheta_n = \phi$ locally uniformly imply

$$u(\bar{x}) \geq \limsup_{n \to \infty} u(\bar{x}_n) \geq \liminf_{n \to \infty} u(\bar{x}_n) \geq \liminf_{n \to \infty} \left(u(\bar{x}) - \vartheta_n(\bar{x}) + \vartheta_n(\bar{x}_n)\right) = u(\bar{x})$$

and therefore $\lim_{n \to \infty} u(\bar{x}_n) = u(\bar{x})$. Combining all of these findings leads to

$$F(\bar{x}, u(\bar{x}), D\phi(\bar{x}), D^2\phi(\bar{x}), \phi(\cdot)) = \lim_{n \to \infty} F(\bar{x}_n, u(\bar{x}_n), D\vartheta_n(\bar{x}_n), D^2\vartheta_n(\bar{x}_n), \vartheta_n(\cdot)) \leq 0,$$

using the continuity assumption (A4). $\square$

It is worth mentioning that the arguments to restrict the class of scanning functions to functions $\phi : \mathbb{R}^d \to \mathbb{R}$, for which $u - \phi$ has a strict global extremum in $x \in \Omega$ with $\phi(x) = u(x)$, are more or less standard (at least for bounded solutions). The restriction to functions with higher smoothness, however, appears not to be covered explicitly by the standard literature. Nevertheless, it seems to be very useful, since the current literature uses several different smoothness assumptions for scanning functions, especially in articles dealing with applications of viscosity solution theory. The following remark shows that, in case of bounded solutions, we can restrict our attention to the even smaller class of scanning functions with bounded derivatives of all orders.

2.1. Viscosity Solutions

Remark 2.5 (Bounded Derivatives of Scanning Functions). It is possible to restrict the class of scanning functions for bounded viscosity solutions (i.e. $p=0$) to

$$\phi \in C_b^\infty(\mathbb{R}^d) = \{f \in C^\infty(\mathbb{R}^d) \mid \forall k \in \mathbb{N}_0 : \|D^k f\|_\infty < \infty\}$$

under the assumptions in Remark 2.3. This can be implied from a thorough inspection of the proof of Lemma 2.4: For $p=0$, the approximations $\vartheta_n \in C^\infty(\mathbb{R}^d)$ are sums of smooth functions with compact support (which always have bounded derivatives of all orders), constants and $\phi_n \in C^\infty(\mathbb{R}^d)$ with

$$\phi_n(x) := (f * \eta_{\delta_n})(x) = \int_{\mathbb{R}^d} f(y) \eta_{\delta_n}(x-y) dy$$

for $x \in \mathbb{R}^d$ and $\delta_n \in (0,1)$, where $\eta_\delta \in C^\infty(\mathbb{R}^d)$ for $\delta > 0$ is the standard mollifier with compact support from Lemma 1.1. Due to the compact support and the identity

$$D^k \phi_n = D^k(\phi * \eta_{\delta_n}) = (\phi * D^k \eta_{\delta_n})$$

for $n \in \mathbb{N}$, the approximations $\phi_n \in C^\infty(\mathbb{R}^d)$ have bounded derivatives of all orders, which implies $\vartheta_n \in C_b^\infty(\mathbb{R}^d)$ for $n \in \mathbb{N}$, as required.

Another delicate problem (for the viscosity solution theory of nonlocal equations) is the choice of domains for solutions and scanning functions. Many articles (such as [1] and [12]) only treat equations on the whole space $\mathbb{R}^d$, where solutions and scanning functions obviously have to be defined on the whole space. However, in articles that consider equations in open subsets $\Omega \subset \mathbb{R}^d$, the appropriate choice of the domain of solutions and scanning functions is an important question. The Dirichlet boundary conditions necessary for the uniqueness of solutions for nonlocal equations in general require the prescription of the solution on the whole complement

$$\Omega^C = \mathbb{R}^d \setminus \Omega$$

instead of only the boundary $\partial \Omega$ (cf. e.g. [63]). It therefore seems natural to assume that solutions and scanning functions for nonlocal equations in an open subset $\Omega \subset \mathbb{R}^d$ are still defined on the whole space $\mathbb{R}^d$. In fact, the corrigendum [4] of [3] shows exemplarily that it can be problematic to assume that viscosity solutions are only defined in $\Omega \subset \mathbb{R}^d$ (or on the closure $\overline{\Omega}$) instead of the whole $\mathbb{R}^d$ for proving uniqueness results: The majority of uniqueness results for general nonlocal equations are based on approximations by supremal convolutions (as introduced in Section 1.2) and related stability arguments (as discussed later in Section 2.2). These stability arguments get more involved as soon as the approximated functions are not defined on the whole space anymore. In the following lemma, we will use our generalization of Tietze's extension theorem for semicontinuous functions from Theorem 1.26, to show that the domain of solutions and scanning functions can be restricted, as long as the nonlocal part only depends on values in an open set $\Omega \subset \mathbb{R}^d$. This naturally solves the difficulties described in [4] (and similar issues in [75], which we will discuss after Lemma 2.8), since it allows us to work (without

2. Integro-Differential Equations

loss of generality) with solutions and scanning functions defined on the whole space. In particular, this will prove itself valuable to apply elliptic results to parabolic problems, where it seems natural to work with solutions, which are only defined on

$$[0, T] \times \mathbb{R}^d \subset \mathbb{R}^{1+d}$$

for some $T > 0$ instead of the whole space $\mathbb{R}^{1+d}$.

Lemma 2.6 (Domain of Solutions). *Suppose that $p \geq 0$, that $\Omega \subset \mathbb{R}^d$ is open and that the consistency (A1) and continuity assumption (A4) are satisfied for two operators*

$$\hat{F} : \Omega \times \mathbb{R} \times \mathbb{R}^d \times \mathbb{S}^{d \times d} \times C_p^2(\Omega) \longrightarrow \mathbb{R}$$
$$F : \mathbb{R}^d \times \mathbb{R} \times \mathbb{R}^d \times \mathbb{S}^{d \times d} \times C_p^2(\mathbb{R}^d) \longrightarrow \mathbb{R}$$

with $F(x, r, q, X, \phi) = \hat{F}(x, r, q, X, \phi|_\Omega)$ for all $(x, r, q, X, \phi) \in \Omega \times \mathbb{R} \times \mathbb{R}^d \times \mathbb{S}^{d \times d} \times C_p^2(\mathbb{R}^d)$. A function $\hat{u} \in USC_p(\overline{\Omega})$ ($\hat{u} \in LSC_p(\overline{\Omega})$) satisfies

$$\hat{F}(x, \hat{u}(x), D\hat{\phi}(x), D^2\hat{\phi}(x), \hat{\phi}(\cdot)) \leq 0 \ (\geq 0) \tag{2.2}$$

for every $\hat{\phi} \in C_p^2(\Omega)$, for which $\hat{u} - \hat{\phi}$ has a global maximum (global minimum) in $x \in \Omega$, if and only if there exists a viscosity subsolution (supersolution) $u \in SC_p(\mathbb{R}^d)$ in Ω of

$$F(x, u(x), Du(x), D^2u(x), u(\cdot)) = 0$$

in terms of Definition 2.1 with $u|_\Omega = \hat{u}$.

Proof. Assume that $\hat{u} \in USC_p(\overline{\Omega})$ such that Equation (2.2) holds for each $\hat{\phi} \in C_p^2(\Omega)$ and every global maximum point $x \in \Omega$ of $\hat{u} - \hat{\phi}$. Define $u : \mathbb{R}^d \to \mathbb{R}$ by $u|_{\overline{\Omega}} := \hat{u}$ and

$$u(x) := \inf_{y \in \Omega} \left\{ \frac{\hat{u}(y)}{1 + |y|^p} + \frac{|x - y|}{d(x, \Omega)} - 1 \right\} \cdot (1 + |x|^p)$$

for all $x \in \mathbb{R}^d \setminus \overline{\Omega}$. Theorem 1.26 shows that $u \in USC_p(\mathbb{R}^d)$. If $\phi \in C_p^2(\mathbb{R}^d)$ such that $u - \phi$ has a global maximum in $x \in \Omega$, then $\hat{u} - \hat{\phi}$ with

$$\hat{\phi} := \phi|_\Omega \in C_p^2(\Omega)$$

has a global maximum in $x \in \Omega$ as well. Therefore, by assumption

$$\hat{F}(x, \hat{u}(x), D\hat{\phi}(x), D^2\hat{\phi}(x), \hat{\phi}(\cdot)) \leq 0.$$

The relationship between the operators F and $\hat{F}$, $u(x) = \hat{u}(x)$ and $D^k\phi(x) = D^k\hat{\phi}(x)$ for $k \in \{1, 2\}$ (by definition of $u \in USC_p(\mathbb{R}^d)$ and $\hat{\phi} \in C_p^2(\Omega)$) finally lead to

$$F(x, u(x), D\phi(x), D^2\phi(x), \phi(\cdot)) = \hat{F}(x, \hat{u}(x), D\hat{\phi}(x), D^2\hat{\phi}(x), \hat{\phi}(\cdot)) \leq 0,$$

which shows that $u \in USC_p(\mathbb{R}^d)$ is a viscosity subsolution in Ω of $F = 0$.

In order to prove the converse, suppose that $u \in \mathrm{USC}_p(\mathbb{R}^d)$ is a viscosity subsolution in Ω of $F = 0$ in terms of Definition 2.1. Obviously, by definition

$$\hat{u} = u|_{\overline{\Omega}} \in \mathrm{USC}_p(\overline{\Omega}).$$

Assume that $\hat{\phi} \in C_p^2(\Omega)$ such that $\hat{u} - \hat{\phi}$ has a global maximum in $\bar{x} \in \Omega$. Proposition 1.7 implies the existence of $\chi_n \in C^\infty(\mathbb{R}^d)$ for every $n \in \mathbb{N}$ such that $0 \leq \chi_n \leq 1$, $\chi_n \equiv 1$ on

$$\{x \in B[0,n] : d(x, \mathbb{R}^d \setminus \Omega) \geq n^{-1}\}$$

and $\chi_n \equiv 0$ outside Ω. Moreover, Lemma 1.17 shows the existence of $\psi \in C^\infty(\mathbb{R}^d)$ with

$$2^{-1}(1+|x|^p) \leq \psi(x) \leq (1+|x|^p)$$

for all $x \in \mathbb{R}^d$. Let $C > \|u\|_p = \sup_{x \in \mathbb{R}^d}(1+|x|^p)^{-1}|u(x)|$ and define $\phi_n \in C_p^2(\mathbb{R}^d)$ by

$$\phi_n(x) := \chi_n(x)\hat{\phi}(x) + \big(1 - \chi_n(x)\big)\big(2C\psi(x) + \hat{\phi}(\bar{x}) - \hat{u}(\bar{x})\big)$$

for all $x \in \mathbb{R}^d$ and $n \in \mathbb{N}$. Since $\chi_n(\bar{x}) = 1$ for large $n \in \mathbb{N}$, and since $\bar{x} \in \Omega$ is a global maximum point of $\hat{u} - \hat{\phi}$, we have $\phi_n(\bar{x}) = \hat{\phi}(\bar{x})$ as well as

$$\hat{\phi}(x) \geq \hat{u}(x) - (\hat{u}(\bar{x}) - \hat{\phi}(\bar{x})) = u(x) - (u(\bar{x}) - \phi_n(\bar{x}))$$

for all $x \in \Omega$ and large $n \in \mathbb{N}$. But since $2C\psi(x) \geq C(1+|x|^p) \geq u(x)$ for all $x \in \mathbb{R}^d$ by construction, we also have

$$2C\psi(x) + \hat{\phi}(\bar{x}) - \hat{u}(\bar{x}) \geq u(x) + \hat{\phi}(\bar{x}) - \hat{u}(\bar{x}) = u(x) - (u(\bar{x}) - \phi_n(\bar{x}))$$

for all $x \in \mathbb{R}^d$ and large $n \in \mathbb{N}$. In particular, a combination of these inequality implies

$$\phi_n(x) = \chi_n(x)\hat{\phi}(x) + \big(1 - \chi_n(x)\big)\big(2C\psi(x) + \hat{\phi}(\bar{x}) - \hat{u}(\bar{x})\big) \geq u(x) - (u(\bar{x}) - \phi_n(\bar{x}))$$

for all $x \in \mathbb{R}^d$ and large $n \in \mathbb{N}$. In other words, $u - \phi_n$ has a global maximum in $\bar{x} \in \Omega$ for large $n \in \mathbb{N}$. Therefore, by assumption

$$F(\bar{x}, u(\bar{x}), D\phi_n(\bar{x}), D^2\phi_n(\bar{x}), \phi_n(\cdot)) \leq 0$$

for large $n \in \mathbb{N}$. Since $\phi_n \equiv \hat{\phi}$ in a neighborhood of $\bar{x} \in \Omega$ for large $n \in \mathbb{N}$, we find $D^k\phi_n(\bar{x}) = D^k\hat{\phi}(\bar{x})$ for $k \in \{1,2\}$. Similarly, $\phi_n \equiv \hat{\phi}$ locally for large $n \in \mathbb{N}$ implies that $\lim_{n \to \infty} D^k\phi_n|_\Omega = D^k\hat{\phi}$ locally uniformly for $k \in \{0,1,2\}$. Finally, since $\sup_{n \in \mathbb{N}} \|\phi_n\|_p \leq \|\hat{\phi}\|_p + \|2C\psi + \hat{\phi}(\bar{x}) - \hat{u}(\bar{x})\|_p < \infty$ by definition, we find

$$\hat{F}(\bar{x}, \hat{u}(\bar{x}), D\hat{\phi}(\bar{x}), D^2\hat{\phi}(\bar{x}), \hat{\phi}(\cdot)) = \lim_{n \to \infty} \hat{F}(\bar{x}, u(\bar{x}), D\phi_n(\bar{x}), D^2\phi_n(\bar{x}), \phi_n|_\Omega(\cdot)) \leq 0,$$

using the continuity assumption (A4) and the relationship between F and $\hat{F}$. □

Since the extension of $\hat{u}$ to u in the proof of Lemma 2.6 is the same for viscosity sub- and supersolutions, we also proved that $\hat{u} \in C_p(\overline{\Omega})$ satisfies

$$\hat{F}(x, \hat{u}(x), D\hat{\phi}(x), D^2\hat{\phi}(x), \hat{\phi}(\cdot)) \leq 0 \ (\geq 0)$$

for each $\hat{\phi} \in C_p^2(\Omega)$ such that $\hat{u} - \hat{\phi}$ has a global maximum (global minimum) in $x \in \Omega$ if and only if there exists a viscosity solution $u \in C_p(\mathbb{R}^d)$ in Ω of

$$F(x, u(x), Du(x), D^2u(x), u(\cdot)) = 0$$

in terms of Definition 2.1 with $u|_\Omega = \hat{u}$.

As discussed earlier, we separated the singular part of the nonlocal part from its rest in the generalized operators F^κ (compared to the original operator F). In Section 2.2, we will assume that the influence of the singular part vanishes as κ tends to zero, in order to develop a uniqueness theory. Some articles (such as [9]) use this generalized version F^κ of the operator F to define a related notion of viscosity solutions directly. In the following lemma, we extend a standard argument, which can already be found in [118] in a simple form. It shows that these two (a-priorily different) notions coincide in our set-up under an additional continuity assumption.

Lemma 2.7 (Characterization of Solutions). *Suppose that $p \geq 0$, that $\Omega \subset \mathbb{R}^d$ is open and that all assumptions from Remark 2.3 hold. A continuous function $u \in C_p(\mathbb{R}^d)$ is a viscosity subsolution (viscosity supersolution) of $F = 0$ in Ω if and only if*

$$F^\kappa(x, u(x), D\phi(x), D^2\phi(x), u(\cdot), \phi(\cdot)) \leq 0 \ (\geq 0) \tag{2.3}$$

for all $\phi \in C^2(\mathbb{R}^d)$ such that $u - \phi$ has a global maximum (global minimum) in $x \in \Omega$.

Proof. Suppose that $u \in C_p(\mathbb{R}^d)$ is a viscosity subsolution of $F = 0$ as in Definition 2.1, and that $\phi \in C_p^2(\mathbb{R}^d)$ such that $u - \phi$ has a global maximum at $x \in \Omega$. Without loss of generality, we can assume that $\phi(x) = u(x)$ holds. Otherwise, replace $\phi : \mathbb{R}^d \to \mathbb{R}$ by

$$y \longmapsto \phi(y) - (\phi(x) - u(x))$$

and use translation invariance assumption (A3). According to Theorem 1.20, there exists $(\phi_n)_{n \in \mathbb{N}} \subset C_p^2(\mathbb{R}^d)$ such that $u \leq \phi_n \leq \phi$ holds, $u - \phi_n$ still has a global maximum in $x \in \Omega$ and $\phi_n \downarrow u$ pointwise as $n \to \infty$. In particular, this implies that $\phi_n - \phi$ also has a global maximum in $x \in \Omega$, and hence $D\phi_n(x) = D\phi(x)$ as well as $D^2\phi_n(x) \leq D^2\phi(x)$ for all $n \in \mathbb{N}$. Moreover, the degenerate ellipticity assumption (A2), the consistency assumption (A1) and Definition 2.1 imply that

$$F^\kappa(x, u(x), D\phi(x), D^2\phi(x), \phi_n(\cdot), \phi(\cdot)) \leq F^\kappa(x, u(x), D\phi_n(x), D^2\phi_n(x), \phi_n(\cdot), \phi_n(\cdot))$$
$$= F(x, u(x), D\phi_n(x), D^2\phi_n(x), \phi_n(\cdot)) \leq 0$$

for all $n \in \mathbb{N}$. Since $u \in C_p(\mathbb{R}^d)$ is continuous, Dini's theorem (cf. Theorem 1.27) implies that the convergence $\lim_{n\to\infty} \phi_n = u$ is locally uniform, and hence Equation (2.3) follows from the continuity assumption (A4).

2.1. Viscosity Solutions

In order to prove the converse, suppose that Equation (2.3) holds for all $\phi \in C^2(\mathbb{R}^d)$ such that $u - \phi$ has a global maximum in $x \in \Omega$. Now, let $\phi \in C_p^2(\mathbb{R}^d)$ such that $x \in \Omega$ is a global maximum point of $u - \phi$. The consistency assumption (A1), the degenerate ellipticity assumption (A2) and Equation (2.3) imply

$$F(x, u(x), D\phi(x), D^2\phi(x), \phi(\cdot)) = F^\kappa(x, u(x), D\phi(x), D^2\phi(x), \phi(\cdot), \phi(\cdot))$$
$$\leq F^\kappa(x, u(x), D\phi(x), D^2\phi(x), u(\cdot), \phi(\cdot)) \leq 0.$$

In particular, $u \in C_p(\mathbb{R}^d)$ is a viscosity subsolution of $F = 0$ in terms of Definition 2.1. □

A careful inspection shows that we only used the continuity of $u : \mathbb{R}^d \to \mathbb{R}$ in the preceding proof once, in order to show that the approximation $\phi_n \to u$ is locally uniform instead of pointwise (by employing Dini's theorem, cf. Theorem 1.27). In particular, if our operators were even continuous for pointwise convergence as in the existing literature, we would get the same result for semicontinuous functions as well. However, in our generalized set-up, the continuity of $u : \mathbb{R}^d \to \mathbb{R}$ is necessary to prove the alternative characterization in Lemma 2.7: Consider the equation related to

$$F^\kappa(x, r, p, X, u, \phi) = -\sup_{|h| \in (1,2)} (u(x+h) - u(x)) = \inf_{|h| \in (1,2)} (u(x) - u(x+h))$$

as in the discussion after Remark 2.3, which corresponds to a classical Poisson process under uncertainty in its jump-height. It is easy to check that $u = \mathbf{1}_{B[0,1]} \in \mathrm{USC}(\mathbb{R}^d)$ is a viscosity subsolution of $F = 0$ in $\mathbb{R}^d$ in terms of Definition 2.1, but

$$F^\kappa(0, u(0), D\phi(0), D^2\phi(0), u, \phi) = \inf_{|h| \in (1,2)} (u(0) - u(0+h)) = 1 - 0 > 0$$

for every $\phi \in C^2(\mathbb{R}^d)$ and $\kappa \in (0,1)$. This counterexample illustrates the difficulties one has to overcome in order to develop a viscosity solution theory for our weaker continuity assumptions. Nevertheless, we will prove a comparison principle for semicontinuous functions (and not only continuous functions) in Section 2.2, which is important for Perron's method in Remark 2.24.

Some articles (such as [1]) not only separate the singular from the remaining nonlocal part as in Lemma 2.7, but also the local part from the nonlocal part by employing second-order sub- and superjets (which play an important role in the viscosity solution theory for local equations, cf. [35]): Instead of $(D\phi(x), D^2\phi(x)) \in \mathbb{R}^d \times \mathbb{S}^{d \times d}$ in Equation (2.3), one can use $(p, X) \in \mathbb{R}^d \times \mathbb{S}^{d \times d}$ that satisfy

$$u(x) + \langle p, y - x \rangle + \tfrac{1}{2} \langle X(y-x), (y-x) \rangle + o(|y-x|^2) \geq u(x) \ (\leq u(x))$$

as $y \to x$, in order to define another notion of viscosity solutions for nonlocal equations. Using a similar approach as in the proof of Lemma 2.7, the approximation result in Theorem 1.23 allows us to show that this notion is also equivalent to Definition 2.1. In fact, we will use (the trivial direction of) this alternative characterization implicitly in Section 2.2, in order to simplify certain compactness arguments.

2. Integro-Differential Equations

It is worth mentioning that the results in this section can be extended to treat generalized boundary conditions as well (i.e. boundary conditions that are only satisfied in the viscosity sense, cf. e.g. [35, Section 7] or [11]), as long as the solutions and scanning functions are defined on the whole space $\mathbb{R}^d$. Since our motivation in this thesis is to study the sublinear generator equations of sublinear Markov semigroups, we are will restrict our attention to classical Dirichlet boundary conditions in this chapter, and leave the treatment of generalized boundary conditions for future work.

2.2. Uniqueness of Solutions

One intrinsic and important property of nonlocal degenerate elliptic equations (as introduced in Section 2.1) is the so-called comparison principle. Broadly speaking, it shows that solutions comparable at the boundary are already comparable everywhere, which immediately implies the uniqueness of related Dirichlet problems. In order to motivate this idea, suppose that $u, v \in C^2(\mathbb{R}^d)$ are classical solutions of

$$F(x, u(x), Du(x), D^2u(x), u) \leq 0$$
$$F(x, v(x), Dv(x), D^2v(x), u) \geq 0$$

in $x \in \Omega$ for a bounded and open set $\Omega \subset \mathbb{R}^d$, and that $u(x) \leq v(x)$ for all $x \in \mathbb{R}^d \setminus \Omega$. Under some mild additional assumptions, e.g. there exists $\lambda > 0$ such that

$$F(x, r, p, X, \phi) - F(x, s, p, X, \phi) \geq \lambda(r - s) \tag{2.4}$$

holds for all $r > s$ and $(x, p, X, \phi) \in \mathbb{R}^d \times \mathbb{R}^d \times \mathbb{S}^{d \times d} \times C_p(\mathbb{R}^d)$, this already implies that

$$u(x) \leq v(x)$$

for all $x \in \mathbb{R}^d$: Suppose that $\sup_{x \in \mathbb{R}^d} (u(x) - v(x)) > 0$, then

$$\sup_{x \in \mathbb{R}^d} (u(x) - v(x)) = \sup_{x \in \overline{\Omega}} (u(x) - v(x)) = u(\bar{x}) - v(\bar{x}) > 0 \tag{2.5}$$

for some $\bar{x} \in \Omega$, using $u \leq v$ in $\mathbb{R}^d \setminus \Omega$ and the extreme value theorem together with the compactness of $\overline{\Omega} \subset \mathbb{R}^d$. Since $u - v$ has a global maximum in $\bar{x} \in \Omega$, we have $Du(\bar{x}) = Dv(\bar{x})$ as well as $D^2u(\bar{x}) \leq D^2v(\bar{x})$ (according to the classical maximum principle for twice differentiable functions). The degenerate ellipticity thus implies

$$0 \geq F(\bar{x}, u(\bar{x}), Du(\bar{x}), D^2u(\bar{x}), u) - F(\bar{x}, v(\bar{x}), Dv(\bar{x}), D^2v(\bar{x}), v)$$
$$\geq F(\bar{x}, u(\bar{x}), Dv(\bar{x}), D^2v(\bar{x}), v) - F(\bar{x}, v(\bar{x}), Dv(\bar{x}), D^2v(\bar{x}), v)$$
$$\geq \lambda(u(\bar{x}) - v(\bar{x})) > 0,$$

which is obviously a contradiction and therefore shows that $\sup_{x \in \mathbb{R}^d} (u(x) - v(x)) \leq 0$.

2.2. Uniqueness of Solutions

In the parabolic case, which is our main interest due to its relation to sublinear Markov semigroups (cf. Chapter 4), we do not even need the additional assumption from (2.4): Suppose that $u, v \in C^2((0,T) \times \mathbb{R}^d) \cap C([0,T] \times \mathbb{R}^d)$ are solutions of

$$\partial_t u(t,x) + G(t,x,u(t,x),Du(t,x),D^2u(t,x),u(t,\cdot)) \leq 0$$
$$\partial_t v(t,x) + G(t,x,v(t,x),Dv(t,x),D^2v(t,x),v(t,\cdot)) \geq 0$$

in $(t,x) \in (0,T) \times \Omega$ and $u(t,x) \leq v(t,x)$ for $(t,x) \in \{0\} \times \mathbb{R}^d \cup (0,T) \times \Omega^C$, where $G : (0,T) \times \Omega \times \mathbb{R}^d \times \mathbb{S}^{d \times d} \times C^2(\mathbb{R}^d) \to \mathbb{R}$ is a nonlocal degenerate elliptic, translation invariant and monotone operator (as in Remark 2.3) and $\Omega \subset \mathbb{R}^d$ is open and bounded. If we assume that $\sup_{(t,x) \in [0,T] \times \mathbb{R}^d} (u(t,x) - v(t,x)) > 0$ holds, then

$$\sup_{(t,x) \in [0,T) \times \mathbb{R}^d} \left(u(t,x) - v(t,x) - \frac{\delta}{T-t} \right) = u(\bar{t},\bar{x}) - v(\bar{t},\bar{x}) - \frac{\delta}{T-\bar{t}} > 0 \quad (2.6)$$

for small $\delta > 0$ and some $(\bar{t},\bar{x}) \in (0,T) \times \Omega$, using $u \leq v$ on the parabolic boundary, the extreme value theorem and the growth of the penalization term as t tends to T. Due to the global maximum, we find $\partial_t u(\bar{t},\bar{x}) - \partial_t v(\bar{t},\bar{x}) = \delta(T-\bar{t})^{-1}$, $Du(\bar{t},\bar{x}) = Dv(\bar{t},\bar{x})$ and $D^2 u(\bar{t},\bar{x}) \leq D^2 v(\bar{t},\bar{x})$ (according to the classical maximum principle). The degenerate ellipticity, translation variance and monotonicity thus imply

$$\begin{aligned}
0 =\ & G(\bar{t},\bar{x},u(\bar{t},\bar{x}),Du(\bar{t},\bar{x}),D^2u(\bar{t},\bar{x}),u(\bar{t},\cdot)) - \delta(T-\bar{t})^{-1}) \\
& - G(\bar{t},\bar{x},u(\bar{t},\bar{x}),Du(\bar{t},\bar{x}),D^2u(\bar{t},\bar{x}),u(\bar{t},\cdot)) \\
\geq\ & G(\bar{t},\bar{x},v(\bar{t},\bar{x}),Dv(\bar{t},\bar{x}),D^2v(\bar{t},\bar{x}),v(\bar{t},\cdot)) \\
& - G(\bar{t},\bar{x},u(\bar{t},\bar{x}),Du(\bar{t},\bar{x}),D^2u(\bar{t},\bar{x}),u(\bar{t},\cdot)) \\
\geq\ & \partial_t u(\bar{t},\bar{x}) - \partial_t v(\bar{t},\bar{x}) = \delta(T-\bar{t})^{-1} > 0,
\end{aligned}$$

which is obviously a contradiction and therefore shows that $u \leq v$ in $[0,T] \times \mathbb{R}^d$. In this section, we show that these ideas can be generalized to (unbounded) viscosity solutions in possibly unbounded domains $\Omega \subset \mathbb{R}^d$. The main difficulties in this approach are: First, to generalize the classical maximum principle for twice differentiable to merely semicontinuous functions, which can be achieved by approximating the (semi-)solutions by supremal convolutions (as introduced in Section 1.2) and a technique called variable doubling. Second, to ensure that the suprema in Equation (2.5) and (2.6) are still attained for unbounded domains, which can be accomplished by introducing penalization terms (as demonstrated exemplarily in the parabolic case to ensure $\bar{t} < T$).

The origin of comparison principles for elliptic equations goes back to Carl Friedrich Gauss, who already used a weak maximum principle for the Laplace equation in [58]. Broadly speaking, the weak maximum principle (which is not to be confused with the maximum principle for twice differentiable functions used earlier) says that subsolutions are bounded above by their values on the boundary, which is a special form of the comparison principle for linear equations. It was later generalized by Eberhard Hopf in [64] to a large class of linear, second-order local equations. The first comparison result for viscosity solutions of nonlinear, first-order local equations was developed by

2. Integro-Differential Equations

Michael Crandall and Pierre-Louis Lions in [36]. Even though Pierre-Louis Lions already provided a comparison principle for the second-order case of local Hamilton-Jacobi-Bellman equations in [88] employing stochastic control methods, the first comparison principle for a general class of fully nonlinear, second-order local equations was obtained by Robert Jensen in [77] and is based on a generalization of the classical maximum principle for twice differentiable functions to semicontinuous functions. In the nonlocal setting, Awatif Sayah derived the first comparison principle for viscosity solutions of fully nonlinear, first-order nonlocal equations in [118], which was later generalized to second-order equations by Olivier Alvarez and Agnès Tourin in [1] for bounded measures and by Espen Jakobsen and Kenneth Karlsen in [75] for unbounded measures.

This section is structured as follows: In the first half, we refine the ideas of [75] to develop a maximum principle for nonlocal equations of the general form

$$F(x, u(x), Du(x), D^2u(x), u(\cdot)) = 0 \tag{E1}$$

in open domains $\Omega \subset \mathbb{R}^d$ for our generalized set-up from Remark 2.3. Our set-up covers operators F with significantly weaker continuity assumptions and viscosity solutions with arbitrary polynomial growth at infinity. In the second half, we apply this general maximum principle in the spirit of [67] and [108] (which only cover bounded viscosity solutions for either spatially homogeneous equations or special Hamilton-Jacobi-Bellman equations with stronger continuity assumptions), to obtain a comparison principle for a general class of parabolic nonlocal equations

$$\partial_t u(t, x) + G(t, x, u(t, x), Du(t, x), D^2u(t, x), u(t, \cdot)) = 0 \tag{E2}$$

in $(0, T) \times \mathbb{R}^d$ for viscosity (semi-)solutions $u : [0, T] \times \mathbb{R}^d \to \mathbb{R}$ with arbitrary polynomial growth at infinity. It is worth mentioning that, even though we restrict our attention to comparison principles for parabolic problems, our maximum principle is much more versatile and can also be used to obtain other comparison, continuous dependence, regularity or existence results as in [12], [74], [10] or [1].

As a start, we present a result that is concerned with the stability of viscosity solutions with respect to the approximation by supremal convolutions. This result was already anticipated in Section 1.2 as a motivation and underpins the importance of supremal convolutions to viscosity solution theory. In particular, remember that supremal convolutions are locally semiconvex according to Lemma 1.12 and therefore twice differentiable almost everywhere by Alexandrov's theorem from Theorem A.7. This argument indicates the suitability of supremal convolutions for viscosity solutions of second-order equations. The stability idea of the following result was first obtained by Robert Jensen, Pierre-Louis Lions and Panagiotis Souganidis in [78] for bounded solutions of second-order local equations and evolved into an essential tool for the standard viscosity solution theory over the years. Note that, due to our generalization of supremal convolutions in Section 1.2, we are able to treat viscosity solutions with arbitrary polynomial growth at infinity, whereas [75] only covers subquadratic growth and the authors claim that this assumption "does not seem so easy to remove".

2.2. Uniqueness of Solutions

Lemma 2.8 (Stability under Supremal Convolutions). *Suppose that $p \geq 0$, that $\Omega \subset \mathbb{R}^d$ is open, that the monotonicity assumption (A5) holds for operator F, and that*

$$\varphi = \varphi_{p \vee 2} \in C^\infty(\mathbb{R}^d)$$

is the quasidistance with $(p \vee 2)$-polynomial growth from Lemma 1.14. If $u \in SC_p(\mathbb{R}^d)$ is a viscosity subsolution in Ω of

$$F(x, u(x), Du(x), D^2 u(x), u(\cdot)) = 0$$

with $C > 0$ such that $|u(x)| \leq C(1 + |x|^p)$ for all $x \in \mathbb{R}^d$, and $0 < \varepsilon \leq (2^{3(p \vee 2) - 2} C)^{-1}$, then the supremal convolution (cf. Lemma 1.12)

$$u^\varepsilon(x) := \Delta_\varphi^\varepsilon[u](x) = \sup_{y \in \mathbb{R}^d} \left(u(y) - \frac{\varphi(x - y)}{\varepsilon} \right) \leq 2^{p + p \vee 2 - 1} C(1 + |x|^p)$$

is a viscosity subsolution in $\Omega_\varepsilon := \{ x \in \mathbb{R}^d : \inf_{y \in \mathbb{R}^d \setminus \Omega} \varphi(y - x) > \varepsilon \cdot \delta(x) \}$ of

$$\Delta_\varphi^\varepsilon[F](x, u^\varepsilon(x), Du^\varepsilon(x), D^2 u^\varepsilon(x), u^\varepsilon(\cdot)) = 0.$$

The smudged operator $\Delta_\varphi^\varepsilon[F] : \mathbb{R}^d \times \mathbb{R} \times \mathbb{R}^d \times \mathbb{S}^{d \times d} \times C_p^2(\mathbb{R}^d) \to \mathbb{R}$ is defined by

$$\Delta_\varphi^\varepsilon[F](x, r, q, X, \phi) := \inf \left\{ F(y, r, q, X, \phi \circ \tau_{x-y}) \;\middle|\; y \in \Omega \text{ with } \varphi(y - x) \leq \varepsilon \cdot \delta(x) \right\}$$

for $(x, r, q, X, \phi) \in \mathbb{R}^d \times \mathbb{R} \times \mathbb{R}^d \times \mathbb{S}^{d \times d} \times C_p^2(\mathbb{R}^d)$, where $\delta(x) := 2^{3(p \vee 2)} C(1 + \varphi(x))$ and $\tau_h : \mathbb{R}^d \to \mathbb{R}^d$ is the translation operator by $h \in \mathbb{R}^d$, i.e. $\tau_h(x) = x + h$ for all $x \in \mathbb{R}^d$.

Proof. We first show that the supremal convolution $u^\varepsilon : \mathbb{R}^d \to \mathbb{R}$ has p-polynomial growth: A case-by-case analysis for $x \in \mathbb{R}^d$ with $|x| \leq 1$ and $|x| \geq 1$ together with the inequalities from Lemma 1.14 imply that

$$(1 + |x|^p) \leq 2(1 + |x|^{p \vee 2}) \leq 2(1 + |x|^2 \vee |x|^{p \vee 2}) \leq 2^{p \vee 2 - 1}(1 + \varphi(x)) \quad (2.7)$$

holds for all $x \in \mathbb{R}^d$. According to Lemma 1.12 with

$$|u(x)| \leq C(1 + |x|^p) \leq 2^{p \vee 2 - 1} C(1 + \varphi(x))$$

for all $x \in \mathbb{R}^d$ and $0 < \varepsilon \leq (2 \cdot 2^{2(p \vee 2) - 2} \cdot 2^{p \vee 2 - 1} C)^{-1} = (2^{3(p \vee 2) - 2} C)^{-1}$, we have $u^\varepsilon \in C(\mathbb{R}^d)$ and there exists $\overline{x} \in \mathbb{R}^d$ for each $x \in \mathbb{R}^d$ such that

$$u^\varepsilon(x) = u(\overline{x}) - \varepsilon^{-1} \varphi(x - \overline{x})$$

and $\varphi(x - \overline{x}) \leq \varepsilon \cdot \delta(x)$ hold. Therefore, for all $x \in \mathbb{R}^d$

$$\begin{aligned}
-C(1 + |x|^p) \leq u(x) \leq u^\varepsilon(x) &= u(\overline{x}) - \varepsilon^{-1} \varphi(x - \overline{x}) \\
&\leq C(1 + |\overline{x}|^p) - \varepsilon^{-1} \varphi(x - \overline{x}) \\
&\leq C(1 + 2^p(|\overline{x} - x|^p + |x|^p)) - \varepsilon^{-1} \varphi(x - \overline{x}) \\
&\leq 2^p 2^{p \vee 2 - 1} C(1 + \varphi(x - \overline{x})) + 2^p C |x|^p - \varepsilon^{-1} \varphi(x - \overline{x}) \\
&\leq 2^{p + p \vee 2 - 1} C(1 + |x|^p) + (2^{p + p \vee 2 - 1} C - \varepsilon^{-1}) \varphi(x - \overline{x}) \\
&\leq 2^{p + p \vee 2 - 1} C(1 + |x|^p),
\end{aligned}$$

using the inequality from Equation (2.7) and $\varepsilon \leq (2^{3(p \vee 2) - 2} C)^{-1} \leq (2^{p + p \vee 2 - 1} C)^{-1}$.

2. Integro-Differential Equations

Suppose that $\phi \in C_p^2(\mathbb{R}^d)$ such that $u^\varepsilon - \phi$ has a global maximum in $x \in \Omega_\varepsilon$. As before, Lemma 1.12 implies the existence of $\overline{x} \in \mathbb{R}^d$ such that $u^\varepsilon(x) = u(\overline{x}) - \varepsilon^{-1}\varphi(x - \overline{x})$,

$$\varphi(\overline{x} - x) = \varphi(x - \overline{x}) \leq \varepsilon \cdot \delta(x) < \inf_{y \in \mathbb{R}^d \setminus \Omega} \varphi(y - x)$$

and thus $\overline{x} \in \Omega$. Since $x \in \Omega_\varepsilon$ is a global maximum point of $u^\varepsilon - \phi$,

$$u(y) - \varepsilon^{-1}\varphi(x - \overline{x}) - \phi(y + (x - \overline{x})) \leq u^\varepsilon(y + (x - \overline{x})) - \phi(y + (x - \overline{x}))$$
$$\leq u^\varepsilon(x) - \phi(x)$$
$$= u(\overline{x}) - \varepsilon^{-1}\varphi(x - \overline{x}) - \phi(\overline{x} + (x - \overline{x}))$$

holds for all $y \in \mathbb{R}^d$. This shows that $\overline{x} \in \Omega$ is a global maximum point of

$$y \longmapsto u(y) - (\phi \circ \tau_{x-\overline{x}})(y) = u(y) - \phi(y + (x - \overline{x})).$$

The chain rule implies that $D^k(\phi \circ \tau_{x-\overline{x}})(\overline{x}) = D^k\phi(x)$ for $k \in \{1, 2\}$. Therefore,

$$\begin{aligned}\Delta_\varphi^\varepsilon[F](x, u^\varepsilon(x)) &\quad, D\phi(x), D^2\phi(x), \phi(\cdot)) \\ \leq \quad F\left(\overline{x},\ u(\overline{x}) - \varepsilon^{-1}|x - \overline{x}|^2, D\phi(x), D^2\phi(x), (\phi \circ \tau_{x-\overline{x}})(\cdot)\right) \\ \leq \quad F\left(\overline{x},\ u(\overline{x})\right.&\quad\left., D\phi(x), D^2\phi(x), (\phi \circ \tau_{x-\overline{x}})(\cdot)\right) \leq 0,\end{aligned}$$

due to the definition of the smudged operator $\Delta^\varepsilon[G]$, the monotonicity assumption (A5) and the fact that $u \in \mathrm{SC}_p(\mathbb{R}^d)$ is a viscosity subsolution of $F = 0$ in Ω. □

A careful inspection of the preceding proof shows that the result of Lemma 2.8 still holds, if we only convolve the solution (and smudge the related operator accordingly) for some of the coordinates (e.g. only with respect to space in case of a parabolic problem). Since a general formulation of this observation would complicate the presentation here, we will restate the result for parabolic problems (that only convolves space coordinates) later in this section.

The construction of the smudged operator $\Delta_\varphi^\varepsilon[F]$ in the preceding Lemma 2.8 suggests once more, why it is reasonable to work with scanning functions in Definition 2.1, which are defined on the whole space $\mathbb{R}^d$ and not only $\Omega \subset \mathbb{R}^d$: In fact, even if the nonlocal part of F only considered values in Ω (e.g. Markov generators of censored jump processes), the scanning function ϕ would have to be defined at least on a neighborhood of Ω, so that the nonlocal part can be evaluated for

$$x \longmapsto (\tau_h \circ \phi)(x) = \phi(x + h)$$

with $h \in \mathbb{R}^d$ close to zero. For that reason, it is unclear to us how to formally interpret the related statement in [75, Lemma 7.3], since the article works with scanning functions ϕ, which are only defined on Ω.

Due to the duality $\Delta_\varphi^\varepsilon[-u] = -\nabla_\varphi^\varepsilon[u]$ between supremal and infimal convolutions, we immediately obtain the following analogue of Lemma 2.8 for the approximation of viscosity supersolutions by infimal convolutions:

2.2. Uniqueness of Solutions

Corollary 2.9 (Stability under Infimal Convolutions). *Suppose that $p \geq 0$, that $\Omega \subset \mathbb{R}^d$ is open, that the monotonicity assumption* (A5) *holds for operator F, and that*

$$\varphi = \varphi_{p \vee 2} \in C^\infty(\mathbb{R}^d)$$

is the quasidistance with $(p \vee 2)$-polynomial growth from Lemma 1.14. If $v \in SC_p(\mathbb{R}^d)$ is a viscosity supersolution in Ω of

$$F(x, v(x), Dv(x), D^2v(x), v(\cdot)) = 0$$

with $C > 0$ such that $|v(x)| \leq C(1 + |x|^p)$ for all $x \in \mathbb{R}^d$, and $0 < \varepsilon \leq (2^{3(p \vee 2) - 2}C)^{-1}$, then the infimal convolution (cf. Lemma 1.12)

$$v_\varepsilon(x) := \nabla^\varepsilon_\varphi[v](x) = \inf_{y \in \mathbb{R}^d} \left(v(y) + \frac{\varphi(x - y)}{\varepsilon} \right) \leq 2^{p + p \vee 2 - 1} C(1 + |x|^p)$$

is a viscosity supersolution in $\Omega_\varepsilon := \{x \in \mathbb{R}^d : \inf_{y \in \mathbb{R}^d \setminus \Omega} \varphi(y - x) > \varepsilon \cdot \delta(x)\}$ of

$$\nabla^\varepsilon_\varphi[F](x, v_\varepsilon(x), Dv_\varepsilon(x), D^2 v_\varepsilon(x), v_\varepsilon(\cdot)) = 0.$$

The smudged operator $\nabla^\varepsilon_\varphi[F] : \mathbb{R}^d \times \mathbb{R} \times \mathbb{R}^d \times \mathbb{S}^{d \times d} \times C^2_p(\mathbb{R}^d) \to \mathbb{R}$ *is defined by*

$$\nabla^\varepsilon_\varphi[F](x, r, q, X, \phi) := \sup \left\{ F(y, r, q, X, \phi \circ \tau_{x - y}) \,\middle|\, y \in \Omega \text{ with } \varphi(y - x) \leq \varepsilon \cdot \delta(x) \right\}$$

for $(x, r, q, X, \phi) \in \mathbb{R}^d \times \mathbb{R} \times \mathbb{R}^d \times \mathbb{S}^{d \times d} \times C^2_p(\mathbb{R}^d)$, where $\delta(x) := 2^{3(p \vee 2)} C(1 + \varphi(x))$ and $\tau_h : \mathbb{R}^d \to \mathbb{R}^d$ is the translation operator by $h \in \mathbb{R}^d$, i.e. $\tau_h(x) = x + h$ for all $x \in \mathbb{R}^d$.

We will now follow an idea from Robert Jensen in [77], which led to the breakthrough in extending the viscosity solution theory for first-order to second-order local equations: Instead of proving a maximum principle directly, we use the stability result in Lemma 2.8 to develop an approximate maximum principle for the smudged operators $\Delta^\varepsilon_\varphi[F]$ first, and then send $\varepsilon > 0$ to zero to obtain a maximum principle for the original operator F. As a result of our weaker continuity assumptions, the related limiting arguments are more involved compared to the existing literature. We therefore start by checking that, if the original operator F satisfies all assumptions from Remark 2.3, so do the smudged operators $\Delta^\varepsilon_\varphi[F]$ for $\varepsilon > 0$.

Lemma 2.10 (Continuity of Smudged Operators). *Suppose that $p \geq 0$ and $\varepsilon > 0$, that $\Omega \subset \mathbb{R}^d$ is open, that $\delta : \mathbb{R}^d \to [0, \infty)$ is continuous, that $\varphi : \mathbb{R}^d \to [0, \infty)$ is a continuous quasidistance such that $\{x \in \mathbb{R}^d : \varphi(x) \leq C\}$ are compact for all $C > 0$, and that $\tau_h : \mathbb{R}^d \to \mathbb{R}^d$ is the translation operator by $h \in \mathbb{R}^d$. If all assumptions from Remark 2.3 hold for the operator F, then they also hold for the smudged operators*

$$\Delta^\varepsilon_\varphi[F](x, r, q, X, \phi) = \inf \left\{ F(y, r, q, X, \phi \circ \tau_{x - y}) \,\middle|\, y \in \Omega \text{ with } \varphi(y - x) \leq \varepsilon \cdot \delta(x) \right\}$$

$$\nabla^\varepsilon_\varphi[F](x, r, q, X, \phi) = \sup \left\{ F(y, r, q, X, \phi \circ \tau_{x - y}) \,\middle|\, y \in \Omega \text{ with } \varphi(y - x) \leq \varepsilon \cdot \delta(x) \right\}$$

with $\Delta^\varepsilon_\varphi[F]^\kappa := \Delta^\varepsilon_\varphi[F^\kappa]$ and $\nabla^\varepsilon_\varphi[F]^\kappa := \nabla^\varepsilon_\varphi[F^\kappa]$ for $0 < \kappa < 1$.

2. Integro-Differential Equations

Proof. It is easy to check that the consistency (A1), the translation invariance (A3) and the monotonicity assumption (A5) still hold for the smudged operators. Due to the duality of $\Delta_\varphi^\varepsilon[F]$ and $\nabla_\varphi^\varepsilon[F]$, it suffices to prove that the degenerate ellipticity (A2) and the continuity assumption (A4) hold for

$$\Delta_\varphi^\varepsilon[F^\kappa](x,r,q,X,u,\phi) = \inf_{\varphi(y-x) \leq \varepsilon \cdot \delta(x)} F^\kappa(y,r,q,X, u \circ \tau_{x-y}, \phi \circ \tau_{x-y})$$

with $\varepsilon > 0$ and fixed $0 < \kappa < 1$:

First, we check the degenerate ellipticity assumption (A2): Suppose that $u-v$ and $\phi - \psi$ have global maxima at $x \in \Omega$, and that $X \leq Y$. The first assumption implies that $u \circ \tau_{x-y} - v \circ \tau_{x-y}$ and $\phi \circ \tau_{x-y} - \psi \circ \tau_{x-y}$ have global maxima in $y \in \Omega$. Therefore,

$$F^\kappa(y,r,q,X, u \circ \tau_{x-y}, \phi \circ \tau_{x-y}) \leq F^\kappa(y,r,q,Y, v \circ \tau_{x-y}, \psi \circ \tau_{x-y})$$

for all $y \in \Omega$, using the degenerate ellipticity assumption (A2) for the original operator F. Taking the infimum over all $y \in \Omega$ with $\varphi(y-x) \leq \varepsilon \cdot \delta(x)$ leads to

$$\Delta_\varphi^\varepsilon[F^\kappa](x,r,q,X,u,\phi) \leq \Delta_\varphi^\varepsilon[F^\kappa](x,r,q,Y,v,\psi),$$

which proves the ellipticity assumption (A2) for the smudged operator $\Delta_\varphi^\varepsilon[F^\kappa]$.

Second, we check the continuity assumption (A4): Suppose that

$$(x_n, r_n, q_n, X_n, u_n, \phi_n)_{n \in \mathbb{N}} \subset \mathbb{R}^d \times \mathbb{R} \times \mathbb{R}^d \times \mathbb{S}^{d \times d} \times \mathrm{SC}_p(\mathbb{R}^d) \times C^2(\mathbb{R}^d)$$

such that $\lim_{n\to\infty}(x_n, r_n, q_n, X_n) = (\hat{x}, \hat{r}, \hat{q}, \hat{X})$, $\lim_{n\to\infty} D^k \phi_n = D^k \hat{\phi}$ locally uniformly for $k \in \{0, 1, 2\}$, and $\lim_{n\to\infty} u_n = \hat{u}$ locally uniformly with $\sup_{n \in \mathbb{N}} \|u_n\|_p < \infty$ and $\hat{u} \in C_p(\mathbb{R}^d)$. According to the continuity assumption (A4) for F, the map

$$y \longmapsto F^\kappa(y, \hat{r}, \hat{q}, \hat{X}, \hat{u} \circ \tau_{\hat{x}-y}, \hat{\phi} \circ \tau_{\hat{x}-y})$$

is continuous, since $\lim_{y \to \bar{y}} f \circ \tau_{\hat{x}-y} = f \circ \tau_{\hat{x}-\bar{y}}$ locally uniformly for continuous $f : \mathbb{R}^d \to \mathbb{R}$. The continuous dependence on $y \in \mathbb{R}^d$ implies that

$$\Delta_\varphi^\varepsilon[F^\kappa](\hat{x}, \hat{r}, \hat{q}, \hat{X}, \hat{u}, \hat{\phi}) = \inf_{\varphi(y-x) < \varepsilon \cdot \delta(x)} F^\kappa(y, \hat{r}, \hat{q}, \hat{X}, \hat{u} \circ \tau_{\hat{x}-y}, \hat{\phi} \circ \tau_{\hat{x}-y}).$$

Furthermore, the definition of $\Delta_\varphi^\varepsilon[F^\kappa]$ as an infimum shows the existence of $(y_n)_{n \in \mathbb{N}} \subset \mathbb{R}^d$ with $\varphi(y_n - x_n) \leq \varepsilon \cdot \delta(x_n)$ such that

$$\Delta_\varphi^\varepsilon[F^\kappa](x_n, r_n, q_n, X_n, u_n, \phi_n) \leq F^\kappa(y_n, r_n, q_n, X_n, u_n \circ \tau_{x_n-y_n}, \phi_n \circ \tau_{x_n-y_n})$$
$$\leq \Delta_\varphi^\varepsilon[F^\kappa](x_n, r_n, q_n, X_n, u_n, \phi_n) + n^{-1}.$$

Since $\varphi(y_n - \hat{x}) \leq \sup_{n \in \mathbb{N}} \rho(\varepsilon \cdot \delta(x_n) + \varphi(x_n - \hat{x})) < \infty$, the compactness assumption for the quasidistance $\varphi : \mathbb{R}^d \to [0, \infty)$ therefore implies that it suffices to prove

$$\lim_{k \to \infty} F^\kappa(y_{n_k}, r_{n_k}, q_{n_k}, X_{n_k}, u_{n_k} \circ \tau_{x_{n_k}-y_{n_k}}, \phi_{n_k} \circ \tau_{x_{n_k}-y_{n_k}}) = \Delta_\varphi^\varepsilon[F^\kappa](\hat{x}, \hat{r}, \hat{q}, \hat{X}, \hat{u}, \hat{\phi})$$

for every convergent subsequence $(y_{n_k})_{k \in \mathbb{N}} \subset (y_n)_{n \in \mathbb{N}}$:

Suppose that $(y_{n_k})_{k \in \mathbb{N}} \subset (y_n)_{n \in \mathbb{N}}$ is a convergent subsequence with $\hat{y} := \lim_{k \to \infty} y_{n_k}$. If $y \in \mathbb{R}^d$ with $\varphi(y - \hat{x}) < \varepsilon \cdot \delta(\hat{x})$, then there exists $K_y \in \mathbb{N}$ such that

$$\varphi(y - x_{n_k}) \leq \varepsilon \delta(x_{n_k})$$

for all $k \geq K_y$, due to the continuity of $\delta : \mathbb{R}^d \to [0, \infty)$ and $\varphi : \mathbb{R}^d \to [0, \infty)$. Therefore,

$$F^\kappa(y, r_{n_k}, q_{n_k}, X_{n_k}, u_{n_k} \circ \tau_{x_{n_k}-y}, \phi_{n_k} \circ \tau_{x_{n_k}-y}) \geq \Delta_\varphi^\varepsilon[F^\kappa](x_{n_k}, r_{n_k}, q_{n_k}, X_{n_k}, u_{n_k}, \phi_{n_k})$$

for all $y \in \mathbb{R}^d$ with $\varphi(y - \hat{x}) < \varepsilon \cdot \delta(\hat{x})$ and $k \geq K_y$, by definition of $\Delta_\varphi^\varepsilon[F^\kappa]$. Note that

$$\sup_{n \in \mathbb{N}} \|u_n \circ \tau_{h_n}\|_p \leq 2^p \sup_{n \in \mathbb{N}} (1 + |h_n|^p) \|u_n\|_p < \infty,$$

$\lim_{n \to} u_n \circ \tau_{h_n} = \hat{u}$ and $\lim_{n \to} D^k(\phi_n \circ \tau_{h_n}) = D^k \hat{\phi}$ locally uniform for $k \in \{0, 1, 2\}$ and $(h_n)_{n \in \mathbb{N}} \subset \mathbb{R}^d$ with $\lim_{n \to 0} h_n = 0$, where we used that $u : \mathbb{R}^d \to \mathbb{R}$ is continuous. Hence, the continuity assumption (A4) for F and the construction of $\Delta_\varphi^\varepsilon[F^\kappa]$ imply that

$$F^\kappa(y, \hat{r}, \hat{q}, \hat{X}, \hat{u} \circ \tau_{\hat{x}-y}, \hat{\phi} \circ \tau_{\hat{x}-y})$$
$$\geq \lim_{k \to \infty} \Delta_\varphi^\varepsilon[F^\kappa](x_{n_k}, r_{n_k}, q_{n_k}, X_{n_k}, u_{n_k}, \phi_{n_k})$$
$$= \lim_{k \to \infty} F^\kappa(y_{n_k}, r_{n_k}, q_{n_k}, X_{n_k}, u_{n_k} \circ \tau_{x_{n_k}-y_{n_k}}, \phi_{n_k} \circ \tau_{x_{n_k}-y_{n_k}})$$
$$= F^\kappa(\hat{y}, \hat{r}, \hat{q}, \hat{X}, \hat{u} \circ \tau_{\hat{x}-\hat{y}}, \hat{\phi} \circ \tau_{\hat{x}-\hat{y}})$$
$$\geq \Delta_\varphi^\varepsilon[F^\kappa](\hat{x}, \hat{r}, \hat{q}, \hat{X}, \hat{u}, \hat{\phi}) = \inf_{|y-x| < \sqrt{\varepsilon} \cdot \delta(x)} F^\kappa(y, \hat{r}, \hat{q}, \hat{X}, \hat{u} \circ \tau_{\hat{x}-y}, \hat{\phi} \circ \tau_{\hat{x}-y})$$

for all $y \in \mathbb{R}^d$ with $\varphi(y - \hat{x}) < \varepsilon \cdot \delta(\hat{x})$. Finally, taking the infimum over all $y \in \mathbb{R}^d$ with $\varphi(y - \hat{x}) < \varepsilon \cdot \delta(\hat{x})$ shows that all limits exist and coincide, which finishes the proof. □

The maximum principle we will develop in the sequel, can be viewed as a generalization of the classical maximum principle for twice differentiable to semicontinuous functions. Recall that the classical maximum principle states that $D\phi(x) = 0$ and $D^2\phi(x) \leq 0$ if $\phi \in C^2(\Omega)$ for some open set $\Omega \subset \mathbb{R}^d$ has a (local) maximum in $x \in \Omega$. In fact, as soon as the viscosity solutions in the following statements are twice differentiable, the results are a straightforward application of the classical maximum principle together with the properties of the related operators from Remark 2.3. The original proof of the maximum principle for semicontinuous functions (in the context of second-order local equations) from Robert Jensen in [77] is based on an application of Alexandrov's theorem (cf. Theorem A.7) from convex analysis. His main argument (often referred to as Jensen's lemma) is isolated in Lemma A.8 and forms the basis for our generalization to nonlocal equations. We will formulate all of the following statements for $k \in \mathbb{N}$ instead of only two viscosity solutions, which allows us to compare more than two viscosity solutions at a time – e.g. to obtain the subadditivity and convexity of solutions with respect to their initial values in Corollary 2.25 and Corollary 2.26.

2. Integro-Differential Equations

Proposition 2.11 (Maximum Principle for Smudged Operators). *Suppose that $p \geq 0$, that $\Omega \subset \mathbb{R}^d$ is open, and that $u_i \in USC_p(\mathbb{R}^d)$ are viscosity solutions in Ω of*

$$F_i(x, u_i(x), Du_i(x), D^2 u_i(x), u_i(\cdot)) \leq 0$$

for $i \in \{1, \ldots, k\}$ and operators F_i, which satisfy all assumptions from Remark 2.3. Furthermore, suppose that $\phi \in C_p^2(\mathbb{R}^{kd})$, that $C > \|u\|_p$ and $0 < \varepsilon \leq (2^{3(p \vee 2)-2} C)^{-1}$, and that $u^\varepsilon : \mathbb{R}^d \to \mathbb{R}$, $\Delta_\varphi^\varepsilon[F_i]$ and $\Omega_\varepsilon \subset \mathbb{R}^d$ are defined as in Lemma 2.8. If

$$x = (x_1, \ldots, x_k) \longmapsto \sum_{i=1}^{k} u_i^\varepsilon(x_i) - \phi(x)$$

has a global maximum in $\overline{x} = (\overline{x}_1, \ldots, \overline{x}_k) \in \Omega_\varepsilon^k = \Omega_\varepsilon \times \ldots \times \Omega_\varepsilon$, and if

$$x = (x_1, \ldots, x_k) \longmapsto \sum_{i=1}^{k} u_i^\varepsilon(x_i) + \frac{|x|^2}{\varepsilon} = \sum_{i=1}^{k} \left(u_i^\varepsilon(x_i) + \frac{|x_i|^2}{\varepsilon} \right)$$

is convex for $x = (x_1, \ldots, x_k) \in \mathbb{R}^{kd}$ in a neighborhood of $\overline{x} \in \Omega_\varepsilon^k$, then there exist symmetric matrices $X_1, \ldots, X_k \in \mathbb{S}^{d \times d}$ satisfying

$$-\frac{2}{\varepsilon} I \leq \operatorname{diag}(X_1, \ldots, X_k) \leq D^2 \phi(\overline{x})$$

such that the following inequalities hold in the ordinary sense

$$\Delta_\varphi^\varepsilon[F_i](\overline{x}_i, u_i^\varepsilon(\overline{x}_i), D_{x_i} \phi(\overline{x}), X_i, \phi(\overline{x}_1, \ldots, \overline{x}_{i-1}, \cdot, \overline{x}_{i+1}, \ldots, \overline{x}_k)) \leq 0$$

for all $i \in \{1, \ldots, k\}$.

Proof. The proof is divided into three parts: In the first part, we will apply our nonlocal generalization of Jensen's lemma from Lemma 1.25, which will approximate the global maximum point $\overline{x} \in \Omega_\varepsilon^k$ by differentiability points of the objective function. In the second and third part, we will use the properties of the operators F_i to pass to the limit, which will yield the desired result.

First, we show how to apply Lemma 1.25: Analogously to the proof of Lemma 2.4, we can construct a function $\psi \in C_p^2(\mathbb{R}^{kd})$ such that $\psi(x) > 0$ for $x \neq \overline{x}$ and

$$\psi(x) = |x - \overline{x}|^4$$

for all $x \in \mathbb{R}^{kd}$ in a neighborhood of $\overline{x} \in \Omega_\varepsilon^k$. If we define $\phi^{(m)} \in C_p^2(\mathbb{R}^{kd})$ by

$$\phi^{(m)}(x) := \phi(x) + m^{-1} \psi(x) + (w(\overline{x}) - \phi(\overline{x})) \tag{2.8}$$

for $x \in \mathbb{R}^{kd}$ and $m \in \mathbb{N}$, and $w(x) := \sum_{i=1}^{k} u_i^\varepsilon(x_i)$ for $x = (x_1, \ldots, x_k) \in \mathbb{R}^{kd}$, then $w - \phi^{(m)}$ has a strict global maximum in $\overline{x} \in \Omega_\varepsilon^k$ with $w(\overline{x}) = \phi^{(m)}(\overline{x})$. Consequently, our generalization of Jensen's lemma in Lemma 1.25 implies the existence of $(x^{(m,n)})_{n \in \mathbb{N}}$ and $(\phi_n^{(m)})_{n \in \mathbb{N}} \subset C_p^2(\mathbb{R}^{kd})$ for every $m \in \mathbb{N}$ such that $\lim_{n \to \infty} x^{(m,n)} = \overline{x}$,

$$x = (x_1, \ldots, x_k) \longmapsto w(x) = \sum_{i=1}^{k} u_i^\varepsilon(x_i)$$

is twice differentiable in $x^{(m,n)} \in \mathbb{R}^{kd}$ for all $n \in \mathbb{N}$, $\lim_{n \to \infty} D^\alpha \phi_n^{(m)} = D^\alpha \phi^{(m)}$ locally uniformly for $\alpha \in \{0, 1, 2\}$, $\sup_{n \in \mathbb{N}} \|\phi_n^{(m)}\|_p < \infty$,

$$\sum_{i=1}^{k} u_i^\varepsilon(x_i) = w(x) \leq \phi_n^{(m)}(x)$$

for all $x = (x_1, \ldots, x_k) \in \mathbb{R}^{kd}$ and $n \in \mathbb{N}$, and

$$\sum_{i=1}^{k} u_i^\varepsilon\left(x_i^{(m,n)}\right) = w\left(x^{(m,n)}\right) = \phi_n^{(m)}\left(x^{(m,n)}\right)$$

for all $n \in \mathbb{N}$. Furthermore, Corollary 1.24 implies the existence of $\psi_n^{(m)} \in C_p^2(\mathbb{R}^{kd})$ for every $n \in \mathbb{N}$ such that $D^\alpha w(x^{(m,n)}) = D^\alpha \psi_n^{(m)}(x^{(m,n)})$ for $\alpha \in \{0, 1, 2\}$, and

$$w(x) \leq \psi_n^{(m)}(x) \leq \phi_n^{(m)}(x)$$

for all $x \in \mathbb{R}^{kd}$. Since $\Omega_\varepsilon \subset \mathbb{R}^d$ is open, there exists an $N_m \in \mathbb{N}$ such that for all $n \geq N_m$

$$(w - \psi_n^{(m)})\left(x_1^{(m,n)}, \ldots, x_{i-1}^{(m,n)}, \cdot, x_{i+1}^{(m,n)}, \ldots, x_k^{(m,n)}\right)$$
$$= u_i^\varepsilon(\cdot) - \left(\sum_{j \neq i} u_j^\varepsilon\left(x_j^{(m,n)}\right) + \psi_n^{(m)}\left(x_1^{(m,n)}, \ldots, x_{i-1}^{(m,n)}, \cdot, x_{i+1}^{(m,n)}, \ldots, x_k^{(m,n)}\right)\right)$$

has a global maximum in $x_i^{(m,n)} \in \Omega_\varepsilon$ for each $i \in \{1, \ldots, k\}$. According to Lemma 2.8, the supremal convolutions $u_i^\varepsilon : \mathbb{R}^d \to \mathbb{R}$ are viscosity solutions of $\Delta_\varphi^\varepsilon[F_i] \leq 0$ in Ω_ε. Therefore, we find together with the translation invariance assumption (A3)

$$\Delta_\varphi^\varepsilon[F_i]\left(x_i^{(m,n)}, u_i^\varepsilon(x_i^{(m,n)}), D_{x_i}\psi_n^{(m)}(x^{(m,n)}), D_{x_i}^2\psi_n^{(m)}(x^{(m,n)}),\right.$$
$$\left.\psi_n^{(m)}\left(x_1^{(m,n)}, \ldots, x_{i-1}^{(m,n)}, \cdot, x_{i+1}^{(m,n)}, \ldots, x_k^{(m,n)}\right)\right) \leq 0$$

for all $m \in \mathbb{N}$ and $n \geq N_m$. The classical maximum principle implies

$$D^j w(x^{(m,n)}) = D^j \psi_n^{(m)}(x^{(m,n)})$$

for all $m \in \mathbb{N}$, $n \geq N_m$ and $j \in \{0, 1, 2\}$. Since $\psi_n^{(m)} - \phi_n^{(m)}$ has a global maximum in $x^{(m,n)} \in \Omega_\varepsilon$ for all $m \in \mathbb{N}$ and $n \geq N_m$, we obtain that $D\psi_n^{(m)}(x^{(m,n)}) = D\phi_n^{(m)}(x^{(m,n)})$.

2. Integro-Differential Equations

Moreover, the degenerate ellipticity assumption (A2) for $\Delta_\varphi^\varepsilon[F_i]$ (cf. Lemma 2.10) implies

$$\Delta_\varphi^\varepsilon[F_i]\left(x_i^{(m,n)}, u_i^\varepsilon(x_i^{(m,n)}), D_{x_i}\phi_n^{(m)}(x^{(m,n)}), X_i^{(m,n)},\right.$$
$$\left.\phi_n^{(m)}\left(x_1^{(m,n)}, \ldots, x_{i-1}^{(m,n)}, \cdot\, , x_{i+1}^{(m,n)}, \ldots, x_k^{(m,n)}\right)\right) \leq 0 \quad (2.9)$$

for all $m \in \mathbb{N}$ and $n \geq N_m$, where

$$X_i^{(m,n)} := D_{x_i}^2 \psi_n^{(m)}(x^{(m,n)}) = D_{x_i}^2 w(x^{(m,n)}) = D^2 u_i^\varepsilon(x_i^{(m,n)})$$

for all $m \in \mathbb{N}$, $n \geq N_m$ and $i \in \{1, \ldots, k\}$.

We now show how to send $n \to \infty$ for fixed $m \in \mathbb{N}$ in Equation (2.9): By assumption,

$$x = (x_1, \ldots, x_k) \longmapsto w(x) + \frac{|x|^2}{\varepsilon} = \sum_{i=1}^k \left(u_i^\varepsilon(x_i) + \frac{|x_i|^2}{\varepsilon}\right)$$

is convex in a neighborhood of $\overline{x} \in \Omega_\varepsilon$. Hence, due to $\lim_{n\to\infty} x^{(m,n)} = \overline{x}$, there exists $N_m' \in \mathbb{N}$ with $N_m' \geq N_m$ for every $m \in \mathbb{N}$ such that

$$D^2\left(w + \frac{|\cdot|^2}{\varepsilon}\right)(x^{(m,n)}) = \operatorname{diag}\left(X_1^{(m,n)}, \ldots, X_k^{(m,n)}\right) + \frac{2}{\varepsilon}I \geq 0$$

for all $m \in \mathbb{N}$ and $n \geq N_m'$. Moreover, since $w - \phi_n^{(m)}$ has a global maximum in $x^{(m,n)} \in \mathbb{R}^{kd}$ and w is twice differentiable in $x^{(m,n)} \in \mathbb{R}^{kd}$ for all $m \in \mathbb{N}$ and $n \in \mathbb{N}$, the classical maximum principle implies

$$D^2(w - \phi_n^{(m)})(x^{(m,n)}) = \operatorname{diag}\left(X_1^{(m,n)}, \ldots, X_k^{(m,n)}\right) - D^2\phi_n^{(m)}(x^{(m,n)}) \leq 0$$

for all $m \in \mathbb{N}$ and $n \in \mathbb{N}$. Combining the last two inequalities leads to

$$-\frac{2}{\varepsilon}I \leq \operatorname{diag}\left(X_1^{(m,n)}, \ldots, X_k^{(m,n)}\right) \leq D^2\phi_n^{(m)}(x^{(m,n)}) \leq \left\|D^2\phi_n^{(m)}(x^{(m,n)})\right\|I$$

for all $m \in \mathbb{N}$ and $n \geq N_m'$, where we used that $\langle A\xi, \xi\rangle \leq \|A\| \cdot |\xi|^2 = \|A\|\langle I\xi, \xi\rangle$ for all $A \in \mathbb{S}^{kd \times kd}$ and $\xi \in \mathbb{R}^{kd}$. Note that for all $m \in \mathbb{N}$

$$\sup_{n \in \mathbb{N}} \left\|D^2\phi_n^{(m)}(x^{(m,n)})\right\| < \infty,$$

because $\lim_{n\to\infty} x^{(m,n)} = \overline{x}$ and $\lim_{n\to\infty} D^2\phi_n^{(m)} = D^2\phi^{(m)}$ locally uniformly for all $m \in \mathbb{N}$. Since sets of the form $\{A \in \mathbb{S}^{d \times d} : -cI \leq A \leq cI\}$ with $C > 0$ are compact (by the Heine-Borel theorem), there exist convergent subsequences $(X_i^{(m,n_j)})_{j \in \mathbb{N}}$ of $(X_i^{(m,n)})_{n \in \mathbb{N}}$ for all $m \in \mathbb{N}$ and $i \in \{1, \ldots, k\}$ with $X_i^{(m)} := \lim_{j\to\infty} X_i^{(m,n_j)}$, such that for every $m \in \mathbb{N}$

$$-\frac{2}{\varepsilon}I \leq \operatorname{diag}\left(X_1^{(m)}, \ldots, X_k^{(m)}\right) = \lim_{j\to\infty} \operatorname{diag}\left(X_1^{(m,n_j)}, \ldots, X_k^{(m,n_j)}\right)$$
$$\leq \lim_{j\to\infty} D^2\phi_n^{(m)}(x^{(m,n_j)}) = D^2\phi^{(m)}(\overline{x}) = D^2\phi(\overline{x}).$$

2.2. Uniqueness of Solutions

Furthermore, since $x^{(m,n)} \in \arg\max(w - \psi_n^{(m)})$ for all $m \in \mathbb{N}$ and $n \in \mathbb{N}$, we find

$$u_i^\varepsilon(x_i^{(m,n)}) = w(x^{(m,n)}) - \psi_n^{(m)}(x^{(m,n)}) + \left(\psi_n^{(m)}(x^{(m,n)}) - \sum_{j \neq i} u_j^\varepsilon\left(x_j^{(m,n)}\right)\right)$$

$$\geq w(\ \overline{x}\ ) - \psi_n^{(m)}(\ \overline{x}\ ) + \left(\psi_n^{(m)}(x^{(m,n)}) - \sum_{j \neq i} u_j^\varepsilon\left(x_j^{(m,n)}\right)\right)$$

for all $m \in \mathbb{N}$, $n \in \mathbb{N}$ and $i \in \{1, \ldots, k\}$. This implies $\lim_{n \to \infty} u_i^\varepsilon(x_i^{(m,n)}) = u_i^\varepsilon(\overline{x})$ for all $m \in \mathbb{N}$ and $i \in \{1, \ldots, k\}$, due to the (semi-)continuity of the functions u_i and ϕ. Finally, the continuity assumption (A4) for $\Delta_\varphi^\varepsilon[F_i]$ (cf. Lemma 2.10) implies

$$\Delta^\varepsilon[F_i]\left(\overline{x}_i, u_i^\varepsilon(\overline{x}_i), D_{x_i}\phi^{(m)}(\overline{x}), X_i^{(m)}, \phi^{(m)}(\overline{x}_1, \ldots, \overline{x}_{i-1}, \cdot, \overline{x}_{i+1}, \ldots, \overline{x}_k)\right) \leq 0 \quad (2.10)$$

for all $m \in \mathbb{N}$ and $i \in \{1, \ldots, k\}$, since $\lim_{n \to \infty} D^\alpha \phi_n^{(m)} = D^\alpha \phi^{(m)}$ locally uniformly for $\alpha \in \{0, 1, 2\}$ and $\sup_{n \in \mathbb{N}} \|\phi_n^{(m)}\|_p < \infty$ for all $m \in \mathbb{N}$ by construction.

At last, we show how to send $m \to \infty$ in Equation (2.10): A similar compactness arguments as before shows that there exist convergent subsequences $(X_i^{(m_j)})_{j \in \mathbb{N}}$ of $(X_i^{(m)})_{m \in \mathbb{N}}$ for all $i \in \{1, \ldots, k\}$ with $X_i := \lim_{j \to \infty} X_i^{(m_j)}$ such that

$$-\frac{2}{\varepsilon} I \leq \mathrm{diag}(X_1, \ldots, X_k) = \lim_{j \to \infty} \mathrm{diag}\left(X_1^{(m_j)}, \ldots, X_k^{(m_j)}\right) \leq D^2 \phi(\overline{x}).$$

From Equation (2.8), we obtain $\sup_{m \in \mathbb{N}} \|\phi^{(m)}\|_p \leq \|\phi + (w(\overline{x}) - \phi(\overline{x}))\|_p + \|\psi\|_p < \infty$,

$$\lim_{j \to \infty} D^\alpha \phi^{(m_j)} = D^\alpha \left(\phi + (w(\overline{x}) - \phi(\overline{x}))\right)$$

locally uniformly for $\alpha \in \{0, 1, 2\}$, and $D\phi^{(m_j)}(\overline{x}) = D\phi(\overline{x})$. The continuity assumption (A4) and the translation invariance (A3) for $\Delta_\varphi^\varepsilon[F_i]$ (cf. Lemma 2.10) hence imply

$$\Delta_\varphi^\varepsilon[F_i](\overline{x}_i, u_i^\varepsilon(\overline{x}_i), D_{x_i}\phi(\overline{x}), X_i, \phi(\overline{x}_1, \ldots, \overline{x}_{i-1}, \cdot, \overline{x}_{i+1}, \ldots, \overline{x}_k)) \leq 0,$$

as required. $\square$

The natural approach to obtain a maximum principle for the original operator F is to use the stability of viscosity solutions in Lemma 2.8 for small $\varepsilon > 0$, apply the maximum principle in Proposition 2.11 to obtain $X_1^\varepsilon, \ldots, X_k^\varepsilon \in \mathbb{S}^{d \times d}$ for the smudged operators $\Delta_\varphi^\varepsilon[F]$, and then send ε to zero. Unfortunately, the lower bound

$$-\frac{2}{\varepsilon} I \leq \mathrm{diag}(X_1^\varepsilon, \ldots, X_k^\varepsilon)$$

explodes as $\varepsilon \to 0$, which makes it complicated to extract a subsequential limit from $(X_1^\varepsilon, \ldots, X_k^\varepsilon)$ using a compactness argument. To circumvent this issue, Michael Crandall introduced the following matrix compactification procedure in [33]. The extended version for $k \in \mathbb{N}$ instead of two matrices presented below is due to Mingshang Hu and Shige Peng in [67]. We include the proof for the convenience of the reader.

2. Integro-Differential Equations

Lemma 2.12 (Matrix Transformation). *For symmetric $X, Y \in \mathbb{S}^{d \times d}$ with*
$$X \leq Y < \gamma^{-1} I$$
for $\gamma > 0$, the matrices $(I - \gamma X)$ and $(I - \gamma Y)$ are invertible. Hence,
$$X^\gamma := X(I - \gamma X)^{-1}$$
$$Y^\gamma := Y(I - \gamma Y)^{-1}$$
are well-defined and again symmetric. Furthermore, they satisfy
$$-\gamma^{-1} I \leq X^\gamma \leq Y^\gamma$$
as well as $X \leq X^\gamma$.

Proof. First of all, note that every $A \in \mathbb{R}^{d \times d}$ with $A > 0$ is invertible, since its kernel only contains the origin, i.e. $\{x \in \mathbb{R}^d : Ax = 0\} = \{0\}$. Moreover, if $A \in \mathbb{S}^{d \times d}$ is invertible, then the equality
$$(A^{-1})^T = (A^{-1})^T (AA^{-1}) = ((A^{-1})^T A^T) A^{-1} = (AA^{-1})^T A^{-1} = A^{-1}$$
holds and therefore A^{-1} is also symmetric. Next, it is easy to see that for $A \in \{X, Y\}$ and $x, y \in \mathbb{R}^d$ the transformation
$$\langle Ax, x \rangle - \gamma^{-1} |x - y|^2 - \langle A(I - \gamma A)^{-1} y, y \rangle$$
$$= \langle A \left(x - (I - \gamma A)^{-1} y \right), \left(x - (I - \gamma A)^{-1} y \right) \rangle - \gamma^{-1} \left| x - (I - \gamma A)^{-1} y \right|^2$$
follows from the bilinearity of the scalar product and the symmetry of the matrices. Moreover, the relation $A < \gamma^{-1} I$ implies that the right hand side is strictly negative and equal to zero if and only if $x = (I - \gamma A)^{-1} y$. Hence,
$$\max_{x \in \mathbb{R}^d} \left\{ \langle Ax, x \rangle - \gamma^{-1} |x - y|^2 \right\} = \langle A(I - \gamma A)^{-1} y, y \rangle$$
for all $y \in \mathbb{R}^d$. In particular, the relation $X \leq Y$ yields
$$\langle X^\gamma y, y \rangle = \langle X(I - \gamma X)^{-1} y, y \rangle = \max_{x \in \mathbb{R}^d} \left\{ \langle Xx, x \rangle - \gamma^{-1} |x - y|^2 \right\}$$
$$\leq \max_{x \in \mathbb{R}^d} \left\{ \langle Yx, x \rangle - \gamma^{-1} |x - y|^2 \right\}$$
$$= \langle Y(I - \gamma Y)^{-1} y, y \rangle$$
$$= \langle Y^\gamma y, y \rangle$$
for all $y \in \mathbb{R}^d$, i.e. $X^\gamma \leq Y^\gamma$ as required. Lastly, the symmetry of X implies
$$\langle X^2 x, x \rangle = \langle X^T X x, x \rangle = \langle Xx, Xx \rangle = |Xx|^2 \geq 0$$
for all $x \in \mathbb{R}^d$ and therefore $X^2 \geq 0$, which shows $X \geq X - \gamma X^2 = X(I - \gamma X)$ as well as $X^\gamma = X(I - \gamma X)^{-1} \geq X$. Analogously,
$$X \geq X - \gamma^{-1} I = -\gamma^{-1} I (I - \gamma X)$$
and thus $X^\gamma = X(I - \gamma X)^{-1} \geq -\gamma^{-1} I$, as required. $\square$

2.2. Uniqueness of Solutions

Corollary 2.13 (Matrix Compactification). *Suppose that $\gamma \in (0, k^{-1})$ with $k \in \mathbb{N}$,*

$$J_{k,d} := \begin{pmatrix} (k-1)I & -I & \cdots & -I \\ -I & \ddots & \ddots & \vdots \\ \vdots & \ddots & \ddots & -I \\ -I & \cdots & -I & (k-1)I \end{pmatrix} \in \mathbb{R}^{kd \times kd}$$

and $X_i \in \mathbb{S}^{d \times d}$ for $i \in \{1, \ldots, k\}$. If the inequality

$$\operatorname{diag}(X_1, \ldots, X_k) \leq J_{k,d}$$

is satisfied, then $X_i^\gamma := X_i(I - \gamma X_i)^{-1} \in \mathbb{S}^{d \times d}$ exists for every $i \in \{1, \ldots, k\}$ such that

$$-\gamma^{-1} I \leq \operatorname{diag}(X_1^\gamma, \ldots, X_k^\gamma) \leq (1 - k\gamma)^{-1} J_{k,d}$$

and $X_i \leq X_i^\gamma$ hold for all $i \in \{1, \ldots, k\}$.

Proof. It is easy to verify by using block matrix multiplication that $J_{k,d}^2 = k J_{k,d}$ holds, since the diagonal blocks of $J_{k,d}^2 = J_{k,d} \cdot J_{k,d}$ are of the form

$$((k-1)I)^2 + (k-1)(-I)^2 = \left((k-1)^2 + (k-1)\right) I = k\left((k-1)I\right),$$

whereas the off-diagonal blocks of $J_{k,d}^2 = J_{k,d} \cdot J_{k,d}$ look like

$$2\left((k-1)I\right)(-I) + (k-2)(-I)^2 = -2(k-1)I + (k-2)I = k(-I).$$

Hence, one can rewrite

$$J_{k,d}(I - \gamma J_{k,d}) = J_{k,d} - \gamma J_{k,d}^2 = (1 - \gamma k) J_{k,d}$$

in order to obtain $J_{k,d}^\gamma := J_{k,d}(I - \gamma J_{k,d})^{-1} = (1 - \gamma k)^{-1} J_{k,d}$. Furthermore, a simple calculation shows that

$$\langle (kI - J_{k,d})x, x \rangle = \left\langle \begin{pmatrix} I & \cdots & I \\ \vdots & \ddots & \vdots \\ I & \cdots & I \end{pmatrix} x, x \right\rangle = \left\langle \begin{pmatrix} x^{(1)} + \ldots + x^{(k)} \\ \vdots \\ x^{(1)} + \ldots + x^{(k)} \end{pmatrix}, \begin{pmatrix} x^{(1)} \\ \vdots \\ x^{(k)} \end{pmatrix} \right\rangle$$

$$= \sum_{i=1}^k \left\langle x^{(1)} + \ldots + x^{(k)}, x^{(i)} \right\rangle$$

$$= \left| x^{(1)} + \ldots + x^{(k)} \right|^2 \geq 0$$

for all $x = (x^{(1)}, \ldots, x^{(k)}) \in \mathbb{R}^{kd}$ and therefore $\gamma^{-1} I > kI \geq J_{k,d}$, using $\gamma < k^{-1}$. In particular, Lemma 2.12 with $X = \operatorname{diag}(X_1, \ldots, X_k)$ and $Y = J_{k,d}$ together with

$$X^\gamma = \operatorname{diag}(X_1, \ldots, X_k)^\gamma = \operatorname{diag}(X_1^\gamma, \ldots, X_k^\gamma),$$

which follows from the block structure of $(I - \gamma X)^{-1}$, implies the desired result. □

2. Integro-Differential Equations

Theorem 2.14 (Maximum Principle). *Suppose that $p \geq 0$, that $\Omega \subset \mathbb{R}^d$ is open, and that $u_i \in USC_p(\mathbb{R}^d)$ are viscosity solutions in Ω of*

$$F_i(x, u_i(x), Du_i(x), D^2 u_i(x), u_i(\cdot)) \leq 0$$

for $i \in \{1, \ldots, k\}$ and operators F_i, which satisfy all assumptions from Remark 2.3. Furthermore, suppose that $\phi \in C_p^2(\mathbb{R}^{kd})$, $g_0 \in C(\mathbb{R}^{kd})$ and $g_i \in C(\mathbb{R}^{kd}; \mathbb{S}^{d \times d})$ such that

$$D^2 \phi(x) \leq g_0(x) J_{k,d} + \mathrm{diag}(g_1(x), \ldots, g_k(x))$$

for all $x \in \mathbb{R}^{kd}$, where $J_{k,d} \in \mathbb{R}^{kd \times kd}$ is defined as in Corollary 2.13. If

$$x = (x_1, \ldots, x_k) \longmapsto \sum_{i=1}^{k} u_i(x_i) - \phi(x)$$

has a global maximum in $\overline{x} = (\overline{x}_1, \ldots, \overline{x}_k) \in \Omega^k = \Omega \times \ldots \times \Omega$ with $g_0(\overline{x}) > 0$, then there exist symmetric matrices $X_1^\gamma, \ldots, X_k^\gamma \in \mathbb{S}^{d \times d}$ for every $\gamma \in (0, k^{-1})$ satisfying

$$-\gamma^{-1} g_0(\overline{x}) I \leq \mathrm{diag}(X_1^\gamma - g_1(\overline{x}), \ldots, X_k^\gamma - g_k(\overline{x})) \leq (1 - k\gamma)^{-1} g_0(\overline{x}) J_{k,d}$$

such that the following inequalities hold in the ordinary sense

$$F_i(\overline{x}_i, u_i(\overline{x}_i), D_{x_i} \phi(\overline{x}), X_i^\gamma, \phi(\overline{x}_1, \ldots, \overline{x}_{i-1}, \cdot, \overline{x}_{i+1}, \ldots, \overline{x}_k)) \leq 0$$

for all $i \in \{1, \ldots, k\}$.

Proof. The proof is divided into four parts: Suppose that $\check{\varphi} = \check{\varphi}_{p \vee 2} \in C^\infty(\mathbb{R}^d)$ and $\hat{\varphi} = \hat{\varphi}_{p \vee 2} \in C^\infty(\mathbb{R}^{kd})$ denote the quasidistances with $(p \vee 2)$-polynomial growth in $\mathbb{R}^d$ and $\mathbb{R}^{kd}$ from Lemma 1.14. In the first part, we will prove the existence of an approximating sequence $(\phi_n)_{n \in \mathbb{N}} \subset C^2(\mathbb{R}^{kd})$ for $\phi \in C_p^2(\mathbb{R}^{kd})$ such that

$$x = (x_1, \ldots, x_k) \longmapsto \sum_{i=1}^{k} u_i(x_i) - \phi_n(x) + n^{-1} \hat{\varphi}(x - \overline{x})$$

has a global maximum in $\overline{x} \in \Omega^k$ for all $n \in \mathbb{N}$. In the second part, we will prove the existence of $(\varepsilon_n)_{n \in \mathbb{N}}$ and $(\overline{x}^n)_{n \in \mathbb{N}} \subset \Omega^k$ such that $\lim_{n \to \infty} \varepsilon_n = 0$, $\lim_{n \to \infty} \overline{x}^n = \overline{x}$ and

$$x = (x_1, \ldots, x_k) \longmapsto \sum_{i=1}^{k} u_i^{\varepsilon_n}(x_i) - \phi_n(x)$$

has a global maximum in $\overline{x}^n \in \Omega^k$ for all $n \in \mathbb{N}$, where $u_i^{\varepsilon_n} = \Delta_{\check{\varphi}}^{\varepsilon_n}[u_i] \in C(\mathbb{R}^d)$ is the supremal convolution from Definition 1.8. In the third part, we will show how to apply the maximum principle for smudged operators from Proposition 2.11 and the matrix compactification from Corollary 2.13 to this approximation. In the forth part, we will send $n \to \infty$ to obtain the desired result.

2.2. Uniqueness of Solutions

First, we will construct the approximations $(\phi_n)_{n\in\mathbb{N}} \subset C^2(\mathbb{R}^{kd})$ for $\phi \in C^2_p(\mathbb{R}^{kd})$: We can assume without loss of generality that

$$\sum_{i=1}^{k} u_i(\overline{x}_i) = \phi(\overline{x}).$$

Otherwise, we replace $\phi(\cdot)$ by $\phi(\cdot) + (\sum_{i=1}^{k} u_i(\overline{x}_i) - \phi(\overline{x}))$ and use the translation invariance assumption (A3) for the operators F_i with $i \in \{1, \ldots, k\}$. According to Lemma 1.6, there exist $(\chi_n)_{n\in\mathbb{N}} \subset C^\infty(\mathbb{R}^{kd})$ such that

$$\mathbb{1}_{B[0,n]}(x) \leq \chi_n(x) \leq \mathbb{1}_{B(0,2n)}(x)$$

for all $x \in \mathbb{R}^d$ and $n \in \mathbb{N}$. Suppose that $C > \|u_i\|_p \vee \|\phi\|_p$ for all $i \in \{1, \ldots, k\}$, and define the approximations $\phi_n \in C^2(\mathbb{R}^{kd})$ by

$$\phi_n(x) := \chi_n(x-\overline{x})\phi(x) + (1-\chi_n(x-\overline{x}))kC(1+|x|^p) + n^{-1}\hat{\varphi}(x-\overline{x}) \qquad (2.11)$$

for all $x \in \mathbb{R}^{kd}$ and large $n \in \mathbb{N}$. Since $\chi_n(x) = 0$ for $x \in \mathbb{R}^{kd} \setminus B(0,2n)$ and thus

$$\sup_{x\in\mathbb{R}^{kd}} |D\chi_n(x-\overline{x})| = \sup_{y\in B[0,2n]} |D\chi_n(y)| < \infty$$

for all $n \in \mathbb{N}$, there exists $C_n > 0$ such that

$$\begin{aligned}|\phi_n(x)| &\leq C_n\left(1+\hat{\varphi}(x)\right) \\ |D\phi_n(x)| &\leq C_n(1+|D\hat{\varphi}(x)|)\end{aligned} \qquad (2.12)$$

for all $x \in \mathbb{R}^{kd}$ and large $n \in \mathbb{N}$. Moreover, we have by assumption $\phi(x) \geq \sum_{i=1}^{k} u_i(x_i)$ and $C(1+|x_i|^p) \geq u_i(x_i)$ for all $x \in \mathbb{R}^{kd}$, which implies

$$\begin{aligned}\phi_n(x) &= \chi_n(x-\overline{x})\phi(x) \quad\quad + (1-\chi_n(x-\overline{x}))kC(1+|x|^p) + n^{-1}\hat{\varphi}(x-\overline{x}) \\ &\geq \chi_n(x-\overline{x})\sum_{i=1}^{k} u_i(x_i) + (1-\chi_n(x-\overline{x}))\sum_{i=1}^{k} u_i(x_i) \quad + n^{-1}\hat{\varphi}(x-\overline{x}) \\ &= \sum_{i=1}^{k} u_i(x_i) + n^{-1}\hat{\varphi}(x-\overline{x})\end{aligned}$$

for all $x \in \mathbb{R}^{kd}$. Similarly, $\phi_n(\overline{x}) = \phi(\overline{x})$ implies $\phi_n(\overline{x}) = \sum_{i=1}^{k} u_i(\overline{x}_i)$. In particular,

$$x = (x_1, \ldots, x_k) \longmapsto \sum_{i=1}^{k} u_i(x_i) - \phi_n(x) + n^{-1}\hat{\varphi}(x-\overline{x})$$

has a global maximum in $\overline{x} \in \Omega^k$ for all $n \in \mathbb{N}$.

2. Integro-Differential Equations

As a next step, we will construct $(\varepsilon_n)_{n\in\mathbb{N}}$ and $(\overline{x}^n)_{n\in\mathbb{N}} \subset \Omega^k$ such that $\lim_{n\to\infty} \varepsilon_n = 0$, $\lim_{n\to\infty} \overline{x}^n = \overline{x}$ and $\overline{x}^n \in \arg\max(\sum_{i=1}^k u_i^{\varepsilon_n} - \phi_n)$ for all $n \in \mathbb{N}$: Due to their continuity and common growth at zero and infinity, there exists some $\lambda \geq 1$ such that

$$\lambda^{-1}\hat{\varphi}(x) \leq \sum_{i=1}^{k} \check{\varphi}(x_i) \leq \lambda\hat{\varphi}(x) \tag{2.13}$$

for $x = (x_1, \ldots, x_k) \in \mathbb{R}^{kd}$. Lemma 1.12 shows the existence of $x^\varepsilon = (x_1^\varepsilon, \ldots, x_k^\varepsilon) \in \mathbb{R}^{kd}$ for $x = (x_1, \ldots, x_k) \in \mathbb{R}^{kd}$ and $0 < \varepsilon \leq (2^{3(p\vee 2)-2}C)^{-1}$ with

$$\check{\varphi}(x_i^\varepsilon - x_i) \leq \varepsilon \cdot 2^{3(p\vee 2)}C(1 + \check{\varphi}(x_i)) \tag{2.14}$$

such that the following identity holds

$$\sum_{i=1}^{k} u_i^\varepsilon(x_i) = \sum_{i=1}^{k} \Delta_{\check{\varphi}}^\varepsilon[u_i](x_i) = \sum_{i=1}^{k} u_i(x_i^\varepsilon) - \varepsilon^{-1}\check{\varphi}(x_i - x_i^\varepsilon).$$

Moreover, Lemma 1.15 and Equation (2.12) imply with $C_n' := 2^{3((p\vee 2)+1)}(p\vee 2)C_n^{3/2} < \infty$

$$0 \leq \phi_n(x) - \nabla_{\hat{\varphi}}^{\lambda^{-1}\varepsilon}[\phi_n](x) \leq \varepsilon^{1/(p\vee 2)} \cdot C_n' \cdot (1 + \hat{\varphi}(x^\varepsilon))$$

for all $x \in \mathbb{R}^{kd}$, $0 < \varepsilon \leq (2^{3(p\vee 2)+1}10C_n)^{-1}$ and large $n \in \mathbb{N}$. Combining these two findings with $\overline{x} \in \arg\max(\sum_{i=1}^k u_i - \phi_n + n^{-1}\hat{\varphi}(\cdot - \overline{x}))$ yields

$$\sum_{i=1}^{k} u_i^\varepsilon(x_i) - \phi_n(x) \leq \sum_{i=1}^{k} u_i(x_i^\varepsilon) - \left(\phi_n(x) + \lambda\varepsilon^{-1}\hat{\varphi}(x - x^\varepsilon)\right)$$

$$\leq \left(\sum_{i=1}^{k} u_i(x_i^\varepsilon) - \phi_n(x^\varepsilon)\right) + \left(\phi_n(x^\varepsilon) - \nabla_{\hat{\varphi}}^{\lambda^{-1}\varepsilon}[\phi_n](x^\varepsilon)\right)$$

$$\leq \left(\sum_{i=1}^{k} u_i(\overline{x}) - \phi_n(\overline{x}) - n^{-1}\hat{\varphi}(x^\varepsilon - \overline{x})\right) + \varepsilon^{1/(p\vee 2)} \cdot C_n' \cdot (1 + \hat{\varphi}(x^\varepsilon))$$

$$\leq \left(\sum_{i=1}^{k} u_i^\varepsilon(\overline{x}) - \phi_n(\overline{x})\right) + \rho_{n,\varepsilon}(x)$$

for all $x \in \mathbb{R}^{kd}$, $0 < \varepsilon \leq (2^{3(p\vee 2)+1}10C_n)^{-1} \wedge (2^{3(p\vee 2)-2}C)^{-1}$ and large $n \in \mathbb{N}$, where

$$\rho_{n,\varepsilon}(x) := \varepsilon^{1/(p\vee 2)}C_n'\left(1 + \hat{\varphi}(x^\varepsilon)\right) - n^{-1}\hat{\varphi}(x^\varepsilon - \overline{x}).$$

Suppose that $0 < r < 1$ such that $B[\overline{x}, r] \subset \Omega^k = \Omega \times \ldots \times \Omega$. We will now show that for every $n \in \mathbb{N}$ there exists $\varepsilon_n > 0$ small enough such that $\rho_{n,\varepsilon_n}(x) < 0$ for all $x \in \mathbb{R}^{kd} \setminus B[\overline{x}, n^{-1}r]$. In particular, this will imply that

$$x = (x_1, \ldots, x_k) \longmapsto \sum_{i=1}^{k} u_i^{\varepsilon_n}(x_i) - \phi_n(x)$$

has a global maximum in some $\overline{x}^n \in B[\overline{x}, n^{-1}r] \subset \Omega^k$ for large $n \in \mathbb{N}$, as required:

2.2. Uniqueness of Solutions

For all $x \in \mathbb{R}^{kd}$ and small $\varepsilon > 0$, Equation (2.13), Equation (2.14) and the generalized triangle inequality of $\hat{\varphi} = \hat{\varphi}_{p\vee 2} \in C^{\infty}(\mathbb{R}^{kd})$ with multiplier $\rho = 2^{2(p\vee 2)-2} \leq 2^{2(p\vee 2)}$ from Lemma 1.14 show that

$$\hat{\varphi}(x^{\varepsilon} - x) \leq \lambda \sum_{i=1}^{k} \check{\varphi}(x_i^{\varepsilon} - x_i) \leq \lambda \sum_{i=1}^{k} \varepsilon \cdot 2^{3(p\vee 2)} C\left(1 + \check{\varphi}(x_i)\right) \leq \varepsilon C'\left(1 + \hat{\varphi}(x)\right)$$

with $C' := k\lambda^2 2^{3(p\vee 2)} C < \infty$, that $\hat{\varphi}(x^{\varepsilon}) \leq 2^{2(p\vee 2)} \hat{\varphi}(x) + \varepsilon\, 2^{2(p\vee 2)} C'(1 + \hat{\varphi}(x))$ and that

$$\hat{\varphi}(x - \overline{x}) \leq 2^{2(p\vee 2)} \left(\hat{\varphi}(x - x^{\varepsilon}) + \hat{\varphi}(x^{\varepsilon} - \overline{x})\right) \leq \varepsilon\, 2^{2(p\vee 2)} C'\left(1 + \hat{\varphi}(x)\right) + 2^{2(p\vee 2)} \hat{\varphi}(x^{\varepsilon} - \overline{x}).$$

This implies that $\hat{\varphi}(x^{\varepsilon} - \overline{x}) \geq 2^{-2(p\vee 2)} \hat{\varphi}(x - \overline{x}) - \varepsilon C'\left(1 + \hat{\varphi}(x)\right)$ and therefore that

$$\begin{aligned}
\rho_{n,\varepsilon}(x) &\leq \varepsilon^{1/(p\vee 2)} C_n'''\left(1 + \hat{\varphi}(x)\right) - n^{-1} 2^{-2(p\vee 2)} \hat{\varphi}(x - \overline{x}) \\
&\leq \varepsilon^{1/(p\vee 2)} 2^{2(p\vee 2)} C_n'''\left(1 + \hat{\varphi}(\overline{x})\right) + \varepsilon^{1/(p\vee 2)} 2^{2(p\vee 2)} C_n''' \hat{\varphi}(x - \overline{x}) - n^{-1} 2^{-2(p\vee 2)} \hat{\varphi}(x - \overline{x}) \\
&= \varepsilon^{1/(p\vee 2)} 2^{2(p\vee 2)} C_n'''\left(1 + \hat{\varphi}(\overline{x})\right) - n^{-1} 2^{-2(p\vee 2)-1} \hat{\varphi}(x - \overline{x}) \\
&\quad + \varepsilon^{1/(p\vee 2)} 2^{2(p\vee 2)} C_n''' \, \hat{\varphi}(x - \overline{x}) - n^{-1} 2^{-2(p\vee 2)-1} \hat{\varphi}(x - \overline{x})
\end{aligned}$$

for all $x \in \mathbb{R}^d$, $n \in \mathbb{N}$ and small $\varepsilon > 0$, where $C_n''' < \infty$ is a constant only depending on $p \geq 0$ and $C_n < \infty$. Hence, there exists $(\varepsilon_n)_{n \in \mathbb{N}} \subset (0, \infty)$ with $\lim_{n \to \infty} \varepsilon_n = 0$ such that $\rho_{n, \varepsilon_n}(x) < 0$ for all $x \in \mathbb{R}^{kd} \setminus B[\overline{x}, n^{-1}r]$ and $n \in \mathbb{N}$. In particular,

$$x = (x_1, \ldots, x_k) \longmapsto \sum_{i=1}^{k} u_i^{\varepsilon_n}(x_i) - \phi_n(x)$$

has a global maximum in some $\overline{x}^n \in B[\overline{x}, n^{-1}r] \subset \Omega^k$ for large $n \in \mathbb{N}$. Moreover, the properties of supremal convolutions from Lemma 1.12 imply that

$$x = (x_1, \ldots, x_k) \longmapsto \sum_{i=1}^{k} u_i^{\varepsilon_n}(x_i) + \frac{|x|^2}{\varepsilon_n} = \sum_{i=1}^{k} \left(u_i^{\varepsilon_n}(x_i) + \frac{|x_i|^2}{\varepsilon_n}\right)$$

is a convex function on $B[\overline{x}, r]$ and that $\overline{x}^n \in \Omega_{\varepsilon_n}^k$ for large $n \in \mathbb{N}$.

Now, we will show how to apply the maximum principle for smudged operators from Proposition 2.11 and the matrix compactification from Corollary 2.13: Lemma 1.18 and

$$|\phi_n(x)| \leq kC(1 + |x|^p) + n^{-1} \hat{\varphi}(x - \overline{x})$$

for all $x \in \mathbb{R}^d$ and $n \in \mathbb{N}$ imply the existence of $(\psi_n)_{n \in \mathbb{N}} \subset C_p^2(\mathbb{R}^{kd})$ such that

$$\begin{aligned}
\{x \in \mathbb{R}^{kd} : \psi_n(x) = \phi_n(x)\} &\supset \{x \in \mathbb{R}^{kd} : |\phi_n(x)| \leq 2kC\left(1 + |x|^p\right)\} \\
&\supset \{x \in \mathbb{R}^{kd} : n^{-1} \hat{\varphi}(x - \overline{x}) \leq kC\left(1 + |x|^p\right)\}
\end{aligned} \quad (2.15)$$

for all $n \in \mathbb{N}$, such that $\sup_{n \in \mathbb{N}} \|\psi_n\|_p \leq 8kC < \infty$, and such that

$$x = (x_1, \ldots, x_k) \longmapsto \sum_{i=1}^{k} u_i^{\varepsilon_n}(x_i) - \psi_n(x)$$

has a global maximum in $\overline{x}^n \in \Omega_{\varepsilon_n}^k$ for large $n \in \mathbb{N}$. Note that the right-hand side of Equation (2.15) increases to $\mathbb{R}^{kd}$ as $n \to \infty$ due to the continuity of $\hat{\varphi} \in C^{\infty}(\mathbb{R}^{kd})$.

2. Integro-Differential Equations

Finally, Proposition 2.11 implies the existence of $\overline{X}_1^{(n)}, \ldots, \overline{X}_k^{(n)} \in \mathbb{S}^{d \times d}$ satisfying

$$-\frac{2}{\varepsilon} I \leq \mathrm{diag}(\overline{X}_1^{(n)}, \ldots, \overline{X}_k^{(n)}) \leq D^2 \psi_n(\overline{x}^n)$$

such that the following inequalities hold in the ordinary sense

$$\Delta^{\varepsilon_n}[F_i](\overline{x}_i^n, u_i^{\varepsilon_n}(\overline{x}_i^n), D_{x_i} \psi_n(\overline{x}^n), \overline{X}_i^{(n)}, \psi_n(\overline{x}_1^n, \ldots, \overline{x}_{i-1}^n, \cdot, \overline{x}_{i+1}^n, \ldots, \overline{x}_k^n)) \leq 0$$

for large $n \in \mathbb{N}$ and $i \in \{1, \ldots, k\}$. Since $\psi_n \equiv \phi_n$ in a neighborhood of $\overline{x} \in \Omega^k$ for large $n \in \mathbb{N}$, and since $\hat{\varphi}(x) \equiv |x|^2$ in a neighborhood of $x = 0$, we find

$$D_{x_i} \psi_n(\overline{x}^n) = D_{x_i} \phi_n(\overline{x}^n) = D_{x_i} \phi(\overline{x}^n) + 2n^{-1}(\overline{x}_i^n - \overline{x}_i)$$
$$D_{x_i}^2 \psi_n(\overline{x}^n) = D_{x_i}^2 \phi_n(\overline{x}^n) = D_{x_i}^2 \phi(\overline{x}^n) + 2n^{-1} I$$

for large $n \in \mathbb{N}$. Therefore,

$$\mathrm{diag}(\overline{X}_1^{(n)}, \ldots, \overline{X}_k^{(n)}) \leq D^2 \psi_n(\overline{x}^n) = D_{x_i}^2 \phi(\overline{x}^n) + 2n^{-1} I$$
$$\leq g_0(\overline{x}^n) J_{k,d} + \mathrm{diag}(g_1(\overline{x}^n), \ldots, g_k(\overline{x}^n)) + 2n^{-1} I$$

for large $n \in \mathbb{N}$, which shows $\mathrm{diag}(Y_1^{(n)}, \ldots, Y_k^{(n)}) \leq J_{k,d}$ with

$$Y_i^{(n)} := g_0(\overline{x}^n)^{-1} \left(\overline{X}_i^{(n)} - g_i(\overline{x}^n) - 2n^{-1} I \right) \in \mathbb{S}^{d \times d}$$

for large $n \in \mathbb{N}$ and $i \in \{1, \ldots, k\}$. Corollary 2.13 now implies that

$$X_i^{(n)} := g_0(\overline{x}^n) \left(Y_i^{(n)} \right)^\gamma + g_i(\overline{x}^n) = g_0(\overline{x}^n) \left(Y_i^{(n)} (I - \gamma Y_i^{(n)})^{-1} \right) + g_i(\overline{x}^n) \in \mathbb{S}^{d \times d}$$

satisfy $\overline{X}_i^{(n)} \leq X_i^{(n)} + 2n^{-1} I$ and

$$-\gamma^{-1} g_0(\overline{x}^n) I \leq \mathrm{diag}(X_1^{(n)} - g_1(\overline{x}^n), \ldots, X_k^{(n)} - g_k(\overline{x}^n)) \leq (1 - k\gamma)^{-1} g_0(\overline{x}^n) J_{k,d} \quad (2.16)$$

for large $n \in \mathbb{N}$ and $i \in \{1, \ldots, k\}$. Hence, the degenerate ellipticity assumption (A2) for the smudged operators $\Delta^{\varepsilon_n}[F_i]$ (cf. Lemma 2.10) leads to

$$\Delta^{\varepsilon_n}[F_i](\overline{x}_i^n, u_i^{\varepsilon_n}(\overline{x}_i^n), D_{x_i} \psi_n(\overline{x}^n), X_i^{(n)} + 2n^{-1} I, \psi_n(\overline{x}_1^n, \ldots, \overline{x}_{i-1}^n, \cdot, \overline{x}_{i+1}^n, \ldots, \overline{x}_k^n)) \leq 0$$

for large $n \in \mathbb{N}$ and $i \in \{1, \ldots, k\}$.

At last, we will use the continuity assumption (A4) to obtain the desired result: Equation (2.16) together with $J_{k,d} \leq kI$, the continuity of $g_0 \in C(\mathbb{R}^{kd})$ and $g_i \in C(\mathbb{R}^{kd}; \mathbb{S}^{d \times d})$ at $\overline{x} \in \Omega^k$, and $g_0(\overline{x}) > 0$ imply (similar to the proof of Proposition 2.11) that the sets

$$\{X_i^{(n)} : n \in \mathbb{N}\}$$

are relatively compact for $i \in \{1, \ldots, k\}$. Thus, there are subsequences $(X_i^{(n_m)})_{m \in \mathbb{N}}$ of $(X_i^{(n)})_{n \in \mathbb{N}}$ and matrices $X_i \in \mathbb{S}^{d \times d}$ such that $\lim_{m \to \infty} X_i^{(n_m)} = X_i$ for all $i \in \{1, \ldots, k\}$.

2.2. Uniqueness of Solutions

By construction of the approximations $(\phi_n)_{n\in\mathbb{N}} \subset C^2(\mathbb{R}^{kd})$ and $(\psi_n)_{n\in\mathbb{N}} \subset C_p^2(\mathbb{R}^{kd})$ (cf. Equation (2.11) and Equation (2.15)), we have

$$\sup_{n\in\mathbb{N}} \|\psi_n\|_p \leq 8kC < \infty$$

and there exists $N_R \in \mathbb{N}$ for every $R > 0$ such that

$$\psi_n(x) = \phi_n(x) = \phi(x) + n^{-1}\hat{\varphi}(x - \overline{x})$$

for all $x \in B[\overline{x}, R]$ and $n \geq N_R$. This shows that

$$\lim_{n\to\infty} D^\alpha \psi_n = D^\alpha \phi$$

locally uniformly for $\alpha \in \{0, 1, 2\}$. Since $\lim_{n\to\infty} \varepsilon_n = 0$, we find $u_i^\varepsilon \geq u_i^{\varepsilon_n} \geq u_i$ for all $\varepsilon > 0$, large $n \in \mathbb{N}$ and $i \in \{1, \ldots, k\}$. Therefore, $\overline{x}^n \in \arg\max(\sum_{i=1}^k u_i^{\varepsilon_n} - \phi_n)$ implies

$$u_j^\varepsilon(\overline{x}_j^n) \geq u_j^{\varepsilon_n}(\overline{x}_j^n) = \left(\sum_{i=1}^k u_i^{\varepsilon_n}(\overline{x}_i^n) - \phi_n(\overline{x}^n)\right) - \left(\sum_{i\neq j} u_i^{\varepsilon_n}(\overline{x}_i^n) - \phi_n(\overline{x}^n)\right)$$

$$\geq \left(\sum_{j=1}^k u_i^{\varepsilon_n}(\overline{x}_i) - \phi_n(\overline{x})\right) - \left(\sum_{i\neq j} u_i^{\varepsilon_n}(\overline{x}_i^n) - \phi_n(\overline{x}^n)\right)$$

$$\geq \left(\sum_{j=1}^k u_i(\overline{x}_i) - \phi_n(\overline{x})\right) - \left(\sum_{i\neq j} u_i^\varepsilon(\overline{x}_i^n) - \phi_n(\overline{x}^n)\right),$$

for large $n \in \mathbb{N}$ and $j \in \{1, \ldots, k\}$. This shows that

$$u_j^\varepsilon(\overline{x}_j) \geq \limsup_{n\to\infty} u_j^{\varepsilon_n}(\overline{x}_j^n) \geq \liminf_{n\to\infty} u_j^{\varepsilon_n}(\overline{x}_j^n)$$

$$\geq \left(\sum_{i=1}^k u_i(\overline{x}_i) - \phi(\overline{x})\right) - \left(\sum_{i\neq j} u_i^\varepsilon(\overline{x}_i) - \phi(\overline{x})\right) = \sum_{i=1}^k u_i(\overline{x}_i) - \sum_{i\neq j} u_i^\varepsilon(\overline{x}_i)$$

for small $\varepsilon > 0$ and $j \in \{1, \ldots, k\}$, using the (semi-)continuity of $u_i \in \mathrm{USC}(\mathbb{R}^d)$ and $u_i^\varepsilon \in C(\mathbb{R}^d)$ for $i \in \{1, \ldots, k\}$, $\lim_{n\to\infty} \overline{x}^n = \overline{x}$ and $\lim_{n\to\infty} \phi_n = \phi$ locally uniformly. In particular, since $\lim_{\varepsilon\to 0} u_j^\varepsilon(x) = u_j(x)$ for $x \in \mathbb{R}^d$ according to Lemma 1.12, this shows

$$\lim_{n\to\infty} u_j^{\varepsilon_n}(\overline{x}_j^n) = u_j(\overline{x}_j)$$

for all $j \in \{1, \ldots, k\}$. Finally, a combination of the preceding convergence results with the continuity assumption (A4) for $\Delta^{\varepsilon_n}[F_i]$ (cf. Lemma 2.10) yields

$$\Delta^\varepsilon[F_i](\overline{x}_i, u_i(\overline{x}_i), D_{x_i}\psi(\overline{x}), X_i, \phi(\overline{x}_1, \ldots, \overline{x}_{i-1}, \cdot, \overline{x}_{i+1}, \ldots, \overline{x}_k)) \leq 0$$

for small $\varepsilon > 0$ and $i \in \{1, \ldots, k\}$. The (pointwise) convergence $\lim_{\varepsilon\to 0} \Delta^\varepsilon[F_i] = F_i$ for $i \in \{1, \ldots, k\}$ (cf. the proof of Lemma 2.10) hence implies the desired result. □

2. Integro-Differential Equations

The preceding lift of the maximum principle for smudged operators to the original problem is considerably more involved compared to the existing literature, such as [33] for local equations or [75] for nonlocal equations: In fact, at the beginning of the proof of [75, Lemma 7.8], the authors claim that since

$$x = (x_1, \ldots, x_k) \longmapsto \sum_{i=1}^{k} u_i(x_i) - \phi(x)$$

has a global maximum $\bar{x} \in \mathbb{R}^d$ by assumption, the approximated problems

$$x = (x_1, \ldots, x_k) \longmapsto \sum_{i=1}^{k} u_i^\varepsilon(x_i) - \phi(x)$$

have global maxima in $\bar{x}^\varepsilon \in \mathbb{R}^d$ for small $\varepsilon > 0$ with $\lim_{\varepsilon \to 0} \bar{x}^\varepsilon = \bar{x}$. Unfortunately, this implication does not seem to hold in this generality: Take e.g. $k=1$ with $u_1(x) = |x|^2$ and $\phi(x) = |x|^2 - (1+|x|^2)^{-1}$ for $x \in \mathbb{R}^d$. A simple calculation shows

$$u_i^\varepsilon(x) = \Delta_{|\cdot|^2}^\varepsilon[u_i](x) = |x|^2 + \varepsilon(1-\varepsilon)^{-1}|x|^2$$

for $\varepsilon > 0$ and $x \in \mathbb{R}^d$, employing the classical maximum principle for differentiable functions. Obviously, $x \mapsto u_1(x) - \phi(x) = (1+|x|^2)^{-1}$ has a global maximum in $\bar{x} = 0$, whereas the functions

$$x \longmapsto u_1^\varepsilon(x) - \phi(x) = \varepsilon(1-\varepsilon)^{-1}|x|^2 + (1+|x|^2)^{-1}$$

are not even bounded for any $\varepsilon > 0$. (Note that, even though we used the parabola $|\cdot|^2$ instead of the quasidistance φ_2 with quadratic growth from Lemma 1.14, the example still works if we would use the latter.) In order to prove a general maximum principle using the stability result from Lemma 2.8 for solutions with p-polynomial growth in possibly unbounded domains Ω, we therefore have to construct approximations ϕ^ε of the original function ϕ as well, which include additional penalization terms to ensure that

$$x = (x_1, \ldots, x_k) \longmapsto \sum_{i=1}^{k} u_i^\varepsilon(x_i) - \phi^\varepsilon(x)$$

stills attains a maximum for small $\varepsilon > 0$. Furthermore, the construction has to be chosen carefully to ensure that the convergence $\lim_{\varepsilon \to 0} \phi^\varepsilon = \phi$ is strong enough to apply our weaker continuity assumptions. Alternatively, one could impose stronger assumptions on the maximum point in the statement of Theorem 2.14, which typically only postpones the additional effort to applications of these maximum principle.

Since we want to use the generalized operators F^κ (which separates the singular nonlocal terms from its rest) to prove a comparison principle later, we now show how to lift the preceding maximum principle from the original operator F to its generalization F^κ. The idea of the following proof essentially corresponds to the one of Lemma 2.7, in which we showed that viscosity solutions (under additional continuity assumptions) can be characterized using F^κ instead of F.

2.2. Uniqueness of Solutions

Lemma 2.15 (Maximum Principle for Generalized Operators). *Suppose that $p \geq 0$, that $\Omega \subset \mathbb{R}^d$ is open, and that $u_i \in C_p(\mathbb{R}^d)$ are viscosity solutions in Ω of*

$$F_i(x, u_i(x), Du_i(x), D^2 u_i(x), u_i(\cdot)) \leq 0$$

for $i \in \{1, \ldots, k\}$ and operators F_i, which satisfy all assumptions from Remark 2.3. Furthermore, suppose that $\phi \in C_p^2(\mathbb{R}^{kd})$, $g_0 \in C(\mathbb{R}^{kd}; \mathbb{R})$ and $g_i \in C(\mathbb{R}^{kd}; \mathbb{S}^{d \times d})$ such that

$$D^2 \phi(x) \leq g_0(x) J_{k,d} + \mathrm{diag}(g_1(x), \ldots, g_k(x))$$

for all $x \in \mathbb{R}^{kd}$, where $J_{k,d} \in \mathbb{R}^{kd \times kd}$ is defined as in Corollary 2.13. If

$$x = (x_1, \ldots, x_k) \longmapsto \sum_{i=1}^{k} u_i(x_i) - \phi(x)$$

has a global maximum in $\overline{x} = (\overline{x}_1, \ldots, \overline{x}_k) \in \Omega^k = \Omega \times \ldots \times \Omega$ with $g_0(\overline{x}) > 0$, then there exist symmetric matrices $X_1^\gamma, \ldots, X_k^\gamma \in \mathbb{S}^{d \times d}$ for every $\gamma \in (0, k^{-1})$ satisfying

$$-\gamma^{-1} g_0(\overline{x}) I \leq \mathrm{diag}(X_1^\gamma - g_1(\overline{x}), \ldots, X_k^\gamma - g_k(\overline{x})) \leq (1 - k\gamma)^{-1} g_0(\overline{x}) J_{k,d}$$

such that the following inequalities hold in the ordinary sense

$$F_i^\kappa(\overline{x}_i, u_i(\overline{x}_i), D_{x_i}\phi(\overline{x}), X_i^\gamma, u_i(\cdot), \phi(\overline{x}_1, \ldots, \overline{x}_{i-1}, \cdot, \overline{x}_{i+1}, \ldots, \overline{x}_k)) \leq 0$$

for all $i \in \{1, \ldots, k\}$ and $0 < \kappa < 1$.

Proof. Similar to the proof of Theorem 2.14, we assume without loss of generality that

$$\sum_{i=1}^{k} u_i(\overline{x}_i) = \phi(\overline{x}).$$

According to Theorem 1.20, there hence exists a sequence $(\phi_n)_{n \in \mathbb{N}} \subset C_p^2(\mathbb{R}^{kd})$ such that

$$\sum_{i=1}^{k} u_i(x_i) \leq \phi_{n+1}(x) \leq \phi_n(x) \leq \phi(x)$$

for all $x = (x_1, \ldots, x_k) \in \mathbb{R}^{kd}$ and $n \in \mathbb{N}$, and such that

$$\lim_{n \to \infty} \phi_n(x) = \sum_{i=1}^{k} u_i(x)$$

for all $x = (x_1, \ldots, x_k) \in \mathbb{R}^{kd}$. In particular, the functions $\phi_n - \phi$ and

$$x = (x_1, \ldots, x_k) \longmapsto \sum_{i=1}^{k} u_i(x_i) - \phi_n(x)$$

have global maxima in $\overline{x} \in \Omega^k$ for every $n \in \mathbb{N}$. Before we can apply Theorem 2.14, we have to obtain bounds for $D^2\phi_n$ with $n \in \mathbb{N}$: A simple calculation shows

$$D^2\phi_n(x) = D^2\phi(x) + D^2(\phi_n - \phi)(x) \le D^2\phi(x) + \lambda_{\max}^+\left[D^2(\phi_n - \phi)(x)\right]I$$
$$\le g_0(x)J_{k,d} + \mathrm{diag}(g_1^{(n)}(x), \ldots, g_k^{(n)}(x))$$

for all $x \in \mathbb{R}^{kd}$ and $n \in \mathbb{N}$, where $\lambda_{\max}^+\left[D^2(\phi_n - \phi)(x)\right]$ denotes the positive part of the largest eigenvalue of $D^2(\phi_n - \phi)(x) \in \mathbb{S}^{kd \times kd}$ and

$$g_i^{(n)}(x) := g_i(x) + \lambda_{\max}^+\left[D^2(\phi_n - \phi)(x)\right]I$$

for $i \in \{1, \ldots, k\}$. Theorem 2.14 now implies for every $\gamma \in (0, k^{-1})$ the existence of symmetric matrices $X_1^{(n)}, \ldots, X_k^{(n)} \in \mathbb{S}^{d \times d}$ satisfying

$$-\gamma^{-1}g_0(\overline{x})I \le \mathrm{diag}(X_1^{(n)} - g_1^{(n)}(\overline{x}), \ldots, X_k^{(n)} - g_k^{(n)}(\overline{x})) \le (1 - k\gamma)^{-1}g_0(\overline{x})J_{k,d}$$

such that the following inequalities hold in the ordinary sense

$$F_i(\overline{x}_i, u_i(\overline{x}_i), D_{x_i}\phi_n(\overline{x}), X_i^{(n)}, \phi_n(\overline{x}_1, \ldots, \overline{x}_{i-1}, \cdot, \overline{x}_{i+1}, \ldots, \overline{x}_k)) \le 0 \qquad (2.17)$$

for all $n \in \mathbb{N}$ and $i \in \{1, \ldots, k\}$.

In order to pass to the limit, recall that $\phi_n - \phi$ have global maxima in $\overline{x} \in \Omega^k$ for every $n \in \mathbb{N}$, which implies $D(\phi_n - \phi)(\overline{x}) = 0$ and $D^2(\phi_n - \phi)(\overline{x}) \le 0$ for all $n \in \mathbb{N}$. Therefore, we obtain $g_j^{(n)}(\overline{x}) = g_j(\overline{x})$ for $j \in \{1, \ldots, k\}$ and hence

$$-\gamma^{-1}g_0(\overline{x})I \le \mathrm{diag}(X_1^{(n)} - g_1(\overline{x}), \ldots, X_k^{(n)} - g_k(\overline{x})) \le (1 - k\gamma)^{-1}g_0(\overline{x})J_{k,d}$$

for all $n \in \mathbb{N}$. Similar to the proof of Corollary 2.13, this implies that

$$\{X_i^{(n)} : n \in \mathbb{N}\} \subset \mathbb{S}^{d \times d}$$

is relatively compact for all $i \in \{1, \ldots, k\}$. Thus, there exist matrices $X_i \in \mathbb{S}^{d \times d}$ and subsequences $(X_i^{(n_m)})_{m \in \mathbb{N}}$ of $(X_i^{(n)})_{n \in \mathbb{N}}$ such that

$$\lim_{m \to \infty} X_i^{(n_m)} = X_i$$

for all $i \in \{1, \ldots, k\}$. According to Dini's theorem from Theorem 1.27, the continuity of $x = (x_1, \ldots, x_k) \mapsto \sum_{i=1}^k u_i(x_i)$ implies that $\lim_{n \to \infty} \phi_n = \sum_{i=1}^k u_i$ locally uniformly. Since $\sum_{i=1}^k u_i(x_i) \le \phi_n(x) \le \phi_1(x)$ for all $x = (x_1, \ldots, x_k) \in \mathbb{R}^{kd}$ and $n \in \mathbb{N}$, we find

$$\sup_{n \in \mathbb{N}} \|\phi_n\|_p \le \|\phi_1\|_p + \sum_{i=1}^k \|u_i\|_p < \infty.$$

Hence, Equation (2.17) together with the degenerate ellipticity (A2) and the continuity assumption (A4) (using $\lim_{m \to \infty} \phi_{n_m}(\overline{x}_1, \ldots, \overline{x}_{i-1}, \cdot, \overline{x}_{i+1}, \ldots, \overline{x}_k) = u_i(\cdot) + \sum_{j \ne i} u_j(\overline{x}_i)$ locally uniformly and $\lim_{m \to \infty} X_i^{(n_m)} = X_i$ for $i \in \{1, \ldots, k\}$) lead to

$$F_i^\kappa(\overline{x}_i, u_i(\overline{x}_i), D_{x_i}\phi(\overline{x}), X_i, u_i(\cdot) + \sum_{j \ne i} u_j(\overline{x}_i), \phi(\overline{x}_1, \ldots, \overline{x}_{i-1}, \cdot, \overline{x}_{i+1}, \ldots, \overline{x}_k)) \le 0$$

for all $i \in \{1, \ldots, k\}$. Finally, an application of the translation invariance assumption (A3) implies the desired result. □

2.2. Uniqueness of Solutions

A careful inspection of the proof of Lemma 2.15 shows (similar to the discussion after Lemma 2.7) that we utilized the continuity of our subsolutions $u_i \in C(\mathbb{R}^d)$ only once, in order to show that the convergence

$$\lim_{n \to \infty} \phi_n = \sum_{i=1}^{k} u_i$$

is locally uniform and not only pointwise (by applying Dini's theorem, cf. Theorem 1.27). This necessity stems from our weaker continuity assumption (A4). In fact, if the operators F_i were continuous with respect to pointwise convergence (cf. the discussion after Remark 2.3), Lemma 2.15 would also hold for discontinuous $u_i \in \text{USC}_p(\mathbb{R}^d)$. Since the preceding maximum principle for generalized operators is the main tool to prove a comparison principle for semicontinuous functions (as we will see later in Theorem 2.22), this shows that our weaker continuity assumptions not only cause more involved proofs for the corresponding maximum principles, but also for their applications.

Before we proceed to show how to apply our preceding results to parabolic problems, let us make another comment on their originality: Although the main line of arguments for proving the maximum principles in Proposition 2.11, Theorem 2.14 and Lemma 2.15 follow the ideas in [75] (which in turn are based on [33] for local equations), we consider it valuable to give all necessary details here. Due to our substantially weaker continuity and growth assumptions (cf. Remark 2.3), some of the approximations at the heart of these arguments have to be significantly generalized and applied with the utmost care. Despite the fact that the technicalities of these generalizations are partly hidden in Chapter 1, their ideas are critical to cover the equations we obtain for sublinear Markov semigroups in Chapter 4.

In the preceding proofs, it became apparent that the approximation of maximum points is an important concept in viscosity solutions theory. The following lemma from [35, Proposition 3.7] presents a very abstract formulation of this concept, and we include its proof here in order to be self-contained.

Lemma 2.16 (Perturbed Maximum Points). *Suppose that $\Omega \subset \mathbb{R}^d$ is an arbitrary set, that $\Phi \in USC(\Omega)$ and $\Psi \in LSC(\Omega)$ with $\Psi(x) \geq 0$ for all $x \in \Omega$, and that*

$$M_\lambda := \sup \{\Phi(x) - \lambda \Psi(x) : x \in \Omega\}$$

for all $\lambda > 0$. If $\lim_{\lambda \to \infty} M_\lambda$ exists in $\mathbb{R}$ and $x^{(\lambda)} \in \Omega$ is chosen such that

$$\lim_{\lambda \to \infty} \left(M_\lambda - \left(\Phi(x^{(\lambda)}) - \lambda \Psi(x^{(\lambda)}) \right) \right) = 0,$$

then $\lim_{\lambda \to \infty} \lambda \Psi(x^{(\lambda)}) = 0$. Moreover,

$$\lim_{\lambda \to \infty} M_\lambda = \Phi(\hat{x}) = \sup \{\Phi(x) : x \in \Omega \text{ with } \Psi(x) = 0\},$$

whenever $\hat{x} \in \Omega$ is a limit point of $(x^{(\lambda)})_{\lambda > 0}$.

Proof. For the first part, it is easy to check that

$$M_{\lambda/2} \geq \Phi(x^{(\lambda)}) - (\lambda/2)\Psi(x^{(\lambda)}) = \left(\Phi(x^{(\lambda)}) - \lambda\Psi(x^{(\lambda)})\right) + (\lambda/2)\Psi(x^{(\lambda)})$$
$$\geq M_\lambda - \delta_\lambda + (\lambda/2)\Psi(x^{(\lambda)})$$

with $\delta_\lambda := M_\lambda - (\Phi(x^{(\lambda)}) - \lambda\Psi(x^{(\lambda)}))$ for all $\lambda > 0$. Therefore, we obtain

$$0 \leq \liminf_{\lambda \to \infty} \left(\lambda\Psi(x^{(\lambda)})\right) \leq \limsup_{\lambda \to \infty} \left(\lambda\Psi(x^{(\lambda)})\right) \leq \limsup_{\lambda \to \infty} \left(2\left(M_{\lambda/2} - M_\lambda + \delta_\lambda\right)\right) = 0,$$

since $\lim_{\lambda \to \infty} \delta_\lambda = 0$ and $\lim_{\lambda \to \infty} M_\lambda = \lim_{\lambda \to \infty} M_{\lambda/2}$ by assumption.

For the second part, suppose that $\hat{x} \in \Omega$ is a limit point of $(x^{(\lambda)})_{\lambda \geq 0}$, i.e. there exists $(\lambda_n)_{n \in \mathbb{N}}$ with $\lim_{n \to \infty} \lambda_n = \infty$ such that $\lim_{n \to \infty} x^{(\lambda_n)} = \hat{x}$. Since $\lim_{n \to \infty} \lambda_n \Psi(x^{(\lambda_n)}) = 0$,

$$0 \leq \Psi(\hat{x}) \leq \liminf_{n \to \infty} \Psi(x^{(\lambda_n)}) = \left(\lim_{n \to \infty} \lambda_n^{-1}\right)\left(\lim_{n \to \infty} \lambda_n \Psi(x^{(\lambda_n)})\right) = 0$$

and hence $\Psi(\hat{x}) = 0$, using the lower semicontinuity of $\Psi \in \text{LSC}(\Omega)$. Moreover,

$$\lim_{n \to \infty} M_{\lambda_n} = \lim_{n \to \infty} (M_{\lambda_n} - \delta_{\lambda_n}) \leq \limsup_{n \to \infty} \left(\Phi(x^{(\lambda_n)}) - \lambda_n \Psi(x^{(\lambda_n)})\right)$$
$$= \limsup_{n \to \infty} \Phi(x^{(\lambda_n)}) \leq \Phi(\hat{x})$$
$$\leq \sup\{\Phi(x) : x \in \Omega \text{ with } \Psi(x) = 0\} \leq \lim_{n \to \infty} M_{\lambda_n}$$

using the upper semicontinuity of $\Phi \in \text{USC}(\Omega)$. Since $\Psi(x) \geq 0$ for all $x \in \Omega$, M_λ decreases as $\lambda > 0$ increases. Finally, this implies

$$\lim_{\lambda \to \infty} M_\lambda = \lim_{n \to \infty} M_{\lambda_n} = \Phi(\hat{x}) = \sup\{\Phi(x) : x \in \Omega \text{ with } \Psi(x) = 0\},$$

and therefore proves the desired result. □

As a special case of the preceding result, we obtain the well-established concept of variable augmentation (also known as variable doubling for $k = 2$): For viscosity subsolutions $u_i \in \text{USC}(\mathbb{R}^d)$ with $i \in \{1, \ldots, k\}$, it is useful to maximize

$$\mathbb{R}^{kd} \ni x = (x_1, \ldots, x_k) \longmapsto \sum_{i=1}^k u_i(x_i) - \lambda\phi(x)$$

instead of $\mathbb{R}^d \ni x \mapsto \sum_{i=1}^k u_i(x)$ for large constants $\lambda > 0$ and some function $\phi \in C^2(\mathbb{R}^d)$, that penalizes $x \in \mathbb{R}^{kd}$ away from the diagonal $\{x = (x_1, \ldots, x_k) \in \mathbb{R}^{kd} : x_1 = \ldots = x_k\}$: Suppose that the augmented function $x \mapsto \sum_{i=1}^k u_i(x_i) - \lambda\phi(x)$ has a global maximum in $\overline{x}^\lambda = (\overline{x}_1^\lambda, \ldots, \overline{x}_k^\lambda) \in \mathbb{R}^{kd}$, then for each $j \in \{1, \ldots, k\}$ the maps

$$\mathbb{R}^d \ni x_j \longmapsto u_j(x_j) + \sum_{i \neq j} u_i(\overline{x}_i^\lambda) - \lambda\phi(\overline{x}^\lambda)$$

have global maxima in $\overline{x}_j^\lambda \in \mathbb{R}^d$. This allows us to apply the definition of viscosity solutions from Definition 2.1 or the degenerate ellipticity assumption (A2) from Remark 2.3. Moreover, the following result shows how we can recover global maxima of the original problem $\mathbb{R}^d \ni x \mapsto \sum_{i=1}^k u_i(x)$ by sending $\lambda > 0$ to infinity.

2.2. Uniqueness of Solutions

Corollary 2.17 (Variable Augmentation). *Suppose that $\Omega \subset \mathbb{R}^d$ is an arbitrary set, and that $u_i \in \mathrm{USC}(\Omega)$ for $i \in \{1, \ldots, k\}$ with $k \in \mathbb{N}$. If $M_\lambda < \infty$ for large $\lambda > 0$, where*

$$M_\lambda := \sup\left\{\sum_{i=1}^k u_i(x_i) - \lambda \sum_{i<j} |x_i - x_j|^2 : x = (x_1, \ldots, x_k) \in \Omega^k\right\},$$

and $x^{(\lambda)} \in \Omega^k$ is chosen such that

$$\lim_{\lambda \to \infty} \left(M_\lambda - \left(\sum_{i=1}^k u_i(x_i^{(\lambda)}) - \lambda \sum_{i<j} |x_i^{(\lambda)} - x_j^{(\lambda)}|^2\right)\right) = 0,$$

then $\lim_{\lambda \to \infty} \lambda \sum_{i<j} |x_i^{(\lambda)} - x_j^{(\lambda)}|^2 = 0$. Moreover,

$$\lim_{\lambda \to \infty} M_\lambda = \sum_{i=1}^k u_i(\hat{x}_i) = \sum_{i=1}^k u_i(\hat{x}_1) = \sup\left\{\sum_{i=1}^k u_i(x) : x \in \Omega\right\}$$

and $\hat{x}_1 = \ldots = \hat{x}_k$, whenever $\hat{x} = (\hat{x}_1, \ldots, \hat{x}_k) \in \Omega^k$ is a limit point of $(x^{(\lambda)})_{\lambda > 0}$.

Let us quickly remark how the variable augmentation from Corollary 2.17 is usually applied for bounded viscosity solutions on bounded domains in the existing literature: First of all, if the domain $\Omega \subset \mathbb{R}^d$ is bounded and closed, then

$$\sup_{x \in \Omega^k}\left(\sum_{i=1}^k u_i(x_i) - \lambda \sum_{i<j} |x_i - x_j|^2\right) = \sum_{i=1}^k u_i(x_i^{(\lambda)}) - \lambda \sum_{i<j} |x_i^{(\lambda)} - x_j^{(\lambda)}|^2$$

for every $\lambda > 0$ with some $x^{(\lambda)} = (x_1^{(\lambda)}, \ldots, x_k^{(\lambda)}) \in \Omega^k$. Furthermore, if

$$\mathbb{R}^d \supseteq \Omega \ni x \longmapsto \sum_{i=1}^k u_i(x)$$

has a strict global maximum in $\hat{x} \in \Omega$, then Corollary 2.17 implies that

$$\lim_{\lambda \to \infty} x^{(\lambda)} = \hat{x}.$$

Since we want to cover unbounded viscosity solutions in unbounded domains, the applications of Corollary 2.17 gets more involved. More precisely, we typically have to introduce penalization terms to the original problem first, to ensure that all related suprema are finite and attained inside a fixed bounded set.

The preceding proof of Corollary 2.17 also allows to augment functions with respect to only some of its coordinates (similar to the discussion after Lemma 2.8). However, since we do not want to complicate our presentation by technical details in this thesis, we leave the more general formulation and proof to the interested reader.

2. Integro-Differential Equations

The main objective for the rest of this section is to obtain a maximum and a comparison principle for general parabolic equations of the form

$$\partial_t u(t,x) + G(t, x, u(t,x), Du(t,x), D^2u(t,x), u(t, \cdot)) = 0, \tag{E2}$$

with solutions $u : [0,T] \times \mathbb{R}^d \to \mathbb{R}$ for some $T > 0$ and degenerate elliptic operators

$$G : (0,T) \times \mathbb{R}^d \times \mathbb{R} \times \mathbb{R}^d \times \mathbb{S}^{d \times d} \times C_p^2(\mathbb{R}^d) \longrightarrow \mathbb{R}$$

satisfying all assumptions in Remark 2.3. Obviously, this a special form of the equations

$$F(x, u(x), Du(x), D^2u(x), u(\cdot)) = 0, \tag{E1}$$

which we have considered up to now. According to Lemma 2.6, we can apply the elliptic maximum principle from Theorem 2.14 to (semi-)solutions that operate on $[0,T] \times \mathbb{R}^d$ (instead of $\mathbb{R}^{1+d}$) without changing any of the arguments, since the operator G only depends on values in $(0,T) \times \mathbb{R}^d$. The proof of the following parabolic maximum principle is based on a sketch in [74] (which in turn originates from [34]). It uses the special form of parabolic equations to obtain a refined statement, which is useful for later applications.

Lemma 2.18 (Parabolic Maximum Principle). *Suppose that $p \geq 0$ and $T > 0$, that $\Omega \subset \mathbb{R}^d$ is open, and that $u_i \in USC_p([0,T] \times \mathbb{R}^d)$ are viscosity solutions in $(0,T) \times \Omega$ of*

$$\partial_t u_i(t,x) + G_i(t, x, u_i(t,x), Du_i(t,x), D^2u_i(t,x), u_i(t, \cdot)) \leq 0$$

for $i \in \{1, \ldots, k\}$ and operators G_i, which satisfy all assumptions from Remark 2.3. Furthermore, suppose that $\phi \in C_p^2((0,T) \times \mathbb{R}^{kd})$ together with $g_0 \in C((0,T) \times \mathbb{R}^{kd}; \mathbb{R})$ and $g_i \in C((0,T) \times \mathbb{R}^{kd}; \mathbb{S}^{d \times d})$ such that

$$D^2 \phi(t,x) \leq g_0(t,x) J_{k,d} + \operatorname{diag}(g_1(t,x), \ldots, g_k(t,x))$$

for all $(t,x) \in (0,T) \times \mathbb{R}^{kd}$, where $J_{k,d} \in \mathbb{R}^{kd \times kd}$ is defined as in Corollary 2.13. If

$$(t,x) = (t, x_1, \ldots, x_k) \longmapsto \sum_{i=1}^{k} u_i(t, x_i) - \phi(t,x)$$

has a global maximum in $(\bar{t}, \bar{x}) = (\bar{t}, \bar{x}_1, \ldots, \bar{x}_k) \in (0,T) \times \Omega^k$ with $g_0(\bar{t}, \bar{x}) > 0$, then there exist symmetric matrices $X_1^\gamma, \ldots, X_k^\gamma \in \mathbb{S}^{d \times d}$ and $b_1, \ldots, b_k \in \mathbb{R}^d$ for every $\gamma \in (0, k^{-1})$ satisfying $\sum_{i=1}^k b_i = \partial_t \phi(\bar{t}, \bar{x})$ and

$$-\gamma^{-1} g_0(\bar{t}, \bar{x}) I \leq \operatorname{diag}(X_1^\gamma - g_1(\bar{t}, \bar{x}), \ldots, X_k^\gamma - g_k(\bar{t}, \bar{x})) \leq (1 - k\gamma)^{-1} g_0(\bar{t}, \bar{x}) J_{k,d}$$

such that the following inequalities hold in the ordinary sense

$$b_i - G_i(\bar{t}, \bar{x}_i, u_i(\bar{t}, \bar{x}_i), D_{x_i} \phi(\bar{t}, \bar{x}), X_i^\gamma, \phi(\bar{t}, \bar{x}_1, \ldots, \bar{x}_{i-1}, \cdot, \bar{x}_{i+1}, \ldots, \bar{x}_k)) \leq 0$$

for all $i \in \{1, \ldots, k\}$.

Proof. Before we begin, note that similar arguments as in the proof of Lemma 2.6 show: We can assume without loss of generality that $\phi \in C_p^2(\mathbb{R} \times \mathbb{R}^{kd})$ and $u_i \in \text{USC}_p(\mathbb{R} \times \mathbb{R}^d)$ for $i \in \{1, \ldots, k\}$, i.e. all functions are defined on the whole space.

The proof is divided into four parts: In the first part, we will construct the approximations $\phi_\lambda^\varepsilon \in C^2(\mathbb{R}^k \times \mathbb{R}^{kd})$ for $\phi \in C^2(\mathbb{R} \times \mathbb{R}^{kd})$ with small $\varepsilon > 0$ and large $\lambda > 0$, introduce the related perturbed problems

$$\mathbb{R}^k \times \mathbb{R}^{kd} \ni (\tau, x) \longmapsto \sum_{i=1}^k u_i(\tau_i, x_i) - \phi_\lambda^\varepsilon(\tau, x)$$

and show that they attain their suprema in some $(\hat{\tau}^{(\varepsilon,\lambda)}, \hat{x}^{(\varepsilon,\lambda)}) \in \mathbb{R}^k \times \mathbb{R}^{kd}$. In the second part, we will use Corollary 2.17 to show that (along a suitable subsequence)

$$\lim_{\varepsilon \to 0} \lim_{\lambda \to \infty} (\hat{\tau}^{(\varepsilon,\lambda)}, \hat{x}^{(\varepsilon,\lambda)}) = (\bar{t}, \ldots, \bar{t}, \bar{x}).$$

In the third part, we will demonstrate how to apply the elliptic maximum principle from Theorem 2.14 to the perturbed problems for fixed $\varepsilon > 0$ and $\lambda > 0$. In the forth part, we will show that the continuity assumption (A4) implies the desired result as $\lambda \to \infty$ and $\varepsilon \to 0$ (along a suitable subsequence).

First, we will introduce the perturbed problems: Define $\phi_\lambda^\varepsilon \in C^2(\mathbb{R}^k \times \mathbb{R}^{kd})$ by

$$\phi_\lambda^\varepsilon(\tau, x) := \phi(\tau_1, x) + \lambda \sum_{i<j} |\tau_i - \tau_j|^2 + \varepsilon \sum_{i=1}^k \left(|x_i - \bar{x}_i|^{p+2} + |\tau_i - \bar{t}|^2\right) \quad (2.18)$$

for $(\tau, x) = (\tau_1, \ldots, \tau_k, x_1, \ldots, x_k) \in \mathbb{R}^k \times \mathbb{R}^{kd}$, $\varepsilon > 0$ and $\lambda \geq 0$, and set

$$M_\lambda^\varepsilon := \sup_{(\tau,x) \in \mathbb{R}^k \times \mathbb{R}^{kd}} \left(\sum_{i=1}^k u_i(\tau_i, x_i) - \phi_\lambda^\varepsilon(\tau, x)\right)$$

for $\varepsilon > 0$ and $\lambda \geq 0$. We want to show that these suprema are attained: By assumption,

$$m_0 := \sup_{(t,x) \in \mathbb{R} \times \mathbb{R}^{kd}} \left(\sum_{i=1}^k u_i(t, x_i) - \phi(t, x)\right) = \sum_{i=1}^k u_i(\bar{t}, \bar{x}_i) - \phi(\bar{t}, \bar{x}) < \infty.$$

Moreover, it is easy to see that for each $(\tau, x) \in \mathbb{R}^k \times \mathbb{R}^{kd}$, $\varepsilon > 0$ and $\lambda \geq 0$

$$\sum_{i=1}^k u_i(\tau_i, x_i) - \phi_\lambda^\varepsilon(\tau, x) \leq \sum_{i=1}^k u_i(\tau_i, x_i) - \phi(\tau_1, x) - \varepsilon \sum_{i=1}^k \left(|x_i - \bar{x}_i|^{p+2} + |\tau_i - \bar{t}|^2\right)$$

$$\leq m_0 + \sum_{i=1}^k (u_i(\tau_i, x_i) - u_i(\tau_1, x_i))$$

$$- \varepsilon \sum_{i=1}^k \left(|x_i - \bar{x}_i|^{p+2} + |\tau_i - \bar{t}|^2\right)$$

$$\leq m_0 + \sum_{i=1}^k 2C(1 + |x_i|^p) - \varepsilon \sum_{i=1}^k \left(|x_i - \bar{x}_i|^{p+2} + |\tau_i - \bar{t}|^2\right).$$

Since the right-hand side of this inequality converges to minus infinity as $|(\tau,x)| \to \infty$ for fixed $\varepsilon > 0$ (independent of $\lambda \geq 0$), there hence exist constants $R_\varepsilon > 0$ such that

$$M_\lambda^\varepsilon = \sup\left\{ \sum_{i=1}^k u_i(\tau_i, x_i) - \phi_\lambda^\varepsilon(\tau, x) \,\bigg|\, (\tau, x) \in B[0, R_\varepsilon] \subset \mathbb{R}^k \times \mathbb{R}^{kd} \right\}$$

for all $\lambda \geq 0$. In particular, there exist

$$(\hat\tau^{(\varepsilon,\lambda)}, \hat x^{(\varepsilon,\lambda)}) \in B[0, R_\varepsilon] \subset \mathbb{R}^k \times \mathbb{R}^{kd}$$

for all $\varepsilon > 0$ and $\lambda \geq 0$ such that

$$M_\lambda^\varepsilon = \sum_{i=1}^k u_i(\hat\tau_i^{(\varepsilon,\lambda)}, \hat x_i^{(\varepsilon,\lambda)}) - \phi_\lambda^\varepsilon(\hat\tau^{(\varepsilon,\lambda)}, \hat x^{(\varepsilon,\lambda)}),$$

due to the (semi-)continuity of $\phi_\lambda^\varepsilon \in C^2(\mathbb{R} \times \mathbb{R}^{kd})$ and $u_i \in \mathrm{USC}(\mathbb{R}^d)$ for $i \in \{1, \ldots, k\}$.

As a next step, we will apply the variable augmentation result from Corollary 2.17: Since $B[0, R_\varepsilon] \subset \mathbb{R}^k \times \mathbb{R}^{kd}$ is compact for all $\varepsilon > 0$, there exist sequences $(\lambda_n^{(\varepsilon)})_{n \in \mathbb{N}}$ with $\lim_{n \to \infty} \lambda_n^{(\varepsilon)} = \infty$ for every $\varepsilon > 0$ such that

$$(\hat\tau^{(\varepsilon)}, \hat x^{(\varepsilon)}) := \lim_{n \to \infty} \left(\hat\tau^{(\varepsilon,\lambda_n^{(\varepsilon)})}, \hat x^{(\varepsilon,\lambda_n^{(\varepsilon)})} \right)$$

exist in $\mathbb{R}^k \times \mathbb{R}^{kd}$. According to Corollary 2.17, we have

$$\lim_{n \to \infty} \lambda_n^{(\varepsilon)} \sum_{i < j} \left| \hat\tau_i^{(\varepsilon,\lambda_n^{(\varepsilon)})} - \hat\tau_j^{(\varepsilon,\lambda_n^{(\varepsilon)})} \right|^2 = 0,$$

which leads to $\hat\tau_1^{(\varepsilon)} = \ldots = \hat\tau_k^{(\varepsilon)}$. Furthermore, Corollary 2.17 also implies that

$$\lim_{\lambda_n^{(\varepsilon)} \to \infty} M_{\lambda_n^{(\varepsilon)}}^\varepsilon = \sum_{i=1}^k u_i(\hat\tau_1^{(\varepsilon)}, \hat x_i^{(\varepsilon)}) - \phi_0^\varepsilon(\hat\tau^{(\varepsilon)}, \hat x^{(\varepsilon)})$$

$$= \sup\left\{ \sum_{i=1}^k u_i(t, x_i) - \phi_0^\varepsilon((t, \ldots, t), x) \,\bigg|\, (t, x) \in \mathbb{R} \times \mathbb{R}^{kd} \right\}$$

for all $\varepsilon > 0$. In particular, since for all $\varepsilon > 0$

$$\mathbb{R} \times \mathbb{R}^{kd} \ni (t, x) = (t, x_1, \ldots, x_k) \longmapsto \sum_{i=1}^k u_i(t, x_i) - \phi_0^\varepsilon((t, \ldots, t), x)$$

have unique maxima in $(\bar t, \bar x) \in (0, T) \times \Omega^k \subset \mathbb{R} \times \mathbb{R}^{kd}$, we obtain

$$\lim_{n \to \infty} (\hat\tau^{(\varepsilon,\lambda_n^{(\varepsilon)})}, \hat x^{(\varepsilon,\lambda_n^{(\varepsilon)})}) = (\bar t, \ldots, \bar t, \bar x)$$

for all $\varepsilon > 0$.

2.2. Uniqueness of Solutions

We will now show how to apply the elliptic maximum principle from Theorem 2.14: Suppose that $(\varepsilon_n)_{n \in \mathbb{N}}$ is a decreasing sequence with $\lim_{n \to \infty} \varepsilon_n = 0$. According to the last paragraph, there exists a sequence $(\lambda_n)_{n \in \mathbb{N}}$ with $\lim_{n \to \infty} \lambda_n = \infty$ such that

$$(0, T) \times \Omega^k \ni (\hat{\tau}^{(n)}, \hat{x}^{(n)}) := (\hat{\tau}^{(\varepsilon_n, \lambda_n)}, \hat{x}^{(\varepsilon_n, \lambda_n)}) \longrightarrow (\bar{t}, \ldots, \bar{t}, \bar{x})$$

as $n \to \infty$ and $\lim_{n \to \infty} \lambda_n \sum_{i<j} |\hat{\tau}_i^{(n)} - \hat{\tau}_j^{(n)}|^2 = 0$, using a diagonal argument. Moreover,

$$\mathbb{R}^k \times \mathbb{R}^{kd} \ni (\tau, x) \longmapsto \sum_{i=1}^{k} u_i(\tau_i, x_i) - \phi_n(\tau, x)$$

with $\phi_n := \phi_{\lambda_n}^{\varepsilon_n} \in C^2(\mathbb{R}^k \times \mathbb{R}^{kd})$ has a global maximum in $(\hat{\tau}^{(n)}, \hat{x}^{(n)}) \in (0, T) \times \Omega^k$ for every $n \in \mathbb{N}$. The definition of $\phi_{\lambda_n}^{\varepsilon_n} \in C^2(\mathbb{R}^k \times \mathbb{R}^{kd})$ from Equation (2.18) together with $\lim_{n \to \infty} \lambda_n \sum_{i<j} |\hat{\tau}_i^{(n)} - \hat{\tau}_j^{(n)}|^2 = 0$ show that

$$|\phi_{\lambda_n}^{\varepsilon_n}(\hat{\tau}^{(n)}, x)| \leq |\phi(\tau_1, x)| + \lambda_n \sum_{i<j} |\hat{\tau}_i^{(n)} - \hat{\tau}_j^{(n)}|^2 + \varepsilon_n \sum_{i=1}^{k} \left(|x_i - \bar{x}_i|^{p+2} + |\hat{\tau}_i^{(n)} - \bar{t}|^2 \right)$$

$$\leq \|\phi\|_p (1 + |x|^p) + \lambda_n \sum_{i<j} |\hat{\tau}_i^{(n)} - \hat{\tau}_j^{(n)}|^2 + \varepsilon_n \sum_{i=1}^{k} |x_i - \bar{x}_i|^{p+2} + 2\varepsilon_n k T^2$$

$$\leq \hat{C}(1 + |x|^p) + \varepsilon_n \sum_{i=1}^{k} |x_i - \bar{x}_i|^{p+2}$$

for all $x \in \mathbb{R}^{kd}$ and $n \in \mathbb{N}$ with some constant $\hat{C} < \infty$ (independent of $n \in \mathbb{N}$). The taming procedure for scanning functions from Lemma 1.18 therefore implies the existence of $(\psi_n)_{n \in \mathbb{N}} \subset C_p^2(\mathbb{R}^k \times \mathbb{R}^{kd})$ with $\sup_{n \in \mathbb{N}} \|\psi_n\|_p \leq 8\hat{C}$ such that

$$\{x \in \mathbb{R}^{kd} : \psi_n(\hat{\tau}^{(n)}, x) = \phi_n(\hat{\tau}^{(n)}, x)\} \supset \{x \in \mathbb{R}^{kd} : |\phi_n(\hat{\tau}^{(n)}, x)| \leq 2\hat{C}(1 + |x|^p)\}$$

$$\supset \left\{ x \in \mathbb{R}^{kd} : \varepsilon_n \sum_{i=1}^{k} |x_i - \bar{x}_i|^{p+2} \leq \hat{C}(1 + |x|^p) \right\}$$

for all $n \in \mathbb{N}$, and such that the functions

$$\mathbb{R}^k \times \mathbb{R}^{kd} \ni (\tau, x) \longmapsto \sum_{i=1}^{k} u_i(\tau_i, x_i) - \psi_n(\tau, x)$$

have global maxima in $(\hat{\tau}^{(n)}, \hat{x}^{(n)}) \in (0, T) \times \Omega^k$ for all $n \in \mathbb{N}$. Note that

$$\left\{ x \in \mathbb{R}^{kd} : \varepsilon_n \sum_{i=1}^{k} |x_i - \bar{x}_i|^{p+2} \leq \hat{C}(1 + |x|^p) \right\} \supset \left\{ x \in \mathbb{R}^{kd} : \varepsilon_n \sum_{i=1}^{k} |x_i - \bar{x}_i|^{p+2} \leq \hat{C} \right\}$$

increases to the whole space $\mathbb{R}^{kd}$ as $n \to \infty$, since $\lim_{n \to \infty} \varepsilon_n = 0$ holds by assumption.

2. Integro-Differential Equations

Furthermore, a simple calculation shows that

$$D^2\psi_n(\tau, x) = D^2\phi(\tau, x) + D^2(\psi_n - \phi)(\tau, x)$$
$$\leq D^2\phi(\tau, x) + \lambda^+_{\max}\left[D^2(\psi_n - \phi)(\tau, x)\right]I$$
$$\leq \hat{g}_0(\tau, x)J_{k,d} + \text{diag}(\hat{g}_1^{(n)}(\tau, x), \ldots, \hat{g}_k^{(n)}(\tau, x))$$

for all $(\tau, x) \in \mathbb{R}^k \times \mathbb{R}^{kd}$, where $\hat{g}_0(\tau, x) := g_0(\tau_1, x)$,

$$\hat{g}_i^{(n)}(\tau, x) := g_i^{(n)}(\tau_1, x) + \lambda^+_{\max}\left[D^2(\psi_n - \phi)(\tau, x)\right]I$$

for $i \in \{1, \ldots, k\}$, and where $\lambda^+_{\max}\left[D^2(\psi_n - \phi)(\tau, x)\right]$ is the positive part of the largest eigenvalue of the symmetric matrix $D^2(\psi_n - \phi)(\tau, x) \in \mathbb{S}^{kd \times kd}$. Finally, Theorem 2.14 implies the existence of $X_1^{(n)}, \ldots, X_k^{(n)} \in \mathbb{S}^{d \times d}$ satisfying

$$-\gamma^{-1}g_0^{(n)}(\hat{\tau}^{(n)}, \hat{x}^{(n)}) \leq \text{diag}(X_1^{(n)} - g_1^{(n)}(\hat{\tau}^{(n)}, \hat{x}^{(n)}), \ldots, X_k^{(n)} - g_k^{(n)}(\hat{\tau}^{(n)}, \hat{x}^{(n)}))$$
$$\leq (1 - k\gamma)^{-1}g_0^{(n)}(\hat{\tau}^{(n)}, \hat{x}^{(n)}) \quad (2.19)$$

such that the following inequalities hold in the ordinary sense

$$\partial_{\tau_i}\psi_n(\hat{\tau}^{(n)}, \hat{x}^{(n)}) + G_i(\hat{\tau}_i^{(n)}, \hat{x}_i^{(n)}, u_i(\hat{\tau}_i^{(n)}, \hat{x}_i^{(n)}), D\psi_n(\hat{\tau}^{(n)}, \hat{x}^{(n)}),$$
$$X_i^{(n)}, \psi_n(\hat{\tau}^{(n)}, \hat{x}_1^{(n)}, \ldots, \hat{x}_{i-1}^{(n)}, \cdot, \hat{x}_{i+1}^{(n)}, \ldots, \hat{x}_k^{(n)})) \leq 0 \quad (2.20)$$

for all $i \in \{1, \ldots, k\}$. Note that we do not need an upper bound for the second derivate of $\psi_n \in C^2_p(\mathbb{R}^k \times \mathbb{R}^{kd})$ with respect to time in order to apply Theorem 2.14, since those derivatives do not occur in our equation.

At last, we will use the continuity assumption (A4) for G_i to obtain the desired result: The constructions of $\psi_n \in C^2_p(\mathbb{R}^k \times \mathbb{R}^{kd})$ and $\phi_n \in C^2(\mathbb{R}^k \times \mathbb{R}^{kd})$ shows that

$$\sum_{i=1}^k \partial_{\tau_i}\psi_n(\hat{\tau}^{(n)}, \hat{x}^{(n)}) = \sum_{i=1}^k \partial_{\tau_i}\phi_n(\hat{\tau}^{(n)}, \hat{x}^{(n)})$$
$$= \partial_t\phi(\hat{\tau}^{(n)}, \hat{x}^{(n)}) + 2\lambda \sum_{i=1}^k \sum_{i<j}\left(\hat{\tau}_i^{(n)} - \hat{\tau}_j^{(n)}\right) = \partial_t\phi(\hat{\tau}^{(n)}, \hat{x}^{(n)}) \quad (2.21)$$

for large $n \in \mathbb{N}$. A combination of Equation (2.20) and Equation (2.21) with the continuity assumption (A4) for G_i with $i \in \{1, \ldots, k\}$ (which is applicable as we will show later), therefore yields an upper and lower bound for the time-derivatives $\partial_{\tau_i}\psi_n(\hat{\tau}^{(n)}, \hat{x}^{(n)}) = \partial_{\tau_i}\phi_n(\hat{\tau}^{(n)}, \hat{x}^{(n)})$ independent of $n \in \mathbb{N}$. Together with Equation (2.19), this implies the existence of a subsequence $(\hat{\tau}^{(n_m)}, \hat{x}^{(n_m)})_{m \in \mathbb{N}}$ such that the limits

$$b_i := \lim_{m \to \infty} \partial_{\tau_i}\phi_n(\hat{\tau}^{(n_m)}, \hat{x}^{(n_m)}) \in \mathbb{R}$$
$$X_i := \lim_{m \to \infty} X_i^{(n_m)} \in \mathbb{S}^{d \times d}$$

exist for all $i \in \{1, \ldots, k\}$. It remains to prove the applicability of assumption (A4): Since

$\lim_{n\to\infty} \lambda_n \sum_{i<j} |\hat{\tau}_i^{(n)} - \hat{\tau}_j^{(n)}|^2 = 0$, since $\lim_{n\to\infty}(\hat{\tau}_1^{(n)}, \hat{x}^{(n)}) = (\bar{t}, \bar{x})$, and since $\psi_n \equiv \phi_n$ locally for large $n \in \mathbb{N}$, we obtain

$$\lim_{n\to\infty} D^\alpha \psi_n(\hat{\tau}^{(n)}, \cdot) = D^\alpha \phi(\bar{t}, \cdot)$$

locally uniformly for $\alpha \in \{0, 1, 2\}$. Moreover, $\sup_{n\in\mathbb{N}} \|\psi_n\|_p \leq 8\hat{C}$ holds by construction. Therefore, the continuity assumption (A4) is applicable and implies

$$b_i - G_i(\bar{t}, \bar{x}_i, u_i(\bar{t}, \bar{x}_i), D_{x_i}\phi(\bar{t}, \bar{x}), X_i^\gamma, \phi(\bar{t}, \bar{x}_1, \ldots, \bar{x}_{i-1}, \cdot, \bar{x}_{i+1}, \ldots, \bar{x}_k)) \leq 0$$

from Equation (2.20), as required. Finally, we find

$$\lim_{n\to\infty} g_i^{(n)}(\hat{\tau}^{(n)}, \hat{x}^{(n)}) = g_i(\bar{t}, \bar{x})$$

for all $i \in \{1, \ldots, k\}$, since $\lim_{n\to\infty} D^2\psi_n = D^2\phi$ locally uniformly and the functions $g_i \in C((0,T) \times \mathbb{R}^{kd}; \mathbb{S}^{d\times d})$ with $i \in \{1, \ldots, k\}$ are continuous. This implies the desired bounds for $X_1, \ldots, X_k$ from Equation (2.19). □

We will now present a lift of the preceding maximum principle from the original operator G to the generalized operators G^κ (that separate the singular parts of the nonlocal part of its rest), in analogy to the elliptic result in Lemma 2.15.

Corollary 2.19 (Parabolic Maximum Principle for Generalized Operators). *Suppose that $p \geq 0$ and $T > 0$, that $\Omega \subset \mathbb{R}^d$ is open, and that $u_i \in USC_p([0,T] \times \mathbb{R}^d)$ with $u(t, \cdot) \in C(\mathbb{R}^d)$ for all $t \in [0,T]$ are viscosity solutions in $(0,T) \times \Omega$ of*

$$\partial_t u_i(t,x) + G_i(t, x, u_i(t,x), Du_i(t,x), D^2 u_i(t,x), u_i(t,\cdot)) \leq 0$$

for $i \in \{1, \ldots, k\}$ and operators G_i, which satisfy all assumptions from Remark 2.3. Furthermore, suppose that $\phi \in C_p^2((0,T) \times \mathbb{R}^{kd})$ together with $g_0 \in C((0,T) \times \mathbb{R}^{kd}; \mathbb{R})$ and $g_i \in C((0,T) \times \mathbb{R}^{kd}; \mathbb{S}^{d\times d})$ such that

$$D^2\phi(t,x) \leq g_0(t,x) J_{k,d} + \mathrm{diag}(g_1(t,x), \ldots, g_k(t,x))$$

for all $(t,x) \in (0,T) \times \mathbb{R}^{kd}$, where $J_{k,d} \in \mathbb{R}^{kd\times kd}$ is defined as in Corollary 2.13. If

$$(t,x) = (t, x_1, \ldots, x_k) \longmapsto \sum_{i=1}^k u_i(t, x_i) - \phi(t,x)$$

has a global maximum in $(\bar{t}, \bar{x}) = (\bar{t}, \bar{x}_1, \ldots, \bar{x}_k) \in (0,T) \times \Omega^k$ with $g_0(\bar{t}, \bar{x}) > 0$, then there exist symmetric matrices $X_1^\gamma, \ldots, X_k^\gamma \in \mathbb{S}^{d\times d}$ and $b_1, \ldots, b_k \in \mathbb{R}^d$ for every $\gamma \in (0, k^{-1})$ satisfying $\sum_{i=1}^k b_i = \partial_t \phi(\bar{t}, \bar{x})$ and

$$-\gamma^{-1} g_0(\bar{t},\bar{x}) I \leq \mathrm{diag}(X_1^\gamma - g_1(\bar{t},\bar{x}), \ldots, X_k^\gamma - g_k(\bar{t},\bar{x})) \leq (1-k\gamma)^{-1} g_0(\bar{t},\bar{x}) J_{k,d}$$

such that the following inequalities hold in the ordinary sense

$$b_i - G_i^\kappa(\bar{t}, \bar{x}_i, u_i(\bar{t}, \bar{x}_i), D_{x_i}\phi(\bar{t}, \bar{x}), X_i, u_i(\bar{t}, \cdot), \phi(\bar{t}, \bar{x}_1, \ldots, \bar{x}_{i-1}, \cdot, \bar{x}_{i+1}, \ldots, \bar{x}_k)) \leq 0$$

for all $i \in \{1, \ldots, k\}$ and $0 < \kappa < 1$.

2. Integro-Differential Equations

Proof. The proof is a straightforward adaptation of the elliptic result in Lemma 2.15. The main difference is that we have to apply the parabolic maximum principle from Lemma 2.18 instead of the elliptic maximum principle from Theorem 2.14. Furthermore, (as mentioned after the proof of Lemma 2.15) the continuity of the subsolutions

$$u_i \in [0,T] \times \mathbb{R}^d \longrightarrow \mathbb{R}$$

is only needed to apply Dini's theorem from Theorem 1.27. Therefore, the continuity of

$$\mathbb{R}^d \ni x \longmapsto u_i(t,x)$$

for each fixed $t \in [0,T]$ suffices to prove the desired result. □

According to Lemma 2.8, the supremal convolution of a viscosity solution is again a viscosity solution of some smudged equation. As discussed after Lemma 2.8, the proof also works if the viscosity solution is only convolved with respect to some of its arguments. In case of parabolic equations, this leads to the following result:

Remark 2.20 (Stability under Parabolic Supremal Convolutions). Suppose that $p \geq 0$, that $\Omega \subset \mathbb{R}^d$ is open, that the monotonicity assumption (A5) holds for operator G, and that $\varphi = \varphi_{p \vee 2} \in C^\infty(\mathbb{R}^d)$ is the quasidistance with $(p \vee 2)$-polynomial growth from Lemma 1.14. If $u \in \mathrm{SC}_p([0,T] \times \mathbb{R}^d)$ is a viscosity subsolution in $(0,T) \times \Omega$ of

$$\partial_t u(t,x) + G(t,x,u(t,x),Du(t,x),D^2u(t,x),u(t,\cdot)) \leq 0$$

with $C > 0$ such that $|u(t,x)| \leq C(1+|x|^p)$ for all $(t,x) \in [0,T] \times \mathbb{R}^d$, and if

$$0 < \varepsilon \leq (2^{3(p \vee 2)-2}C)^{-1},$$

then the supremal convolution (cf. Lemma 1.12)

$$u^\varepsilon(t,x) := \Delta_\varphi^\varepsilon[u(t,\cdot)](x) = \sup_{y \in \mathbb{R}^d}\left(u(t,y) - \frac{\varphi(x-y)}{\varepsilon}\right) \leq 2^{p+p \vee 2-1}C(1+|x|^p)$$

is a viscosity subsolution in $(0,T) \times \Omega_\varepsilon := (0,T) \times \{x \in \mathbb{R}^d : \inf_{y \in \mathbb{R}^d \setminus \Omega} \varphi(y-x) > \varepsilon \cdot \delta(x)\}$ of

$$\partial_t u^\varepsilon(t,x) + \Delta_\varphi^\varepsilon[G](t,x,u^\varepsilon(t,x),Du^\varepsilon(t,x),D^2u^\varepsilon(t,x),u^\varepsilon(t,\cdot)) = 0.$$

The *smudged operator* $\Delta_\varphi^\varepsilon[G] : \mathbb{R}^d \times \mathbb{R} \times \mathbb{R}^d \times \mathbb{S}^{d \times d} \times C_p^2(\mathbb{R}^d) \to \mathbb{R}$ is defined by

$$\Delta_\varphi^\varepsilon[G](t,x,r,q,X,\phi) := \inf\left\{F(t,y,r,q,X,\phi \circ \tau_{x-y}) \;\middle|\; y \in \Omega \text{ with } \varphi(y-x) \leq \varepsilon \cdot \delta(x)\right\}$$

for $(t,x,r,q,X,\phi) \in (0,T) \times \mathbb{R}^d \times \mathbb{R} \times \mathbb{R}^d \times \mathbb{S}^{d \times d} \times C_p^2(\mathbb{R}^d)$, where $\delta(x) := 2^{3(p \vee 2)}C(1+\varphi(x))$ and $\tau_h : \mathbb{R}^d \to \mathbb{R}^d$ is the translation operator by $h \in \mathbb{R}^d$, i.e. $\tau_h(x) = x + h$ for all $x \in \mathbb{R}^d$.

As motivated at the beginning of this section, we usually need additional structure conditions to prove comparison principles. In case of our general parabolic equations, we will impose the following regularity condition, which is essentially an additional continuous dependence assumptions for the related operators.

2.2. Uniqueness of Solutions

Definition 2.21 (Regularity Condition). Suppose that $p \geq 0$ and $T > 0$, that $u_i^\varepsilon \in \mathrm{USC}_p([0,T] \times \mathbb{R}^d)$ is the supremal convolution in space (from Remark 2.20) of some $u_i \in \mathrm{USC}_p([0,T] \times \mathbb{R}^d)$ for $\varepsilon > 0$ and $i \in \{1, \ldots, k\}$ with $k \in \mathbb{N}$, and that

$$\phi_\alpha^\lambda(t, x) := \frac{\delta}{T - t} + \lambda \sum_{i<j} |x_i - x_j|^2 + \alpha e^{\mu t} \sum_{i=1}^{k} |x_i|^q$$

for $(t, x) = (t, x_1, \ldots, x_k) \in [0, T) \times \mathbb{R}^{kd}$ with $\delta, \mu, \alpha, \lambda \geq 0$. A family of operators

$$G_i : (0, T) \times \mathbb{R}^d \times \mathbb{R} \times \mathbb{R}^d \times \mathbb{S}^{d \times d} \times C_p^2(\mathbb{R}^d) \longrightarrow \mathbb{R}$$

for $i \in \{1, \ldots, k\}$ satisfies a *regularity condition* for $\beta_1, \ldots, \beta_k > 0$ and $q > p$ with $q \geq 2$, if there exists $C' > 0$ and $\omega : [0, \infty) \to [0, \infty)$ with $\lim_{t \to 0+} \omega(t) = \omega(0) = 0$ such that the following implication holds: If the function

$$(0, T) \times \mathbb{R}^{kd} \ni (t, x) = (t, x_1, \ldots, x_k) \longmapsto \sum_{i=1}^{k} \beta_i u_i^\varepsilon(t, x_i) - \phi_\alpha^\lambda(t, x),$$

has a global maximum in $(\bar{t}, \bar{x}) = (\bar{t}^{(\alpha, \lambda, \varepsilon)}, \bar{x}^{(\alpha, \lambda, \varepsilon)}) \in (0, T) \times \mathbb{R}^{kd}$ with

$$\sup_{(t,x) \in (0,T) \times \mathbb{R}^{kd}} \left(\sum_{i=1}^{k} \beta_i u_i^\varepsilon(t, x_i) - \phi_\alpha^\lambda(t, x) \right) = \sum_{i=1}^{k} \beta_i u_i^\varepsilon(\bar{t}, \bar{x}_i) - \phi_\alpha^\lambda(\bar{t}, \bar{x}) \geq 0,$$

and if $X_i = X_i^{(\alpha, \lambda, \varepsilon)} \in \mathbb{S}^{d \times d}$ for $i \in \{1, \ldots, k\}$ satisfy

$$\mathrm{diag}(X_1, \ldots, X_k) \leq 4\lambda J_{k,d} + q(q-1)\alpha e^{\mu t} \mathrm{diag}(|\bar{x}_1|^{q-2} I, \ldots, |\bar{x}_k|^{q-2} I),$$

then the following inequality holds in the ordinary sense

$$-\sum_{i=1}^{k} \beta_i \Delta_\varphi^\varepsilon [G_i^\kappa](\bar{t}, \bar{x}_i, u_i^\varepsilon(\bar{t}, \bar{x}_i), \beta_i^{-1} D_{x_i} \phi_\alpha^\lambda(\bar{t}, \bar{x}), X_i,$$

$$u_i^\varepsilon(\bar{t}, \cdot), \beta_i^{-1} \phi_\alpha^\lambda(\bar{t}, \bar{x}_1, \ldots, \bar{x}_{i-1}, \cdot, \bar{x}_{i+1}, \ldots, \bar{x}_k))$$

$$\leq \omega \left(\left(\sum_{i<j} |\bar{x}_i - \bar{x}_j| \right) \left(1 + \sum_{i=1}^{k} |\bar{x}_i|^p \right) \right) + C'\lambda \sum_{i<j} |\bar{x}_i - \bar{x}_j|^2$$

$$+ C'\alpha e^{\mu t} \left(1 + \sum_{i=1}^{k} |\bar{x}_i|^q \right) + \varrho_{\alpha, \lambda, \varepsilon, \kappa}$$

for a remainder $\varrho_{\alpha, \lambda, \varepsilon, \kappa}$ with $\limsup_{\alpha \to 0} \limsup_{\lambda \to \infty} \limsup_{\varepsilon \to 0} \limsup_{\kappa \to 0} \varrho_{\alpha, \lambda, \varepsilon, \kappa} \leq 0$.

2. Integro-Differential Equations

We will now show how to apply our parabolic maximum principle from Corollary 2.19 together with the variable augmentation from Corollary 2.17, in order to obtain a parabolic comparison principle for viscosity solutions. Compared to the existing literature (such as [1], [12], [75] or [108]), our result covers equations with weaker continuity assumptions and viscosity solutions with arbitrary polynomial growths. Furthermore, we will formulate our result for $k \in \mathbb{N}$ instead of only two solutions, and refer to it as "domination principle" instead of "comparison principle" accordingly. This idea can already be found in [35] for local equations, and allows us to obtain additional structure results later (cf. e.g. Corollary 2.26).

Theorem 2.22 (Domination Principle). *Suppose that $p \geq 0$, $T > 0$ and $k \in \mathbb{N}$. Moreover, suppose that $u_i \in USC_p([0,T] \times \mathbb{R}^d)$ are viscosity solutions in $(0,T) \times \mathbb{R}^d$ of*

$$\partial_t u_i(t,x) + G_i(t,x,u_i(t,x),Du_i(t,x),D^2u_i(t,x),u_i(t,\cdot)) \leq 0$$

for $i \in \{1,\ldots,k\}$ and operators G_i, which satisfy all assumptions from Remark 2.3 and a regularity condition for $\beta_1,\ldots,\beta_k > 0$ and $q > p$ with $q \geq 2$ from Definition 2.21. If the initial conditions $u_i(0,\cdot): \mathbb{R}^d \to \mathbb{R}$ are continuous for all $i \in \{1,\ldots,k\}$, and if

$$\sum_{i=1}^k \beta_i u_i(0,x) \leq 0$$

for all $x \in \mathbb{R}^d$, then $\sum_{i=1}^k \beta_i u_i(t,x) \leq 0$ for all $(t,x) \in (0,T) \times \mathbb{R}^d$.

Proof. The key idea of this proof is to show that $\sum_{i=1}^k \beta_i u_i(t_0,x_0) > 0$ for some point $(t_0,x_0) \in (0,T) \times \mathbb{R}^d$ implies the existence of

$$(\bar{t},\bar{x}) = (\bar{t}^{(\alpha,\lambda,\varepsilon)}, \bar{x}^{(\alpha,\lambda,\varepsilon)}) \in (0,T) \times \mathbb{R}^{kd} \tag{2.22}$$

for small $\delta, \varepsilon, \alpha > 0$ and large $\lambda \geq 0$ such that $\sup_{\lambda \geq 0} \sup_{\varepsilon > 0} \sum_{i=1}^k |\bar{x}_i|^p < \infty$,

$$\sup_{(t,x) \in (0,T) \times \mathbb{R}^{kd}} \left(\sum_{i=1}^k \beta_i u_i^\varepsilon(t,x_i) - \phi_\alpha^\lambda(t,x) \right) = \sum_{i=1}^k \beta_i u_i^\varepsilon(\bar{t},\bar{x}_i) - \phi_\alpha^\lambda(\bar{t},\bar{x}) \geq 0 \tag{2.23}$$

and $\liminf_{\lambda \to \infty} \liminf_{\varepsilon \to 0} \lambda \sum_{i<j} |\bar{x}_i - \bar{x}_j|^2 = 0$ (at least along suitable subsequences), where the penalization functions $\phi_\alpha^\lambda : [0,T) \times \mathbb{R}^{kd} \to \mathbb{R}$ are defined as in Definition 2.21. The parabolic maximum principle from Corollary 2.19 with $\gamma = (2k)^{-1}$ then implies the existence of $b_i = b_i^{(\alpha,\lambda,\varepsilon)} \in \mathbb{R}$ and $X_i = X_i^{(\alpha,\lambda,\varepsilon)} \in \mathbb{S}^{d \times d}$ for $i \in \{1,\ldots,k\}$ satisfying

$$\sum_{i=1}^k \beta_i b_i = \partial_t \phi_\alpha^\lambda(\bar{t},\bar{x})$$

$$\text{diag}\left(X_1,\ldots,X_k\right) \leq 4\lambda J_{k,d} + q(q-1)\alpha e^{\mu t} \text{diag}\left(|\bar{x}_1|^{q-2}I,\ldots,|\bar{x}_k|^{q-2}I\right)$$

such that the following inequalities hold in the ordinary sense

$$b_i + \Delta_\varphi^\varepsilon[G_i^\kappa](\bar{t},\bar{x}_i,u_i^\varepsilon(\bar{t},\bar{x}_i),p_i,X_i,u_i^\varepsilon(\bar{t},\cdot),\beta_i^{-1}\phi_\alpha^\lambda(\bar{t},\bar{x}_1,\ldots,\bar{x}_{i-1},\cdot,\bar{x}_{i+1},\ldots,\bar{x}_k)) \leq 0$$

for $i \in \{1,\ldots,k\}$ and $\kappa \in (0,1)$, where $p_i := p_i^{(\alpha,\lambda,\varepsilon)} := \beta_i^{-1} D_{x_i} \phi_\alpha^\lambda(\bar{t},\bar{x})$ for $i \in \{1,\ldots,k\}$.

2.2. Uniqueness of Solutions

Therefore, the given regularity condition for $\beta_1, \ldots, \beta_k > 0$ and $q > p$ with $q \geq 2$ yields

$$\frac{\delta}{(T-\bar{t})^2} + \mu \alpha e^{\mu t} \sum_{i=1}^{k} |\bar{x}_i|^q = \partial_t \phi_\alpha^\lambda(\bar{t}, \bar{x}) = \sum_{i=1}^{k} \beta_i b_i$$

$$\leq \omega \left(\left(\sum_{i<j} |\bar{x}_i - \bar{x}_j| \right) \left(1 + \sum_{i=1}^{k} |\bar{x}_i|^p \right) \right) + C'\lambda \sum_{i<j} |\bar{x}_i - \bar{x}_j|^2$$

$$+ C'\alpha e^{\mu t} \left(1 + \sum_{i=1}^{k} |\bar{x}_i|^q \right) + \varrho_{\alpha,\lambda,\varepsilon,\kappa}$$

for small $\delta, \varepsilon, \alpha > 0$, large $\lambda \geq 0$ and $\kappa \in (0,1)$, with some (independent) $C' > 0$,

$$\limsup_{\alpha \to 0} \limsup_{\lambda \to \infty} \limsup_{\varepsilon \to 0} \limsup_{\kappa \to 0} \varrho_{\alpha,\lambda,\varepsilon,\kappa} \leq 0$$

and $\omega : [0, \infty) \to [0, \infty)$ with $\lim_{t \to 0+} \omega(t) = \omega(0) = 0$. If $\mu \geq C'$ holds, this implies that

$$0 < \frac{\delta}{T^2} \leq \frac{\delta}{(T-\bar{t})^2}$$

$$\leq \omega \left(\left(\sum_{i<j} |\bar{x}_i - \bar{x}_j| \right) \left(1 + \sum_{i=1}^{k} |\bar{x}_i|^p \right) \right) + C'\lambda \sum_{i<j} |\bar{x}_i - \bar{x}_j|^2 + C'\alpha e^{\mu t} + \varrho_{\alpha,\lambda,\varepsilon,\kappa}$$

for small $\delta, \varepsilon, \alpha > 0$, large $\lambda \geq 0$ and $\kappa \in (0,1)$. Sending $\kappa \to 0$, $\varepsilon \to 0$, $\lambda \to \infty$, $\alpha \to 0$ (in this order) along suitable subsequences ultimately leads to a contradiction, since

$$\sup_{\lambda > 0} \sup_{\varepsilon > 0} \sum_{i=1}^{k} |\bar{x}_i|^p < \infty$$

$$\liminf_{\lambda \to \infty} \liminf_{\varepsilon \to 0} \lambda \sum_{i<j} |\bar{x}_i - \bar{x}_j|^2 = 0$$

(2.24)

by construction of $(\bar{t}, \bar{x}) = (\bar{t}^{(\alpha,\lambda,\varepsilon)}, \bar{x}^{(\alpha,\lambda,\varepsilon)})$ from Equation (2.22). Consequently, we have

$$\sum_{i=1}^{k} \beta_i u_i(t, x) \leq 0$$

for all $(t, x) \in (0, T) \times \mathbb{R}^d$. In order to finish the proof, it therefore remains to show that

$$\sup_{(t,x) \in (0,T) \times \mathbb{R}^d} \sum_{i=1}^{k} \beta_i u_i(t, x) > 0$$

implies the existence of $(\bar{t}, \bar{x}) = (\bar{t}^{(\alpha,\lambda,\varepsilon)}, \bar{x}^{(\alpha,\lambda,\varepsilon)}) \in (0, T) \times \mathbb{R}^{kd}$ for small $\delta, \varepsilon, \alpha > 0$ and large $\lambda \geq 0$ such that Equation (2.23) and Equation (2.24) hold (at least along suitable subsequences).

2. Integro-Differential Equations

The remaining proof is divided into three parts: Suppose that $\mu \geq C'$ with $C' > 0$ from our regularity condition, and that there exists some $(t_0, x_0) \in (0, T) \times \mathbb{R}^d$ such that

$$m_0 := \sum_{i=1}^{k} \beta_i u_i(t_0, x_0) > 0. \tag{2.25}$$

In the first part, we will prove that there exist $(\bar{t}, \bar{x}) = (\bar{t}^{(\alpha, \lambda, \varepsilon)}, \bar{x}^{(\alpha, \lambda, \varepsilon)}) \in [0, T) \times \mathbb{R}^{kd}$ for small $\delta, \varepsilon, \alpha > 0$ and large $\lambda \geq 0$ such that

$$\sup_{(t,x) \in (0,T) \times \mathbb{R}^{kd}} \left(\sum_{i=1}^{k} \beta_i u_i^\varepsilon(t, x_i) - \phi_\alpha^\lambda(t, x) \right) = \sum_{i=1}^{k} \beta_i u_i^\varepsilon(\bar{t}, \bar{x}_i) - \phi_\alpha^\lambda(\bar{t}, \bar{x})$$

and $\sup_{\lambda \geq 0} \sup_{\varepsilon > 0} \sum_{i=1}^{k} |\bar{x}_i|^p < \infty$. In the second part, we will demonstrate that

$$\liminf_{\lambda \to \infty} \liminf_{\varepsilon \to 0} \lambda \sum_{i<j} |\bar{x}_i - \bar{x}_j|^2 = 0$$

for small $\delta, \alpha > 0$. In the third part, we will show that $\bar{t} = \bar{t}^{(\alpha, \lambda, \varepsilon)} > 0$ and

$$\sum_{i=1}^{k} \beta_i u_i^\varepsilon(\bar{t}, \bar{x}_i) - \phi_\alpha^\lambda(\bar{t}, \bar{x}) > 0$$

for small $\delta, \varepsilon, \alpha > 0$ and large $\lambda \geq 0$.

First, we will construct the maximum points $(\bar{t}, \bar{x}) = (\bar{t}^{(\alpha, \lambda, \varepsilon)}, \bar{x}^{(\alpha, \lambda, \varepsilon)}) \in [0, T) \times \mathbb{R}^{kd}$: According to Remark 2.20 with $C > \max_{i \in \{1, \ldots, k\}} \|u_i\|_p$, there exists $\varepsilon_0 > 0$ such that

$$\max_{i \in \{1, \ldots, k\}} |u_i^\varepsilon(t, x)| \leq 2^{p+(p \vee 2)-1} C(1 + |x|^p)$$

for all $(t, x) \in [0, T) \times \mathbb{R}^d$ and $\varepsilon \in (0, \varepsilon_0]$. Hence,

$$\sum_{i=1}^{k} \beta_i u_i^\varepsilon(t, x_i) - \phi_\alpha^\lambda(t, x) \leq 2^{p+(p \vee 2)-1} C \sum_{i=1}^{k} \beta_i (1 + |x_i|^p) - \alpha e^{\mu t} \sum_{i=1}^{k} |x_i|^q,$$

for all $(t, x) \in [0, T) \times \mathbb{R}^d$ and $\varepsilon \in (0, \varepsilon_0]$. Since $q > p$ by assumption, the right-hand side tends to $-\infty$ as $|x| \to \infty$ for all $\alpha > 0$. In particular, there exists $R_\alpha > 0$ such that

$$\sum_{i=1}^{k} \beta_i u_i^\varepsilon(t, x_i) - \phi_\alpha^\lambda(t, x) \leq \sum_{i=1}^{k} \beta_i u_i(t_0, x_0) - \left(\frac{1}{T - t_0} + \alpha e^{\mu t_0} \sum_{i=1}^{k} |x_0|^q \right)$$

$$\leq \sum_{i=1}^{k} \beta_i u_i^\varepsilon(t_0, x_0) - \phi_\alpha^\lambda(t_0, (x_0, \ldots, x_0))$$

for all $(t, x) \in [0, T) \times \mathbb{R}^{kd}$ with $|x| \geq R_\alpha$, and all $\lambda \geq 0$, $\varepsilon \in (0, \varepsilon_0]$ and $\delta \in (0, 1]$.

Similarly, for every $\alpha > 0$ and $\delta \in (0,1]$ there exists $T_{\alpha,\delta} \in (0,T)$ such that

$$\sum_{i=1}^{k} \beta_i u_i^\varepsilon(t,x_i) - \phi_\alpha^\lambda(t,x) \le \sum_{i=1}^{k} \beta_i u_i^\varepsilon(t_0,x_0) - \phi_\alpha^\lambda(t_0,(x_0,\ldots,x_0))$$

for all $(t,x) \in [T_{\alpha,\delta},T) \times \mathbb{R}^{kd}$ with $|x| \le R_\alpha$, and all $\lambda \ge 0$ and $\varepsilon \in (0,\varepsilon_0]$. Therefore, the semicontinuity implies for every $\alpha > 0$, $\lambda \ge 0$, $\varepsilon \in (0,\varepsilon_0]$ and $\delta \in (0,1]$ that

$$\sup_{(t,x)\in[0,T)\times\mathbb{R}^{kd}} \left(\sum_{i=1}^{k} \beta_i u_i^\varepsilon(t,x_i) - \phi_\alpha^\lambda(t,x)\right)$$
$$= \sup_{(t,x)\in[0,T_{\alpha,\delta}]\times B[0,R_\alpha]} \left(\sum_{i=1}^{k} \beta_i u_i^\varepsilon(t,x_i) - \phi_\alpha^\lambda(t,x)\right) = \sum_{i=1}^{k} \beta_i u_i^\varepsilon(\bar{t},\bar{x}_i) - \phi_\alpha^\lambda(\bar{t},\bar{x}) \quad (2.26)$$

for some $(\bar{t},\bar{x}) = (\bar{t}^{(\alpha,\lambda,\varepsilon)},\bar{x}^{(\alpha,\lambda,\varepsilon)}) \in [0,T_{\alpha,\delta}] \times \mathbb{R}^{kd}$ with $|\bar{x}| \le R_\alpha$. In particular,

$$\sup_{\lambda \ge 0}\sup_{\varepsilon > 0} \sum_{i=1}^{k} |\bar{x}_i|^p \le R_\alpha < \infty$$

for all $\alpha > 0$ and $\delta \in (0,1]$, where $\bar{x} = \bar{x}^{(\alpha,\lambda,\varepsilon)} := 0 \in \mathbb{R}^{kd}$ for $\varepsilon > \varepsilon_0$.
As a next step, we will show that for every $\alpha > 0$ and $\delta \in (0,1]$

$$\liminf_{\lambda\to\infty}\liminf_{\varepsilon\to 0} \lambda \sum_{i<j} |\bar{x}_i - \bar{x}_j|^2 = 0.$$

Since $[0,T_{\alpha,\delta}] \times B[0,R_\alpha] \subset [0,T) \times \mathbb{R}^{kd}$ are compact for all $\alpha > 0$ and $\delta \in (0,1]$, there exist

$$(\varepsilon_n)_{n\in\mathbb{N}} = (\varepsilon_n^{(\alpha,\lambda)})_{n\in\mathbb{N}} \subset (0,\varepsilon_0]$$

for every $\alpha,\lambda \ge 0$ and $\delta \in (0,1]$ with $\lim_{n\to\infty} \varepsilon_n^{(\alpha,\lambda)} = 0$ such that

$$(\hat{t}^{(\alpha,\lambda)},\hat{x}^{(\alpha,\lambda)}) := \lim_{n\to\infty} (\bar{t}^{(\alpha,\lambda,\varepsilon_n)},\bar{x}^{(\alpha,\lambda,\varepsilon_n)}) \in [0,T_{\alpha,\delta}] \times B[0,R_\alpha] \subset [0,T) \times \mathbb{R}^{kd}.$$

The (semi-)continuity of $u_i^\varepsilon : [0,T] \times \mathbb{R}^d \to \mathbb{R}$ and $\phi_\alpha^\lambda : [0,T) \times \mathbb{R}^{kd} \to \mathbb{R}$ therefore implies

$$\sum_{i=1}^{k} \beta_i u_i^\varepsilon(\hat{t}^{(\alpha,\lambda)},\hat{x}_i^{(\alpha,\lambda)}) - \phi_\alpha^\lambda(\hat{t}^{(\alpha,\lambda)},\hat{x}^{(\alpha,\lambda)})$$
$$\ge \limsup_{n\to\infty} \left(\sum_{i=1}^{k} \beta_i\, u_i^\varepsilon(\bar{t}^{(\alpha,\lambda,\varepsilon_n)},\bar{x}_i^{(\alpha,\lambda,\varepsilon_n)}) - \phi_\alpha^\lambda(\bar{t}^{(\alpha,\lambda,\varepsilon_n)},\bar{x}^{(\alpha,\lambda,\varepsilon_n)})\right) \quad (2.27)$$
$$\ge \limsup_{n\to\infty} \left(\sum_{i=1}^{k} \beta_i u_i^{\varepsilon_n}(\bar{t}^{(\alpha,\lambda,\varepsilon_n)},\bar{x}_i^{(\alpha,\lambda,\varepsilon_n)}) - \phi_\alpha^\lambda(\bar{t}^{(\alpha,\lambda,\varepsilon_n)},\bar{x}^{(\alpha,\lambda,\varepsilon_n)})\right)$$

for all $\alpha > 0$, $\lambda \ge 0$, $\varepsilon \in (0,\varepsilon_0]$ and $\delta \in (0,1]$, using the monotonicity of $(0,\varepsilon_0] \ni \varepsilon \mapsto u_i^\varepsilon$.

2. Integro-Differential Equations

Furthermore, the construction of $(\bar{t}, \bar{x}) = (\bar{t}^{(\alpha,\lambda,\varepsilon)}, \bar{x}^{(\alpha,\lambda,\varepsilon)}) \in [0,T) \times \mathbb{R}^{kd}$ as a maximum point in Equation (2.26) shows that

$$\sum_{i=1}^{k} \beta_i u_i^{\varepsilon_n}(\bar{t}^{(\alpha,\lambda,\varepsilon_n)}, \bar{x}_i^{(\alpha,\lambda,\varepsilon_n)}) - \phi_\alpha^\lambda(\bar{t}^{(\alpha,\lambda,\varepsilon_n)}, \bar{x}^{(\alpha,\lambda,\varepsilon_n)})$$
$$= \sup_{(t,x) \in [0,T) \times \mathbb{R}^{kd}} \left(\sum_{i=1}^{k} \beta_i u_i^{\varepsilon_n}(t, x_i) - \phi_\alpha^\lambda(t, x) \right) \geq \sum_{i=1}^{k} \beta_i u_i^{\varepsilon_n}(t, x_i) - \phi_\alpha^\lambda(t, x) \quad (2.28)$$

for all $(t, x) \in [0, T) \times \mathbb{R}^{kd}$, $n \in \mathbb{N}$, $\alpha > 0$, $\lambda \geq 0$ and $\delta \in (0, 1]$. A combination of Equation (2.27), Equation (2.28) with $(t, x) = (\hat{t}^{(\alpha,\lambda)}, \hat{x}^{(\alpha,\lambda)}) \in [0, T) \times \mathbb{R}^{kd}$ and

$$\lim_{\varepsilon \to 0} u_i^\varepsilon(t, x) = u_i(t, x)$$

for all $(t, x) \in [0, T) \times \mathbb{R}^d$ and $i \in \{1, \ldots, k\}$ from Remark 2.20 leads to

$$\lim_{n \to \infty} \left(\sum_{i=1}^{k} \beta_i u_i^{\varepsilon_n}(\bar{t}^{(\alpha,\lambda,\varepsilon_n)}, \bar{x}_i^{(\alpha,\lambda,\varepsilon_n)}) - \phi_\alpha^\lambda(\bar{t}^{(\alpha,\lambda,\varepsilon_n)}, \bar{x}^{(\alpha,\lambda,\varepsilon_n)}) \right)$$
$$= \sum_{i=1}^{k} \beta_i u_i(\hat{t}^{(\alpha,\lambda)}, \hat{x}_i^{(\alpha,\lambda)}) - \phi_\alpha^\lambda(\hat{t}^{(\alpha,\lambda)}, \hat{x}^{(\alpha,\lambda)}) = \sup_{(t,x) \in [0,T) \times \mathbb{R}^{kd}} \left(\sum_{i=1}^{k} \beta_i u_i(t, x_i) - \phi_\alpha^\lambda(t, x) \right)$$

for all $\alpha > 0$, $\lambda \geq 0$ and $\delta \in (0, 1]$. Finally, Corollary 2.17 shows that

$$0 \leq \liminf_{\lambda \to \infty} \liminf_{\varepsilon \to 0} \lambda \sum_{i<j} |\bar{x}_i - \bar{x}_j|^2 \leq \liminf_{\lambda \to \infty} \lambda \lim_{n \to \infty} \sum_{i<j} |\bar{x}_i^{(\alpha,\lambda,\varepsilon_n)} - \bar{x}_j^{(\alpha,\lambda,\varepsilon_n)}|^2$$
$$= \lim_{\lambda \to \infty} \lambda \sum_{i<j} |\hat{x}_i^{(\alpha,\lambda)} - \hat{x}_j^{(\alpha,\lambda)}|^2 = 0$$

for all $\alpha > 0$ and $\delta \in (0, 1]$. Hence, we obtain

$$\liminf_{\lambda \to \infty} \liminf_{\varepsilon \to 0} \lambda \sum_{i<j} |\bar{x}_i - \bar{x}_j|^2 = 0 \quad (2.29)$$

for all $\alpha > 0$ and $\delta \in (0, 1]$, as required.

At last, we will show that $\bar{t} = \bar{t}^{(\alpha,\lambda,\varepsilon)} > 0$ for an appropriate choice of $\alpha > 0$, $\lambda \geq 0$, $\varepsilon \in (0, \varepsilon_0]$ and $\delta \in (0, 1]$: Since $(t_0, x_0) \in (0, T) \times \mathbb{R}^d$ was chosen in (2.25) such that

$$m_0 = \sum_{i=1}^{k} \beta_i u_i(t_0, x_0) > 0$$

holds, there exist $\alpha_0, \delta_0 > 0$ such that for all $\alpha \in (0, \alpha_0]$, $\lambda \geq 0$, $\varepsilon \in (0, \varepsilon_0]$ and $\delta \in (0, \delta_0]$

$$\sum_{i=1}^{k} \beta_i u_i(t_0, x_0) - \phi_\alpha^\lambda(t_0, (x_0, \ldots, x_0)) \geq m_0 - \left(\frac{\delta_0}{T - t_0} + \alpha_0 e^{\mu t} k |x_0|^q \right) \geq \frac{m_0}{2} > 0.$$

In particular, we obtain for all $\alpha \in (0, \alpha_0]$, $\lambda \geq 0$, $\varepsilon \in (0, \varepsilon_0]$ and $\delta \in (0, \delta_0]$ that

$$\sum_{i=1}^{k} \beta_i u_i^{\varepsilon}(\bar{t}, \bar{x}_i) - \phi_{\alpha}^{\lambda}(\bar{t}, \bar{x}) = \sup_{(t,x) \in [0,T] \times \mathbb{R}^{kd}} \left(\sum_{i=1}^{k} \beta_i u_i^{\varepsilon}(t, x_i) - \phi_{\alpha}^{\lambda}(t, x) \right) \qquad (2.30)$$

$$\geq \sum_{i=1}^{k} \beta_i u_i(t_0, x_0) - \phi_{\alpha}^{\lambda}(t, (x_0, \ldots, x_0)) \geq \frac{m_0}{2} > 0.$$

We will now prove the remaining result by contradiction: Fix $\delta \in (0, \delta_0]$ and $\alpha \in (0, \alpha_0]$, and suppose that there exist $(\varepsilon_n)_{n \in \mathbb{N}} \subset (0, \varepsilon_0]$ with $\lim_{n \to \infty} \varepsilon_n = 0$ and $(\lambda_n)_{n \in \mathbb{N}} \subset [0, \infty)$ with $\lim_{n \to \infty} \lambda_n = \infty$ such that

$$\bar{t} = \bar{t}^{(\alpha, \lambda_n, \varepsilon_n)} = 0$$

for infinitely many $n \in \mathbb{N}$. In fact, we can assume without loss of generality that $\bar{t} = \bar{t}^{(\alpha, \lambda_n, \varepsilon_n)} = 0$ holds for all $n \in \mathbb{N}$ (by proceeding with a subsequence if necessary). The monotone convergence of supremal convolutions from Lemma 1.12, Dini's theorem from Theorem 1.27, and the continuity of $x \mapsto u_i(0, x)$ for all $i \in \{1, \ldots, k\}$ show that there exists $N_\alpha \in \mathbb{N}$ such that

$$\sum_{i=1}^{k} \beta_i u_i^{\varepsilon_n}(\bar{t}, \bar{x}_i) - \sum_{i=1}^{k} \beta_i u_i(\bar{t}, \bar{x}_i) \leq \sum_{i=1}^{k} \beta_i \sup_{|x| \leq R_\alpha} (u_i^{\varepsilon_n}(0, x) - u_i(0, x)) \leq \frac{m_0}{4}.$$

for all $n \geq N_\alpha$. In particular, the positivity $\phi_\alpha^\lambda(\bar{t}, \bar{x}) \geq 0$ and Equation (2.30) yield

$$\sum_{i=1}^{k} \beta_i u_i(\bar{t}, \bar{x}_i) \geq \sum_{i=1}^{k} \beta_i u_i^{\varepsilon_n}(\bar{t}, \bar{x}_i) - \frac{m_0}{4} \geq \left(\sum_{i=1}^{k} \beta_i u_i^{\varepsilon_n}(\bar{t}, \bar{x}_i) - \phi_\alpha^\lambda(\bar{t}, \bar{x}) \right) - \frac{m_0}{4} \geq \frac{m_0}{4} > 0$$

for all $n \geq N_\alpha$. Moreover, Equation (2.29) and the compactness of

$$[0, T_{\alpha, \delta}] \times B[0, R_\alpha] \subset [0, T) \times \mathbb{R}^{kd}$$

imply the existence of subsequences $(\varepsilon_{n_k})_{k \in \mathbb{N}}$ and $(\lambda_{n_k})_{k \in \mathbb{N}}$ such that

$$(\hat{t}, \hat{x}) = (\hat{t}^{(\alpha)}, \hat{x}^{(\alpha)}) := \lim_{k \to \infty} (\bar{t}, \bar{x}) = \lim_{k \to \infty} (\bar{t}^{(\alpha, \lambda_{n_k}, \varepsilon_{n_k})}, \bar{x}^{(\alpha, \lambda_{n_k}, \varepsilon_{n_k})}) \in [0, T_{\alpha, \delta}] \times \mathbb{R}^{kd}$$

with $\hat{t} = 0$ and $\hat{x}_1 = \ldots = \hat{x}_k$. Finally, the upper semicontinuity of $u_i : [0, T] \times \mathbb{R}^d \to \mathbb{R}$ for $i \in \{1, \ldots, k\}$ leads to

$$\sum_{i=1}^{k} \beta_i u_i(\hat{t}, \hat{x}_1) = \sum_{i=1}^{k} \beta_i u_i(\hat{t}, \hat{x}_i) \geq \limsup_{n \to \infty} \sum_{i=1}^{k} \beta_i u_i(\bar{t}, \bar{x}_i) \geq \frac{m_0}{4} > 0,$$

which contradicts our assumption that $\sum_{i=1}^{k} \beta_i u_i(0, x) \leq 0$ for all $x \in \mathbb{R}^d$. Therefore, there exist sequences $(\varepsilon_n)_{n \in \mathbb{N}} \subset (0, \varepsilon_0]$ with $\lim_{n \to \infty} \varepsilon_n = 0$ and $(\lambda_n)_{n \in \mathbb{N}} \subset [0, \infty)$ with $\lim_{n \to \infty} \lambda_n = \infty$ for every $\alpha \in (0, \alpha_0]$ and $\delta \in (0, \delta_0]$ such that

$$\bar{t} = \bar{t}^{(\alpha, \lambda_n, \varepsilon_n)} > 0$$

for all $n \in \mathbb{N}$, as required. □

2. Integro-Differential Equations

Before we continue with some applications and corollaries of our domination principle, let us make a few comments on its proof:

First of all, the proof can be generalized to solutions in $(0,T) \times \Omega$ with open $\Omega \subset \mathbb{R}^d$ instead of the whole space $(0,T) \times \mathbb{R}^d$ under some additional boundary conditions: Similar to $\bar{t} = \bar{t}^{(\alpha,\lambda,\varepsilon)} > 0$ at the end of the proof, we can use

$$\sum_{i=1}^{k} u_i(t,x) \leq 0$$

for $(t,x) \in (0,T) \times \Omega^C = (0,T) \times (\mathbb{R}^d \setminus \Omega)$ to show that $\bar{x} = \bar{x}^{(\alpha,\lambda,\varepsilon)} \in \Omega$ holds for an appropriate choice of $\alpha, \varepsilon, \lambda, \delta > 0$. In order to prove that, we also need that $u_i : [0,T] \times \mathbb{R}^d \to \mathbb{R}$ are continuous on $(0,T) \times \Omega^C$ for $i \in \{1, \ldots, k\}$.

Moreover, we generalized the exponential-scaling-in-time idea (i.e. the $e^{\mu t}$ factor in the definition of $\phi_\alpha^\lambda : [0,T) \times \mathbb{R}^{kd} \to \mathbb{R}$) from [108, Section 4] for solutions with linear growth, in order to cover solutions with arbitrary polynomial growth at infinity. It is interesting to note that we can get rid of the related assumption that $q \geq 2$ holds for $p < 2$: A close inspection shows that it is only used to ensure that ϕ_α^λ is twice differentiable. An elaborate construction for ϕ_α^λ (which maintains the growth of the current definition at infinity and is twice differentiable even if $p < q < 2$) can therefore resolve this issue, while retaining the original line of argument.

Although we used our parabolic maximum principle for generalized operators from Corollary 2.19, which can only be applied to continuous solutions, we obtained a domination principle for semicontinuous solutions (which is important for some applications, such as Perron's method from Remark 2.24). We achieved this by approximating the semicontinuous solutions by supremal convolutions first (similar to [67, Theorem 51]). This additional step leads to our complicated regularity condition from Definition 2.21 and significantly increases the complexity of many related approximations.

Corollary 2.23 (Comparison Principle). *Suppose that $p \geq 0$ and $T > 0$, and that $u, v \in SC_p([0,T] \times \mathbb{R}^d)$ are viscosity solutions in $(0,T) \times \mathbb{R}^d$ of*

$$\partial_t u(t,x) + G(t,x,u(t,x),Du(t,x),D^2u(t,x),u(t,\cdot)) \leq 0$$
$$\partial_t v(t,x) + G(t,x,v(t,x),Dv(t,x),D^2v(t,x),v(t,\cdot)) \geq 0$$

for an operators G, which satisfies all assumptions from Remark 2.3. Moreover, suppose that the two operators G_1 and G_2, which are defined by

$$G_1(t,x,r,p,X,\phi) := G(t,x,\ r,\ p,\ X,\ \phi)$$
$$G_2(t,x,r,p,X,\phi) := -G(t,x,-r,-p,-X,-\phi)$$

for $(t,x,r,p,X,\phi) \in (0,T) \times \mathbb{R}^d \times \mathbb{R} \times \mathbb{R}^d \times \mathbb{S}^{d \times d} \times C_p^2(\mathbb{R}^d)$, satisfy a regularity condition for $\beta_1 = \beta_2 = 1$ and $q > p$ with $q \geq 2$ from Definition 2.21. If the initial conditions $u(0,\cdot) : \mathbb{R}^d \to \mathbb{R}$ and $v(0,\cdot) : \mathbb{R}^d \to \mathbb{R}$ are continuous, and if

$$u(0,x) \leq v(0,x)$$

for all $x \in \mathbb{R}^d$, then $u(t,x) \leq v(t,x)$ for all $(t,x) \in (0,T) \times \mathbb{R}^d$.

Proof. It is easy to see that $u_1 := u$ and $u_2 := -v$ are viscosity solutions in $(0,T) \times \mathbb{R}^d$ of
$$\partial_t u_i(t,x) + G_i(t,x,u_i(t,x),Du_i(t,x),D^2u_i(t,x),u_i(t,\cdot)) \leq 0$$
for $i \in \{1,2\}$. Since $\sum_{i=1}^2 \beta_i u_i(0,x) = u(0,x) - v(0,x) \leq 0$ for all $x \in \mathbb{R}^d$ by assumption, Theorem 2.22 implies the desired result. $\square$

An immediate consequence of the preceding result is that the initial value problems
$$\partial_t u(t,x) + G(t,x,u(t,x),Du(t,x),D^2u(t,x),u(t,\cdot)) = 0 \quad \text{in } (0,T) \times \mathbb{R}^d$$
$$u(t,x) = \varphi(x) \quad \text{on } \{0\} \times \mathbb{R}^d$$
with initial values $\varphi \in C_p(\mathbb{R}^d)$ have unique viscosity solutions $u \in C_p([0,T] \times \mathbb{R}^d)$, if the operator $G : (0,T) \times \mathbb{R}^d \times \mathbb{R} \times \mathbb{R}^d \times \mathbb{S}^{d \times d} \times C_p^2(\mathbb{R}^d) \to \mathbb{R}$ satisfies all assumptions from Remark 2.3 as well as the regularity condition from Corollary 2.23. We will often denote by $u^\varphi \in C_p(\mathbb{R}^d)$ the unique viscosity solution for initial value $\varphi \in C_p(\mathbb{R}^d)$.

Remark 2.24 (Perron's Method). Another important application of our parabolic comparison principle from Corollary 2.23 is Perron's method for proving the existence of viscosity solutions. The original idea is due to Oskar Perron for the Dirichlet problem of the Laplace equation in [107]: Roughly speaking, if
$$\underline{u} : (0,T) \times \mathbb{R}^d \longrightarrow \mathbb{R}$$
$$\overline{u} : (0,T) \times \mathbb{R}^d \longrightarrow \mathbb{R}$$
are continuous viscosity subsolutions and supersolutions, respectively, in $(0,T) \times \mathbb{R}^d$ of
$$\partial_t u(t,x) + G(t,x,u(t,x),Du(t,x),D^2u(t,x),u(t,\cdot)) = 0 \quad \text{(E2)}$$
with $\underline{u}(0,x) \leq \varphi(x) \leq \overline{u}(0,x)$ for $x \in \mathbb{R}^d$ and some initial value $\varphi \in C(\mathbb{R}^d)$, then
$$u(t,x) := \sup\{v(t,x) \mid v \text{ is viscosity subsolution of (E2) with } v \leq \overline{u} \text{ and } v(0,\cdot) \leq \varphi\}$$
for $(t,x) \in [0,T] \times \mathbb{R}^d$ defines a viscosity solution of (E2) in $(0,T) \times \mathbb{R}^d$ with $u(0,\cdot) = \varphi$. This reduces the proof for the existence of viscosity solutions to the construction of continuous viscosity sub- and supersolutions, which can usually be obtained from additional boundedness and growth assumptions for the operator G. Since we prove the existence of viscosity solutions for our equations in Chapter 4 using a probabilistic method, we do not follow this approach here. For a modern proof of Perron's method, which can be adapted to our generalized set-up, we refer the interested reader to [1].

Due to the abstract nature of Perron's method, it is often important for applications to obtain additional results on the structure of the resulting viscosity solutions. In the rest of this section, we will focus on the following two results, which study the subadditivity and convexity of viscosity solutions with respect to their initial values. These results are relevant in the field of sublinear Markov semigroups, since Mingshang Hu and Shige Peng used Perron's method in [67] to construct Lévy processes for sublinear expectations. Although we use a probabilistic approach in Chapter 4 instead, these result show that we could also use Perron's method to construct Lévy-type processes for sublinear expectations.

Corollary 2.25 (Subadditivity of Solutions). *Suppose that $p \geq 0$ and $T > 0$, and that $u^\varphi \in C_p([0,T] \times \mathbb{R}^d)$ are viscosity solutions in $(0,T) \times \mathbb{R}^d$ of*

$$\partial_t u^\varphi(t,x) + G(t,x,u^\varphi(t,x),Du^\varphi(t,x),D^2u^\varphi(t,x),u^\varphi(t,\cdot)) = 0$$

for $u^\varphi(0,\cdot) = \varphi \in C_p(\mathbb{R}^d)$ and an operators G, which satisfies all assumptions from Remark 2.3 and the regularity condition from Corollary 2.23. If G is superadditive, i.e.

$$G(t,x,r_1+r_2,p_1+p_2,X_1+X_2,\phi_1+\phi_2)$$
$$\geq G(t,x,r_1,p_1,X_1,\phi_1) + G(t,x,r_2,p_2,X_2,\phi_2)$$

for $(t,x,r_i,p_i,X_i,\phi_i) \in (0,T) \times \mathbb{R}^d \times \mathbb{R} \times \mathbb{R}^d \times \mathbb{S}^{d \times d} \times C_p^2(\mathbb{R}^d)$ and $i \in \{1,2\}$, then

$$u^{\varphi+\psi}(t,x) \leq u^\varphi(t,x) + u^\psi(t,x)$$

for all $(t,x) \in (0,T) \times \mathbb{R}^d$ and $\varphi, \psi \in C_p(\mathbb{R}^d)$.

Proof. Due to the superadditivity of G, the given regularity condition from Corollary 2.23 implies that the operators $(G_i)_{i \in \{1,2,3\}}$, which are defined by

$$G_1(t,x,r,p,X,\phi) := G(t,x,r,p,X,\phi)$$
$$G_2(t,x,r,p,X,\phi) := G_3(t,x,r,p,X,\phi) := -G(t,x,-r,-p,-X,-\phi)$$

for $(t,x,r,p,X,\phi) \in (0,T) \times \mathbb{R}^d \times \mathbb{R} \times \mathbb{R}^d \times \mathbb{S}^{d \times d} \times C_p^2(\mathbb{R}^d)$, satisfy a regularity condition for $\beta_1 = \beta_2 = \beta_3 = 1$ and $q > p$ with $q \geq 2$ from Definition 2.21. Hence, the result follows from Theorem 2.22 with $u_1 := u^{\varphi+\psi}$, $u_2 := -u^\varphi$ and $u_3 := -u^\psi$. $\square$

Corollary 2.26 (Convexity of Solutions). *Suppose that $p \geq 0$ and $T > 0$, and that $u^\varphi \in C_p([0,T] \times \mathbb{R}^d)$ are viscosity solutions in $(0,T) \times \mathbb{R}^d$ of*

$$\partial_t u^\varphi(t,x) + G(t,x,u^\varphi(t,x),Du^\varphi(t,x),D^2u^\varphi(t,x),u^\varphi(t,\cdot)) = 0$$

for $u^\varphi(0,\cdot) = \varphi \in C_p(\mathbb{R}^d)$ and an operators G, which satisfies all assumptions from Remark 2.3 and the regularity condition from Corollary 2.23. If G is concave, i.e.

$$G(t,x,\lambda r_1+(1-\lambda)r_2,\lambda p_1+(1-\lambda)p_2,\lambda X_1+(1-\lambda)X_2,\lambda\phi_1+(1-\lambda)\phi_2)$$
$$\geq \lambda G(t,x,r_1,p_1,X_1,\phi_1) + (1-\lambda)G(t,x,r_2,p_2,X_2,\phi_2)$$

for $(t,x,r_i,p_i,X_i,\phi_i) \in (0,T) \times \mathbb{R}^d \times \mathbb{R} \times \mathbb{R}^d \times \mathbb{S}^{d \times d} \times C_p^2(\mathbb{R}^d)$, $\lambda \in (0,1)$ and $i \in \{1,2\}$, then

$$u^{\lambda\varphi+(1-\lambda)\psi}(t,x) \leq \lambda u^\varphi(t,x) + (1-\lambda)u^\psi(t,x)$$

for all $(t,x) \in (0,T) \times \mathbb{R}^d$, $\lambda \in (0,1)$ and $\varphi, \psi \in C_p(\mathbb{R}^d)$.

Proof. Due to the convexity of G, the given regularity condition from Corollary 2.23 implies that the operators $(G_i)_{i \in \{1,2,3\}}$, which are defined by

$$G_1(t,x,r,p,X,\phi) := G(t,x,r,p,X,\phi)$$
$$G_2(t,x,r,p,X,\phi) := G_3(t,x,r,p,X,\phi) := -G(t,x,-r,-p,-X,-\phi)$$

for $(t,x,r,p,X,\phi) \in (0,T) \times \mathbb{R}^d \times \mathbb{R} \times \mathbb{R}^d \times \mathbb{S}^{d \times d} \times C_p^2(\mathbb{R}^d)$, satisfy a regularity condition for $\beta_1 = 1$, $\beta_2 = \lambda$, $\beta_3 = 1 - \lambda$ and $q > p$ with $q \geq 2$ from Definition 2.21. Hence, the result follows from Theorem 2.22 with $u_1 := u^{\lambda\varphi+(1-\lambda)\psi}$, $u_2 := -u^\varphi$ and $u_3 := -u^\psi$. $\square$

2.3. Hamilton–Jacobi–Bellman Equations

Hamilton-Jacobi-Bellman equations are second-order equations of the form

$$\partial_t u(t,x) + \inf_{\alpha \in \mathcal{A}} G_\alpha(t,x,u(t,x),Du(t,x),D^2u(t,x),u(t,\cdot)) = 0 \qquad (E3)$$

for a family of linear, nonlocal degenerate elliptic operators

$$G_\alpha : (0,T) \times \mathbb{R}^d \times \mathbb{R} \times \mathbb{R}^d \times \mathbb{S}^{d \times d} \times C^2(\mathbb{R}^d) \longrightarrow \mathbb{R}$$

with $\alpha \in \mathcal{A}$ – linear in the sense that

$$(r,p,X,\phi) \longmapsto G_\alpha(t,x,r,p,X,\phi)$$

are linear functionals for all $(t,x) \in (0,T) \times \mathbb{R}^d$. These equations play an important role in optimal control theory – a connection that was first obtained by Richard Bellman in [14] using a novel dynamic programming approach. Similar equations related to the calculus of variations were already studied earlier by Constantin Carathéodory in [24], generalizing the relation to Hamilton-Jacobi equations in mathematical physics from William R. Hamilton in [60] and Carl G. J. Jacobi in [72]. Bellman's dynamic programming approach was later extended by Rufus Isaacs in [68] to demonstrate a more general connection between (fully nonlinear) equations of the form (often referred to as Isaacs equations or Hamilton-Jacobi-Isaacs equations)

$$\partial_t u(t,x) + \sup_{\alpha \in \mathcal{A}} \inf_{\beta \in \mathcal{B}} G_{\alpha,\beta}(t,x,u(t,x),Du(t,x),D^2u(t,x),u(t,\cdot)) = 0$$

to differential game theory, where

$$G_{\alpha,\beta} : (0,T) \times \mathbb{R}^d \times \mathbb{R} \times \mathbb{R}^d \times \mathbb{S}^{d \times d} \times C(\mathbb{R}^d) \longrightarrow \mathbb{R}$$

with $(\alpha,\beta) \in \mathcal{A} \times \mathcal{B}$ is again a family of linear, nonlocal degenerate elliptic operators. Before Michael Crandall and Pierre-Louis Lions established viscosity solutions for Hamilton-Jacobi-Bellman equations related to optimal control theory in [36], several notions of generalized solutions were proposed and used in the literature. For a detailed account on the history of Hamilton-Jacobi-Bellman equations (including the related notions of generalized solutions) and a short introduction into their connection to optimal control theory, we refer the interested reader to [7, Chapter 1].

This section is structured as follows: At the beginning, we will introduce a general form of Hamilton-Jacobi-Bellman equations and all related assumptions, which are necessary to apply our general theory from Section 2.1 and Section 2.2. Moreover, we will list some typical examples and discuss their connection to Markov processes. Afterwards, we will show that these general Hamilton-Jacobi-Bellman equations satisfy all assumptions from Remark 2.3 and the regularity condition from Definition 2.21. At last, we will formulate the resulting corollaries from Section 2.2, such as the comparison principle and the convexity of viscosity solutions with respect to their initial values.

2. Integro-Differential Equations

Definition 2.27 (Hamilton–Jacobi–Bellman Equations). A nonlinear, parabolic, second-order nonlocal equation in $(0,T) \times \mathbb{R}^d$ with time horizon $T > 0$

$$\partial_t u(t,x) + G(t,x,u(t,x),Du(t,x),D^2u(t,x),u(t,\cdot)) = 0 \tag{E3}$$

is called *Hamilton-Jacobi-Bellman equation* (or HJB equation for short) if

$$G(t,x,r,p,X,\phi) = G^\kappa(t,x,r,p,X,\phi,\phi)$$
$$G^\kappa(t,x,r,p,X,u,\phi) = \inf_{\alpha \in \mathcal{A}} G^\kappa_\alpha(t,x,r,p,X,u,\phi)$$

for a family $(G^\kappa_\alpha(t,x,r,p,X,u,\phi))_{\alpha \in \mathcal{A}}$ of linear operators

$$G^\kappa_\alpha : (0,T) \times \mathbb{R}^d \times \mathbb{R} \times \mathbb{R}^d \times \mathbb{S}^{d \times d} \times \mathrm{SC}_p(\mathbb{R}^d) \times C^2(\mathbb{R}^d) \longrightarrow \mathbb{R}$$

with $p \geq 0$ and $0 < \kappa < 1$ of the form

$$G^\kappa_\alpha(t,x,r,p,X,u,\phi) = f_\alpha(t,x) - \mathcal{L}_\alpha(t,x,r,p,X) - \mathcal{I}^\kappa_\alpha(t,x,u,\phi)$$
$$\mathcal{L}_\alpha(t,x,r,p,X) = c_\alpha(t,x)r + b^T_\alpha(t,x)p + \mathrm{tr}\left(\sigma_\alpha(t,x)\sigma^T_\alpha(t,x)X\right)$$
$$\mathcal{I}^\kappa_\alpha(t,x,u,\phi) = \check{\mathcal{I}}^\kappa_\alpha(t,x,\phi) + \overline{\mathcal{I}}^\kappa_\alpha(t,x,u,\phi) + \hat{\mathcal{I}}^\kappa_\alpha(t,x,u),$$

where $f_\alpha(t,x), c_\alpha(t,x) \in \mathbb{R}$, $b_\alpha(t,x) \in \mathbb{R}^d$, $\sigma_\alpha(t,x) \in \mathbb{S}^{d \times d}$ and

$$\check{\mathcal{I}}^\kappa_\alpha(t,x,\phi) = \int_{|z| \leq \kappa} (\phi(x+j_\alpha(t,x,z)) - \phi(x) - D\phi(x)j_\alpha(t,x,z))\, m_\alpha(dz)$$

$$\overline{\mathcal{I}}^\kappa_\alpha(t,x,u,\phi) = \int_{\kappa < |z| \leq 1} (u(x+j_\alpha(t,x,z)) - u(x) - D\phi(x)j_\alpha(t,x,z))\, m_\alpha(dz)$$

$$\hat{\mathcal{I}}^\kappa_\alpha(t,x,u) = \int_{1 < |z|} (u(x+j_\alpha(t,x,z)) - u(x))\, m_\alpha(dz)$$

with Borel measures $m_\alpha : \mathcal{B}(\mathbb{R}^d) \to [0,\infty]$ and $j_\alpha(t,x,z) \in \mathbb{R}^d$.

Let us briefly discuss the explicit form of the linear nonlocal operators $(G_\alpha)_{\alpha \in \mathcal{A}}$: Suppose that $C^2_c(\mathbb{R}^d)$ is the family of twice-continuously differentiable functions with compact support and that $\mathfrak{B}(\mathbb{R}^d)$ is the family of Borel-measurable functions on $\mathbb{R}^d$. Philippe Courrège showed in [32] that every linear operator

$$A : C^2_c(\mathbb{R}^d) \longrightarrow \mathfrak{B}(\mathbb{R}^d),$$

which satisfies a nonlocal degenerate ellipticity condition (sometimes also referred to as *positive maximum principle* in the literature) such that

$$A\varphi(x) \geq A\psi(x)$$

whenever $\varphi - \psi$ has a global maximum in $x \in \mathbb{R}^d$ for $\varphi, \psi \in C^\infty_c(\mathbb{R}^d)$ with $(\varphi - \psi)(x) \geq 0$, is of the form of $(G_\alpha)_{\alpha \in \mathcal{A}}$ from Definition 2.27 with Borel-measurable coefficients. In particular, the form of $(G_\alpha)_{\alpha \in \mathcal{A}}$ from Definition 2.27 is archetypical, since our theory in Section 2.1 and Section 2.2 only works for nonlocal degenerate elliptic operators. For a modern proof of Courrège's statement in the context of classical Lévy-type processes, we refer the interested reader to [21, Theorem 2.21, p. 56].

2.3. Hamilton–Jacobi–Bellman Equations

Remark 2.28 (Assumptions on Coefficients). Suppose that $(G_\alpha)_{\alpha \in \mathcal{A}}$ is a family of Hamilton-Jacobi-Bellman operators as defined in Definition 2.27 with $p \geq 0$ and $T > 0$. Throughout this section we will assume that the following conditions hold:

(B1) (Boundedness) The coefficients of the local part are bounded

$$\sup_{\alpha \in \mathcal{A}} \left(|f_\alpha(t,x)| + |c_\alpha(t,x)| + |b_\alpha(t,x)| + |\sigma_\alpha(t,x)| \right) < \infty$$

in every $(t,x) \in (0,T) \times \mathbb{R}^d$, and the family of measures of the nonlocal part meet

$$\sup_{\alpha \in \mathcal{A}} \int \left(|z|^2 \mathbf{1}_{|z| \leq 1} + |z|^p \mathbf{1}_{|z| > 1} \right) m_\alpha(dz) < \infty,$$

a uniform integrability condition at zero and infinity.

(B2) (Tightness) The family of measures of the nonlocal part satisfies

$$\lim_{\kappa \to 0} \sup_{\alpha \in \mathcal{A}} \int_{|z| \leq \kappa} |z|^2 \, m_\alpha(dz) = 0 = \lim_{R \to \infty} \sup_{\alpha \in \mathcal{A}} \int_{R < |z|} |z|^p \, m_\alpha(dz),$$

a uniform tightness condition at the origin and infinity.

(B3) (Continuity) There exists some constant $C > 0$ and a function $\omega : \mathbb{R}_+ \to \mathbb{R}_+$ with $\omega(0) = \lim_{h \to 0} \omega(h) = 0$ such that

$$\sup_{\alpha \in \mathcal{A}} \left(|\sigma_\alpha(t,x) - \sigma_\alpha(s,y)| + |b_\alpha(t,x) - b_\alpha(s,y)| \right) \leq \omega(|t-s|) + C|x-y|$$

$$\sup_{\alpha \in \mathcal{A}} \left(|f_\alpha(t,x) - f_\alpha(s,y)| + |c_\alpha(t,x) - c_\alpha(s,y)| \right) \leq \omega(|t-s|) + \omega(|x-y|)$$

$$\sup_{\alpha \in \mathcal{A}} |j_\alpha(t,x,z) - j_\alpha(s,y,z)| \leq |z| \left(\omega(|t-s|) + C|x-y| \right)$$

for all $t, s \in (0,T)$ and $x, y, z \in \mathbb{R}^d$.

(B4) (Growth Condition) The jump-height coefficients of the nonlocal part have at most linear growth in space, i.e. there exists a constant $C > 0$ such that

$$\sup_{\alpha \in \mathcal{A}} |j_\alpha(t,x,z)| \leq C \left(1 + |x| \right) |z|$$

for all $t \in (0,T)$ and $x, z \in \mathbb{R}^d$.

(B5) (Monotonicity) The codomain coefficients are non-positive

$$c_\alpha(t,x) \leq 0$$

for all $t \in (0,T)$ and $x \in \mathbb{R}^d$.

2. Integro-Differential Equations

In order to apply our theory from Section 2.1 and Section 2.2, we have to impose strong continuity assumptions on the coefficients for the Hamilton-Jacobi-Bellman equations in Remark 2.28. Even though there exist comparison principles for viscosity solutions related to certain discontinuous operators such as [69], the example in [1, Remark 4] shows that it is not possible to prove a comparison principle for general Hamilton-Jacobi-Bellman equations with discontinuous coefficients.

Although we restrict our attention to parabolic Hamilton-Jacobi-Bellman equations, the same line of arguments also shows that elliptic Hamilton-Jacobi-Bellman equations (and similarly elliptic Isaacs equations, cf. Remark 2.29)

$$\inf_{\alpha \in \mathcal{A}} F_\alpha(x, u(x), Du(x), D^2u(x), u(\cdot)) = 0$$

are covered by our general theory from Section 2.1, where $(F_\alpha)_{\alpha \in \mathcal{A}}$ are linear operators of the same form as $(G_\alpha)_{\alpha \in \mathcal{A}}$ in Definition 2.27 without the dependence on time. In particular, the maximum principles in Theorem 2.14 and Lemma 2.15 can be applied to elliptic Hamilton-Jacobi-Bellman equations under the assumptions in Remark 2.28.

Remark 2.29 (Isaacs Equations). Almost all of the arguments in the rest of this section (except for the subadditivity and the convexity with respect to the initial values) also work for parabolic equations of the form (E2), where

$$G(t, x, r, p, X, \phi) = G^\kappa(t, x, r, p, X, \phi, \phi)$$
$$G^\kappa(t, x, r, p, X, u, \phi) = \sup_{\alpha \in \mathcal{A}} \inf_{\beta \in \mathcal{B}} G^\kappa_{\alpha,\beta}(t, x, r, p, X, u, \phi)$$

with some operators $(G^\kappa_{\alpha,\beta})_{(\alpha,\beta) \in \mathcal{A} \times \mathcal{B}}$ as in Definition 2.27 and the assumptions from Remark 2.28, where all related suprema have to be taken over all $(\alpha, \beta) \in \mathcal{A} \times \mathcal{B}$. This is essentially due to the fact that we only need

$$\sup_{\alpha \in \mathcal{A}} \inf_{\beta \in \mathcal{B}} G^\kappa_{\alpha,\beta}(t_1, x_1, r_1, p_1, X_1, u_1, \phi_1) - \sup_{\alpha \in \mathcal{A}} \inf_{\beta \in \mathcal{B}} G^\kappa_{\alpha,\beta}(t_2, x_2, r_2, p_2, X_2, u_2, \phi_2)$$
$$\leq \sup_{(\alpha,\beta) \in \mathcal{A} \times \mathcal{B}} \left(G^\kappa_{\alpha,\beta}(t_1, x_1, r_1, p_1, X_1, u_1, \phi_1) - G^\kappa_{\alpha,\beta}(t_2, x_2, r_2, p_2, X_2, u_2, \phi_2) \right)$$

for all $(t_i, x_i, r_i, p_i, X_i, u_i, \phi_i) \in (0, T) \times \mathbb{R}^d \times \mathbb{R} \times \mathbb{R}^d \times \mathbb{S}^{d \times d} \times \mathrm{SC}_p(\Omega) \times C^2(\Omega)$ and $i \in \{1, 2\}$ in the following arguments. Such fully nonlinear equations are often referred to as *Isaacs equations* or *Hamilton-Jacobi-Isaacs equations*, and play an important role in the theory of (stochastic) differential game theory, cf. e.g. [8, Part I.2] or [22].

Remark 2.30 (Connection to Markov Processes). As we will see in Section 4.1, there is a strong connection between Markov processes

$$X : \Omega \times \mathbb{R}_+ \longrightarrow \mathbb{R}$$

for sublinear expectations and viscosity solutions of Hamilton-Jacobi-Bellman equations:

2.3. Hamilton–Jacobi–Bellman Equations

Under some additional boundedness and regularity assumptions, the function
$$u(t, x) = E^x(\varphi(X_t))$$
is a viscosity solution in $(0, \infty) \times \mathbb{R}^d$ of the so-called sublinear generator equation
$$\partial_t u(t, x) - A(u(t, \cdot))(x) = 0$$
for regular $\varphi : \mathbb{R}^d \to \mathbb{R}$, where E^x is the sublinear expectation related to $X = (X_t)_{t \geq 0}$ starting in $x \in \mathbb{R}^d$ and A is the associated sublinear (Markov) generator. This connection generalizes the well-known theory for classical Markov processes (cf. e.g. [83]), where E^x are linear expectations (i.e. Lebesgue integrals with respect to probability measures) and the associated generators A are linear operators. The following examples give an informal overview over a large class of Markov processes including their related generator equations – for a rigorous treatment, see Chapter 4.

(i) (Heat Equation) The linear local equation
$$\partial_t u(t, x) - \tfrac{1}{2} \operatorname{tr}\left(\sigma \sigma^T D^2 u(t, x)\right) = 0$$
corresponds to a classical Brownian motion with volatility (or covariance) $\sigma\sigma^T$. For unit volatility $\sigma = I$, this correspondence was already used by Albert Einstein in [48] to relate the Brownian particle movement to the classical heat equation.

(iii) (Lévy Equation) The linear nonlocal equation
$$\partial_t u(t, x) - (\mathcal{L}u(t, x) + \mathcal{I}u(t, x)) = 0$$
with the spatially homogeneous operators
$$\mathcal{L}u(t, x) = b^T Du(t, x) + \tfrac{1}{2} \operatorname{tr}\left(\sigma \sigma^T D^2 u(t, x)\right)$$
$$\mathcal{I}u(t, x) = \int \left(u(t, x+z) - u(t, x) - Du(t, x)z\mathbb{1}_{|z| \leq 1}\right) m(dz)$$
corresponds to a classical Lévy process with (spatially homogeneous) Lévy-triplet $(b, \sigma\sigma^T, m)$, as shown e.g. in [117, Section 31, p. 205].

(iv) (Lévy-Type Equation) The linear nonlocal equation
$$\partial_t u(t, x) - (\mathcal{L}u(t, x) + \mathcal{I}u(t, x)) = 0$$
with the spatially inhomogeneous operators
$$\mathcal{L}u(t, x) = b(x)^T Du(t, x) + \tfrac{1}{2} \operatorname{tr}\left(\sigma(x)\sigma(x)^T D^2 u(t, x)\right)$$
$$\mathcal{I}u(t, x) = \int \left(u(t, x+z) - u(t, x) - Du(t, x)z\mathbb{1}_{|z| \leq 1}\right) m(x, dz)$$
corresponds to a classical Lévy-type process with (spatially inhomogeneous) Lévy-triplet $(b(x), \sigma(x)\sigma(x)^T, m(x))$. For an introduction and elaborate overview over the recent advances in the field of Lévy-type processes, we refer the interested reader to the monograph [21].

2. Integro-Differential Equations

(v) (Sublinear Heat Equation) The nonlinear local equation

$$\partial_t u(t,x) - \sup_{\alpha \in \mathcal{A}} \left(\tfrac{1}{2} \operatorname{tr} \left(\sigma_\alpha \sigma_\alpha^T D^2 u(t,x) \right) \right) = 0$$

corresponds to a Brownian motion with uncertainty in its volatility $(\sigma_\alpha \sigma_\alpha^T)_{\alpha \in \mathcal{A}}$. This process was introduced by Shige Peng in [102] (under the name G-Brownian motion) as the canonical continuous stochastic process for sublinear expectations with independent, stationary increments, and is now part of many applications in mathematical finance. For an introduction to the topic and a first overview over the related literature, see the draft monograph [105].

(vi) (Sublinear Lévy Equation) The nonlinear nonlocal equation

$$\partial_t u(t,x) - \sup_{\alpha \in \mathcal{A}} \left(\mathcal{L}_\alpha u(t,x) + \mathcal{I}_\alpha u(t,x) \right) = 0$$

with the spatially homogeneous operators

$$\mathcal{L}_\alpha u(t,x) = b_\alpha^T Du(t,x) + \tfrac{1}{2} \operatorname{tr} \left(\sigma_\alpha \sigma_\alpha^T D^2 u(t,x) \right)$$

$$\mathcal{I}_\alpha u(t,x) = \int \left(u(t,x+z) - \phi(t,x) - Du(t,x) z \mathbb{1}_{|z| \leq 1} \right) m_\alpha(dz)$$

correspond to a Lévy process with uncertainty in its (spatially homogeneous) Lévy-triplet $(b_\alpha, \sigma_\alpha \sigma_\alpha^T, m_\alpha)_{\alpha \in \mathcal{A}}$. Spatially homogeneous jump processes for sublinear expectations were introduced by Mingshang Hu and Shige Peng in [67] (under the name G-Lévy processes) and their construction was later generalized by Ariel Neufeld and Marcel Nutz in [93].

(vii) (Sublinear Lévy-Type Equation) The nonlinear nonlocal equation

$$\partial_t u(t,x) - \sup_{\alpha \in \mathcal{A}} \left(\mathcal{L}_\alpha u(t,x) + \mathcal{I}_\alpha u(t,x) \right) = 0$$

with the spatially inhomogeneous operators

$$\mathcal{L}_\alpha u(t,x) = b_\alpha(x)^T Du(t,x) + \tfrac{1}{2} \operatorname{tr} \left(\sigma_\alpha(x) \sigma_\alpha(x)^T D^2 u(t,x) \right)$$

$$\mathcal{I}_\alpha u(t,x) = \int \left(u(t,x+z) - \phi(t,x) - Du(t,x) z \mathbb{1}_{|z| \leq 1} \right) m_\alpha(x,dz)$$

correspond to a Lévy-type process with uncertainty in its (spatially inhomogeneous) Lévy-triplet $(b_\alpha(x), \sigma_\alpha(x)\sigma_\alpha(x)^T, m_\alpha(x))$. In Chapter 4, we will construct and characterize a large class of Lévy-type processes for sublinear expectations with bounded coefficients.

Remark 2.31 (Universal Constant $C \geq 0$). In order to simplify the presentation of the following very technical proofs for Lemma 2.32 and Proposition 2.33, we are going to employ the concept of universal constants: We will always use the same constant $C \geq 0$, which may change from line to line (in a trivial manner), but must not depend on any of the parameters involved in subsequent limiting procedures.

2.3. Hamilton–Jacobi–Bellman Equations

Lemma 2.32 (Admissibility of Equations). *Suppose that G is a nonlinear Hamilton-Jacobi-Bellman operator as defined in Definition 2.27. If all assumptions on the coefficients from Remark 2.28 hold, then the corresponding Hamilton-Jacobi-Bellman equation*

$$\partial_t u(t,x) + G(t,x,u(t,x),Du(t,x),D^2u(t,x),u(t,\cdot)) = 0$$

satisfies all assumptions from Remark 2.3.

Proof. Due to the definition of Hamilton-Jacobi-Bellman operators, it is easy to check that the consistency (A1) and translation invariance assumption (A3) are satisfied. Moreover, monotonicity assumption (B5) implies the monotonicity assumption (A5). It remains to prove that the degenerate ellipticity (A2) and continuity assumption (A4) are satisfied.

First of all, we will show the degenerate ellipticity assumption (A2): Due to the monotonicity of the infimum, it suffices to prove the degenerate ellipticity for each of the linear operators G_α with $\alpha \in \mathcal{A}$ separately. If $X, Y \in \mathbb{S}^{d \times d}$ with $X \leq Y$, then

$$\mathrm{tr}\left(\sigma_\alpha(t,x)\sigma_\alpha^T(t,x)X\right) \leq \mathrm{tr}\left(\sigma_\alpha(t,x)\sigma_\alpha^T(t,x)Y\right),$$

since the trace operator is monotone increasing (cf. [89, Section 4.1.2, p. 43]) together with $\sigma_\alpha(t,x)\sigma_\alpha^T(t,x) \geq 0$ for all $(t,x) \in (0,T) \times \mathbb{R}^d$ and $\alpha \in \mathcal{A}$. In particular,

$$\mathcal{L}_\alpha(t,x,r,p,X) \leq \mathcal{L}_\alpha(t,x,r,p,Y)$$

for all $(t,x,r,p) \in (0,T) \times \mathbb{R}^d \times \mathbb{R} \times \mathbb{R}^d$ and $\alpha \in \mathcal{A}$. Furthermore, if $u - v$ and $\phi - \psi$ have global maxima in $(t,x) \in (0,T) \times \mathbb{R}^d$, then rearranging the terms leads to

$$u(t, x + j_\alpha(t,x,z)) - u(t,x) \leq v(t, x + j_\alpha(t,x,z)) - v(t,x)$$
$$\phi(t, x + j_\alpha(t,x,z)) - \phi(t,x) \leq \psi(t, x + j_\alpha(t,x,z)) - \psi(t,x)$$

for all $z \in \mathbb{R}^d$ and $\alpha \in \mathcal{A}$. Hence, $\mathcal{I}_\alpha^\kappa(t,x,u,\phi) \leq \mathcal{I}_\alpha^\kappa(t,x,v,\psi)$ holds, since the maximum of $\phi - \psi$ in $(t,x) \in (0,T) \times \mathbb{R}^d$ implies $D\phi(t,x) = D\psi(t,x)$. Altogether, this leads to

$$G_\alpha^\kappa(t,x,r,p,X,u,\phi) \geq G_\alpha^\kappa(t,x,r,p,Y,u,\psi)$$

for all $(r,p) \in \mathbb{R} \times \mathbb{R}^d$ and $\alpha \in \mathcal{A}$, as required.

In the remaining proof, we will show the continuity assumption (A4): Suppose that

$$\lim_{n \to \infty}(t_n, x_n, r_n, q_n, X_n) = (t, x, r, q, X)$$

as well as $\lim_{n \to \infty} u_n = u$ and $\lim_{n \to \infty} D^k \phi_n = D^k \phi$ locally uniformly for $k \in \{0,1,2\}$ with $u \in C_p([0,T] \times \mathbb{R}^d)$ and $\sup_{n \in \mathbb{N}} \|u_n\|_p < \infty$. The triangle inequality implies

$$|G^\kappa(t_n, x_n, r_n, p_n, X_n, u_n, \phi_n) - G^\kappa(t,x,r,p,X,u,\phi)|$$
$$\leq \sup_{\alpha \in \mathcal{A}} |G_\alpha^\kappa(t_n, x_n, r_n, p_n, X_n, u_n, \phi_n) - G_\alpha^\kappa(t,x,r,p,X,u,\phi)|$$
$$\leq \sup_{\alpha \in \mathcal{A}} |f_\alpha(t_n,x_n) - f_\alpha(t,x)| + \sup_{\alpha \in \mathcal{A}} |\mathcal{L}_\alpha(t_n, x_n, r_n, p_n, X_n) - \mathcal{L}_\alpha(t,x,r,p,X)|$$
$$+ \sup_{\alpha \in \mathcal{A}} |\mathcal{I}_\alpha^\kappa(t_n, x_n, u_n, \phi_n) - \mathcal{I}_\alpha^\kappa(t,x,u,\phi)|.$$

Thus, it suffices to check (A4) for the inhomogeneous, local and nonlocal part separately:

2. Integro-Differential Equations

For the inhomogeneous part and local part $\mathcal{L}_\alpha$, we find

$$\limsup_{n\to\infty} \sup_{\alpha\in\mathcal{A}} |f_\alpha(t_n, x_n) - f_\alpha(t, x)| \leq \limsup_{n\to\infty} \omega(|t_n - t|) + \omega(|x_n - x|) = 0$$

using the continuity assumption (B3). Similarly, assumption (B3) implies

$$|c_\alpha(t_n, x_n)r_n - c_\alpha(t, x)r| \leq |c_\alpha(t_n, x_n) - c_\alpha(t, x)| \cdot |r_n| + |c_\alpha(t, x)| \cdot |r_n - r|$$
$$\leq (\omega(|t_n - t|) + \omega(|x_n - x|)) \cdot |r_n| + |r_n - r| \sup_{\alpha\in\mathcal{A}} |c_\alpha(t, x)|$$
$$|b_\alpha^T(t_n, x_n)p_n - b_\alpha^T(t, x)p| \leq |b_\alpha^T(t_n, x_n) - b_\alpha^T(t, x)| \cdot |p_n| + |b_\alpha^T(t, x)| \cdot |p_n - p|$$
$$\leq (\omega(|t_n - t|) + C|x_n - x|) \cdot |p_n| + |p_n - p| \sup_{\alpha\in\mathcal{A}} |b_\alpha(t, x)|$$

for the killing and the drift term of the local part $\mathcal{L}_\alpha$. For the diffusion term, we obtain

$$\left| \operatorname{tr}\left(\sigma_\alpha(t_n, x_n)\sigma_\alpha^T(t_n, x_n)X_n\right) - \operatorname{tr}\left(\sigma_\alpha(t, x)\sigma_\alpha^T(t, x)X\right) \right|$$
$$\leq \left| \operatorname{tr}\left((\sigma_\alpha(t_n, x_n) - \sigma_\alpha(t, x))(\sigma_\alpha(t_n, x_n) - \sigma_\alpha(t, x))^T X_n\right) \right|$$
$$+ \left| \operatorname{tr}\left(\sigma_\alpha(t, x)(\sigma_\alpha(t_n, x_n) - \sigma_\alpha(t, x))^T X_n\right) \right|$$
$$+ \left| \operatorname{tr}\left((\sigma_\alpha(t_n, x_n) - \sigma_\alpha(t, x))\sigma_\alpha^T(t, x)X_n\right) \right|$$
$$+ \left| \operatorname{tr}\left(\sigma_\alpha(t, x)\sigma_\alpha^T(t, x)(X_n - X)\right) \right|$$
$$\leq d |(\sigma_\alpha(t_n, x_n) - \sigma_\alpha(t, x)| \cdot |(\sigma_\alpha(t_n, x_n) - \sigma_\alpha(t, x)| \cdot |X_n|$$
$$+ d |\sigma_\alpha(t, x)| \cdot |(\sigma_\alpha(t_n, x_n) - \sigma_\alpha(t, x)| \cdot |X_n|$$
$$+ d |(\sigma_\alpha(t_n, x_n) - \sigma_\alpha(t, x)| \cdot |\sigma_\alpha(t, x)| \cdot |X_n|$$
$$+ d |\sigma_\alpha(t, x)| \cdot |\sigma_\alpha(t, x)| \cdot |X_n - X|$$
$$\leq d \left(\omega(t_n - t) + C|x_n - x|\right)^2 |X_n|$$
$$+ d \left(\omega(t_n - t) + C|x_n - x|\right) |X_n| \sup_{\alpha\in\mathcal{A}} |\sigma_\alpha(t, x)|$$
$$+ d \left(\omega(t_n - t) + C|x_n - x|\right) |X_n| \sup_{\alpha\in\mathcal{A}} |\sigma_\alpha(t, x)|$$
$$+ d |X_n - X| \sup_{\alpha\in\mathcal{A}} |\sigma_\alpha(t, x)|^2,$$

using the continuity assumption (B3) and several properties of the trace operator from [89, Section 4.1.2, p. 43], such as its linearity or the inequality $\operatorname{tr}(A) \leq d\|A\|$ for $A \in \mathbb{S}^{d\times d}$. A combination of the inequalities for the killing, drift and diffusion term leads to

$$\limsup_{n\to\infty} \sup_{\alpha\in\mathcal{A}} |\mathcal{L}_\alpha(t_n, x_n, r_n, p_n, X_n) - \mathcal{L}_\alpha(t, x, r, p, X)| \leq 0.$$

For the nonlocal part $\mathcal{I}_\alpha^\kappa$, the triangle inequality allows us to consider the small jump term $\check{\mathcal{I}}_\alpha^\kappa$, medium jump term $\overline{\mathcal{I}}_\alpha^\kappa$ and large jump term $\hat{\mathcal{I}}_\alpha^\kappa$ separately:

For the small jump term $\check{\mathcal{I}}_\alpha^\kappa$ of the nonlocal part $\mathcal{I}_\alpha^\kappa$, we find

$$\check{\mathcal{I}}_\alpha^\kappa(t,x,\phi) = \int_{|z|\leq\kappa}\int_0^1 (1-\xi) j_\alpha^T(t,x,z) D^2\phi\big(t, x + \xi j_\alpha(t,x,z)\big) j_\alpha(t,x,z)\, d\xi\, m_\alpha(dz)$$

by the multivariate Taylor theorem from [2, Theorem 7.5, p. 278]. Since by assumption $\lim_{n\to\infty} D^2\phi_n = D^2\phi$ locally uniformly and $\lim_{n\to\infty}(t_n,x_n) = (t,x)$, the right-hand side

$$\left| D^2\phi_n\big(t_n, x_n + \xi j_\alpha(t_n,x_n,z)\big) - D^2\phi\big(t_n, x_n + \xi j_\alpha(t_n,x_n,z)\big) \right|$$
$$\leq \sup_{|\zeta|\leq C(1+|x_n|)} \left| D^2\phi_n(t_n, x_n + \zeta) - D^2\phi(t_n, x_n + \zeta) \right|$$

vanishes independently of $\alpha \in \mathcal{A}$ and $|z| \leq \kappa \leq 1$, using the growth assumption (B4). Similarly, the continuity (B3) and growth assumption (B4) imply

$$\Big| j_\alpha^T(t_n,x_n,z) D^2\phi(t_n, x_n + \xi j_\alpha(t_n,x_n,z)) j_\alpha(t_n,x_n,z)$$
$$\qquad - j_\alpha^T(t,x,z) D^2\phi(t, x + \xi j_\alpha(t,x,z)) j_\alpha(t,x,z) \Big|$$
$$\leq \left| D^2\phi(t_n, x_n + \xi j_\alpha(t_n,x_n,z)) - D^2\phi(t, x + \xi j_\alpha(t,x,z)) \right| |j_\alpha(t_n,x_n,z)|^2$$
$$+ \left| D^2\phi(t, x + \xi j_\alpha(t,x,z)) \right| \left| j_\alpha^T(t_n,x_n,z) j_\alpha(t_n,x_n,z) - j_\alpha^T(t,x,z) j_\alpha(t,x,z) \right|$$
$$\leq \Bigg(\sup_{|\zeta| \vee |\zeta'| \leq C(1+|x_n|)} \left| D^2\phi(t_n, x_n + \zeta + \zeta'(\omega(|t_n - t|) + |x_n - x|)) - D^2\phi(t, x + \zeta) \right|$$
$$+ \sup_{|\zeta|\leq C(1+|x_n|)} \left| D^2\phi(t, x + \zeta) \right| (\omega(|t_n - t|) + |x_n - x|) \Bigg) C(1+|x|+|x_n|+|x_n|^2)|z|^2$$

for $|z| \leq \kappa \leq 1$. Since $D^2\phi : (0,T) \times \mathbb{R}^d \to \mathbb{S}^{d\times d}$ is continuous and $\lim_{n\to\infty}(t_n,x_n) = (t,x)$, the preceding right-hand side also vanishes independently of $\alpha \in \mathcal{A}$ and $|z| \leq \kappa \leq 1$. A combination of these two findings with the boundedness assumption (B1) leads to

$$|\check{\mathcal{I}}_\alpha^\kappa(t_n, x_n, \phi_n) - \check{\mathcal{I}}_\alpha^\kappa(t, x, \phi)|$$
$$\leq |\check{\mathcal{I}}_\alpha^\kappa(t_n,x_n,\phi_n) - \check{\mathcal{I}}_\alpha^\kappa(t_n,x_n,\phi)| + |\check{\mathcal{I}}_\alpha^\kappa(t_n,x_n,\phi) - \check{\mathcal{I}}_\alpha^\kappa(t,x,\phi)|$$
$$\leq C\left(1 + \sup_{n\in\mathbb{N}} |x_n|^2 \right) \sup_{\alpha\in\mathcal{A}} \int_{|z|\leq 1} |z|^2\, m_\alpha(dz)$$
$$\Bigg(\sup_{|\zeta|\leq C(1+\sup_n |x_n|)} |D^2\phi_n(t_n, x_n + \zeta) - D^2\phi(t_n, x_n + \zeta)|$$
$$+ \sup_{|\zeta|\vee|\zeta'|\leq C(1+\sup_n |x_n|)} |D^2\phi(t_n, x_n + \zeta + \zeta'(\omega(|t_n-t|) + |x_n - x|)) - D^2\phi(t, x + \zeta)|$$
$$+ \sup_{|\zeta|\leq C(1+\sup_n |x_n|)} |D^2\phi(t, x + \zeta)| (\omega(|t_n - t|) + |x_n - x|) \Bigg)$$

and hence $\limsup_{n\to\infty} \sup_{\alpha\in\mathcal{A}} |\check{\mathcal{I}}_\alpha^\kappa(t_n, x_n, \phi_n) - \check{\mathcal{I}}_\alpha^\kappa(t, x, \phi)| \leq 0$.

2. Integro-Differential Equations

For the large jump term $\hat{\mathcal{I}}_\alpha^\kappa$ of the nonlocal part $\mathcal{I}_\alpha^\kappa$, we find

$$\left|\left(u_n(t_n, x_n + j_\alpha(t_n, x_n, z)) - u_n(t_n, x_n)\right) - \left(u(t, x + j_\alpha(t, x, z)) - u(t, x)\right)\right|$$

$$\leq \left|u_n(t_n, x_n + j_\alpha(t_n, x_n, z)) - u(t, x + j_\alpha(t, x, z))\right|$$
$$+ \left|u_n(t_n, x_n) - u(t, x)\right|$$

$$\leq \left|u(t_n, x_n + j_\alpha(t_n, x_n, z)) - u(t, x + j_\alpha(t, x, z))\right|$$
$$+ \left|u(t_n, x_n) - u(t, x)\right|$$
$$+ 2 \sup_{|\zeta| \leq C(1+|x_n|)R} \left|u_n(t_n, x_n + \zeta) - u(t_n, x_n + \zeta)\right|$$

$$\leq 2\Bigg(\sup_{|\zeta| \vee |\zeta'| \leq C(1+|x_n|)R} \left|u(t_n, x_n + \zeta + \zeta'(\omega(|t_n - t|) + |x_n - x|)) - u(t, x + \zeta)\right|$$
$$+ \sup_{|\zeta| \leq C(1+|x_n|)R} \left|u_n(t_n, x_n + \zeta) - u(t_n, x_n + \zeta)\right|\Bigg)$$

for all $|z| \leq R$, using the continuity (B3) and growth assumption (B4). Hence, we obtain

$$\left|\hat{\mathcal{I}}_\alpha^\kappa(t_n, x_n, u_n) - \hat{\mathcal{I}}_\alpha^\kappa(t, x, u)\right|$$

$$\leq C\left(\sup_{n \in \mathbb{N}} \|u_n\|_p + \|u\|_p\right)\left(1 + |x_n|^p + |x|^p\right) \sup_{\alpha \in \mathcal{A}} \int_{R<|z|} (1 + |z|^p)\, m_\alpha(dz)$$

$$+ 2\Bigg(\sup_{|\zeta| \vee |\zeta'| \leq C(1+\sup_n |x_n|)R} \left|u(t_n, x_n + \zeta + \zeta'(\omega(|t_n - t|) + |x_n - x|)) - u(t, x + \zeta)\right|$$

$$+ \sup_{|\zeta| \leq C(1+\sup_n |x_n|)R} \left|u_n(t_n, x_n + \zeta) - u(t_n, x_n + \zeta)\right|\Bigg) \sup_{\alpha \in \mathcal{A}} \int_{1<|z|\leq R} 1\, m_\alpha(dz)$$

for all $R \geq 1$. Furthermore, the boundedness assumption (B1), $\lim_{n \to \infty} u_n = u$ locally uniformly and $\lim_{n \to \infty}(t_n, x_n) = (t, x)$ lead to

$$\limsup_{n \to \infty} \sup_{\alpha \in \mathcal{A}} \left|\hat{\mathcal{I}}_\alpha^\kappa(t_n, x_n, u_n) - \hat{\mathcal{I}}_\alpha^\kappa(t, x, u)\right|$$
$$\leq C\left(\sup_{n \in \mathbb{N}} \|u_n\|_p + \|u\|_p\right)\left(1 + 2|x|^p\right) \sup_{\alpha \in \mathcal{A}} \int_{R<|z|} (1 + |z|^p)\, m_\alpha(dz) \qquad (2.31)$$

for all $R \geq 1$. Finally, a simple calculation shows that

$$\int_{R<|z|} (1 + |z|^p)\, m_\alpha(dz) \leq 2 \int_{R<|z|} |z|^p\, m_\alpha(dz)$$

for all $R \geq 1$. Therefore, Equation (2.31) together with $\sup_{n \in \mathbb{N}} \|u_n\|_p < \infty$ and the tightness assumption (B2) imply $\limsup_{n \to \infty} \sup_{\alpha \in \mathcal{A}} |\hat{\mathcal{I}}_\alpha^\kappa(t_n, x_n, u_n) - \hat{\mathcal{I}}_\alpha^\kappa(t, x, u)| \leq 0$ as $R \to \infty$, as required.

For the medium jump term $\overline{\mathcal{I}}_\alpha^\kappa$ of the nonlocal part $\mathcal{I}_\alpha^\kappa$, we find

$$\Big|\Big(u_n(t_n, x_n + j_\alpha(t_n, x_n, z)) - u_n(t_n, x_n) - D\phi_n(t_n, x_n)j_\alpha(t_n, x_n, z)\Big)$$
$$- \Big(u(t, x + j_\alpha(t, x, z)) - u(t, x) - D\phi(t, x)j_\alpha(t, x, z)\Big)\Big|$$
$$\leq \Big|\Big(u_n(t_n, x_n + j_\alpha(t_n, x_n, z)) - u_n(t_n, x_n)\Big) - \Big(u(t, x + j_\alpha(t, x, z)) - u(t, x)\Big)\Big|$$
$$+ \Big|D\phi(t, x)\Big| \cdot \Big|j_\alpha(t_n, x_n, z) - j_\alpha(t, x, z)\Big|$$
$$+ \Big|D\phi_n(t_n, x_n) - D\phi(t, x)\Big| \cdot \Big|j_\alpha(t_n, x_n, z)\Big|$$
$$\leq \Big|\Big(u_n(t_n, x_n + j_\alpha(t_n, x_n, z)) - u_n(t_n, x_n)\Big) - \Big(u(t, x + j_\alpha(t, x, z)) - u(t, x)\Big)\Big|$$
$$+ \Big|D\phi_n(t, x)\Big|\Big(\omega(|t_n - t|) + C|x_n - x|\Big)|z|$$
$$+ \Big|D\phi_n(t_n, x_n) - D\phi(t, x)\Big|C\Big(1 + |x_n|\Big)|z|$$

for all $|z| \leq R$, using the continuity (B3) and growth assumption (B4). Therefore,

$$\Big|\overline{\mathcal{I}}_\alpha^\kappa(t_n, x_n, u_n, \phi_n) - \overline{\mathcal{I}}_\alpha^\kappa(t, x, u, \phi)\Big|$$
$$\leq +2\Bigg(\sup_{|\zeta|\vee|\zeta'|\leq C(1+\sup_n|x_n|)R} \Big|u(t, x_n + \zeta + \zeta'(\omega(|t_n - t|) + |x_n - x|)) - u(t, x + \zeta)\Big|$$
$$+ \sup_{|\zeta|\leq C(1+\sup_n|x_n|)R} \Big|u_n(t_n, x_n + \zeta) - u(t_n, x_n + \zeta)\Big|\Bigg)\sup_{\alpha\in\mathcal{A}}\int_{\kappa<|z|\leq 1} 1\, m_\alpha(dz)$$
$$+ C\Bigg(\Big|D\phi_n(t, x)\Big|\Big(\omega(|t_n - t|) + |x_n - x|\Big)$$
$$+ \Big|D\phi_n(t_n, x_n) - D\phi(t, x)\Big|\Big(1 + \sup_{n\in\mathbb{N}}|x_n|\Big)\Bigg)\sup_{\alpha\in\mathcal{A}}\int_{\kappa<|z|\leq 1} |z|\, m_\alpha(dz).$$

Finally, the boundedness assumption (B1) and

$$\int_{\kappa<|z|\leq 1} 1\, m_\alpha(dz) + \int_{\kappa<|z|\leq 1} |z|\, m_\alpha(dz) \leq \big(\kappa^{-2} + \kappa^{-1}\big)\int_{|z|\leq 1} |z|^2\, m_\alpha(dz)$$

together with $\lim_{n\to\infty} u_n = u$ and $\lim_{n\to\infty} D\phi_n = D\phi$ locally uniformly imply

$$\limsup_{n\to\infty}\sup_{\alpha\in\mathcal{A}} |\overline{\mathcal{I}}_\alpha^\kappa(t_n, x_n, u_n, \phi_n) - \overline{\mathcal{I}}_\alpha^\kappa(t, x, u, \phi)| \leq 0,$$

as required. $\square$

2. Integro-Differential Equations

Proposition 2.33 (Regularity Condition). *Suppose that G is a nonlinear Hamilton-Jacobi-Bellman operator (as defined in Definition 2.27) such that all assumptions on the coefficients from Remark 2.28 hold. If $p = 0$, then the two operators $(G_i)_{i \in \{1,2\}}$ with*

$$G_1(t, x, r, p, X, \phi) := G(t, x, r, p, X, \phi)$$
$$G_2(t, x, r, p, X, \phi) := -G(t, x, -r, -p, -X, -\phi)$$

for $(t, x, r, p, X, \phi) \in (0, T) \times \mathbb{R}^d \times \mathbb{R} \times \mathbb{R}^d \times \mathbb{S}^{d \times d} \times C_p^2(\mathbb{R}^d)$ satisfy a regularity condition (as introduced in Definition 2.21) for $\beta_1 = \beta_2 = 1$ and any $q \geq 2$. If, on the other hand, $p > 0$ and $q > p$ with $q \geq 2$ such that the uniform integrability condition

$$\sup_{\alpha \in \mathcal{A}} \int_{|z|>1} |z|^q m_\alpha(dz) < \infty$$

holds, then the operators $(G_i)_{i \in \{1,2\}}$ satisfy a regularity condition for $\beta_1 = \beta_2 = 1$ and that particular $q > p$ with $q \geq 2$.

Proof. Suppose that $u, -v \in \text{USC}_p([0, T] \times \mathbb{R}^d)$ and that for $\delta, \mu, \beta, \lambda \geq 0$

$$\phi_\beta^\lambda(t, x, y) := \frac{\delta}{T-t} + \lambda|x-y|^2 + \beta e^{\mu t}(|x|^q + |y|^q) \tag{2.32}$$

with $(t, x, y) \in (0, T) \times \mathbb{R}^d \times \mathbb{R}^d$. Moreover, suppose that

$$u^\varepsilon : [0, T] \times \mathbb{R}^d \longrightarrow \mathbb{R}$$

is the parabolic supremal convolution of $u : [0, T] \times \mathbb{R}^d \to \mathbb{R}$ (as in Remark 2.20), that

$$v_\varepsilon : [0, T] \times \mathbb{R}^d \longrightarrow \mathbb{R}$$

is the parabolic infimal convolution of $v : [0, T] \times \mathbb{R}^d \to \mathbb{R}$ (i.e. $v_\varepsilon := -(-v)^\varepsilon$), and that

$$\Delta_\varphi^\varepsilon[G^\kappa](t, x, r, q, X, w, \psi) := \inf_{\varphi(y-x) \leq \varepsilon \cdot \delta(x)} G^\kappa(t, y, r, q, X, w \circ \tau_{x-y}, \psi \circ \tau_{x-y}) \tag{2.33}$$

for $(t, x, r, q, X, w, \psi) \in (0, T) \times \mathbb{R}^d \times \mathbb{R} \times \mathbb{R}^d \times \mathbb{S}^{d \times d} \times \text{SC}_p(\mathbb{R}^d) \times C^2(\mathbb{R}^d)$ and $\varepsilon > 0$, where

$$\varphi = \varphi_{p \vee 2} \in C^\infty(\mathbb{R}^d)$$

is the quasidistance with $(p \vee 2)$-polynomial growth from Lemma 1.14, $\tau_h : \mathbb{R}^d \to \mathbb{R}^d$ is the translation operator by $h \in \mathbb{R}^d$ (i.e. $\tau_h(x) = x + h$ for all $x \in \mathbb{R}^d$) and

$$\delta(x) := 2^{3(p \vee 2)} C(1 + \varphi(x))$$

for $x \in \mathbb{R}^d$. According to the definition of the regularity condition in Definition 2.21, we have to show the existence of some $C' > 0$ and some $\omega' : [0, \infty) \to [0, \infty)$ with $\omega'(0) = \lim_{h \to 0+} \omega'(h) = 0$, such that the following implication holds:

2.3. Hamilton–Jacobi–Bellman Equations

If $(\bar{t}, \bar{x}, \bar{y}) \in (0, T) \times \mathbb{R}^d \times \mathbb{R}^d$ is a global maximum point with

$$\sup_{(t,x,y)\in(0,T)\times\mathbb{R}^d\times\mathbb{R}^d} \left(u^\varepsilon(t,x) - v_\varepsilon(t,y) - \phi^\lambda_\beta(t,x,y)\right) = u^\varepsilon(\bar{t},\bar{x}) - v_\varepsilon(\bar{t},\bar{y}) - \phi^\lambda_\beta(\bar{t},\bar{x},\bar{y}) \geq 0$$

and if $X, Y \in \mathbb{S}^{d \times d}$ are symmetric matrices with

$$\begin{pmatrix} X & 0 \\ 0 & -Y \end{pmatrix} \leq 4\lambda \begin{pmatrix} I & -I \\ -I & I \end{pmatrix} + q(q-1)\beta e^{\mu t} \begin{pmatrix} |\bar{x}|^{q-2}I & 0 \\ 0 & |\bar{y}|^{q-2}I \end{pmatrix} \quad (2.34)$$

for some $\delta, \mu, \beta, \lambda, \varepsilon > 0$, then

$$\Delta^\varepsilon_\varphi[G^\kappa]\left(\bar{t}, \bar{y}, v_\varepsilon(\bar{t},\bar{y}), -D_y \phi^\lambda_\beta(\bar{t},\bar{x},\bar{y}), Y, v_\varepsilon(\bar{t},\cdot), -\phi^\lambda_\beta(\bar{t},\bar{x},\cdot)\right)$$
$$- \Delta^\varepsilon_\varphi[G^\kappa]\left(\bar{t}, \bar{x}, u^\varepsilon(\bar{t},\bar{x}), D_x \phi^\lambda_\beta(\bar{t},\bar{x},\bar{y}), X, u^\varepsilon(\bar{t},\cdot), \phi^\lambda_\beta(\bar{t},\cdot,\bar{y})\right)$$
$$\leq \omega'\left(|\bar{x}-\bar{y}|(1+|\bar{x}|^p+|\bar{y}|^p)\right) + C'\lambda|\bar{x}-\bar{y}|^2 + C'\beta e^{\mu t}\left(1+|\bar{x}|^q+|\bar{y}|^q\right) + \varrho_{\beta,\lambda,\varepsilon,\kappa}$$

for a remainder $\varrho_{\beta,\lambda,\varepsilon,\kappa}$ with $\limsup_{\beta\to 0} \limsup_{\lambda\to\infty} \limsup_{\varepsilon\to 0} \limsup_{\kappa\to 0} \varrho_{\beta,\lambda,\varepsilon,\kappa} \leq 0$.

We will now prove this implication: First of all, the construction of $G = \inf_{\alpha \in \mathcal{A}} G_\alpha$ in Definition 2.27 and Equation (2.33) imply

$$\Delta^\varepsilon_\varphi[G^\kappa]\left(\bar{t}, \bar{y}, v_\varepsilon(\bar{t},\bar{y}), -D_y \phi^\lambda_\beta(\bar{t},\bar{x},\bar{y}), Y, v_\varepsilon(\bar{t},\cdot), -\phi^\lambda_\beta(\bar{t},\bar{x},\cdot)\right)$$
$$- \Delta^\varepsilon_\varphi[G^\kappa]\left(\bar{t}, \bar{x}, u^\varepsilon(\bar{t},\bar{x}), D_x \phi^\lambda_\beta(\bar{t},\bar{x},\bar{y}), X, u^\varepsilon(\bar{t},\cdot), \phi^\lambda_\beta(\bar{t},\cdot,\bar{y})\right)$$
$$\leq \sup_{(\alpha,x,y)\in\mathcal{A}_\varepsilon} \Big\{ \left(f_\alpha(\bar{t},y) - f_\alpha(\bar{t},x)\right)$$
$$+ \Big(\mathcal{L}_\alpha(\bar{t}, x, u^\varepsilon(\bar{t},\bar{x}), D_x \phi^\lambda_\beta(\bar{t},\bar{x},\bar{y}), X)$$
$$- \mathcal{L}_\alpha(\bar{t}, y, v_\varepsilon(\bar{t},\bar{y}), -D_y \phi^\lambda_\beta(\bar{t},\bar{x},\bar{y}), Y) \Big)$$
$$+ \Big(\mathcal{I}^\kappa_\alpha(\bar{t}, x, u^\varepsilon(\bar{t},\cdot) \circ \tau_{\bar{x}-x}, \phi^\lambda_\beta(\bar{t},\cdot,\bar{y}) \circ \tau_{\bar{x}-x})$$
$$- \mathcal{I}^\kappa_\alpha(\bar{t}, y, v_\varepsilon(\bar{t},\cdot) \circ \tau_{\bar{y}-y}, \phi^\lambda_\beta(\bar{t},\bar{x},\cdot) \circ \tau_{\bar{y}-y}) \Big) \Big\}$$

with $\mathcal{A}_\varepsilon := \{(\alpha,x,y) \in \mathcal{A} \times \mathbb{R}^d \times \mathbb{R}^d : \varphi(x-\bar{x}) \leq \varepsilon \cdot \delta(\bar{x}), \varphi(y-\bar{y}) \leq \varepsilon \cdot \delta(\bar{y})\}$ for $\varepsilon > 0$. Therefore, we can treat the inhomogeneous, local and nonlocal part separately. Since

$$\sup_{(\alpha,x,y)\in\mathcal{A}_\varepsilon} \left(f_\alpha(\bar{t},y) - f_\alpha(\bar{t},x)\right) \leq \sup_{(\alpha,x,y)\in\mathcal{A}_\varepsilon} \omega(|x-y|)$$
$$\leq \sup_{(\alpha,x,y)\in\mathcal{A}_\varepsilon} \omega(|\bar{x}-\bar{y}| + |x-\bar{x}| + |y-\bar{y}|)$$
$$\leq \sup_{(\alpha,x,y)\in\mathcal{A}_\varepsilon} \omega\left(|\bar{x}-\bar{y}| + 2^{p\vee 2-2}\left(\varphi(x-\bar{x})^{1/2} + \varphi(y-\bar{y})^{1/2}\right)\right)$$
$$\leq \omega\left(|\bar{x}-\bar{y}| + 2^{p\vee 2-2}\varepsilon^{1/2}\left(\delta(\bar{x})^{1/2} + \delta(\bar{y})^{1/2}\right)\right)$$

follows from the continuity assumption (B3) and the bounds for $\varphi \in C^\infty(\mathbb{R}^d)$ from Lemma 1.14, it remains to prove the implication for $\mathcal{L}_\alpha$ and $\mathcal{I}^\kappa_\alpha$:

2. Integro-Differential Equations

For the local part $\mathcal{L}_\alpha$, we find for $(\alpha, x, y) \in \mathcal{A}_\varepsilon$ that

$$\begin{aligned}
&\mathcal{L}_\alpha(\bar{t}, x, u^\varepsilon(\bar{t}, \bar{x}), D_x \phi_\beta^\lambda(\bar{t}, \bar{x}, \bar{y}), X) - \mathcal{L}_\alpha(\bar{t}, y, v_\varepsilon(\bar{t}, \bar{y}), -D_y \phi_\beta^\lambda(\bar{t}, \bar{x}, \bar{y}), Y) \\
&= \Big(c_\alpha(\bar{t}, x) u^\varepsilon(\bar{t}, \bar{x}) - c_\alpha(\bar{t}, y) v_\varepsilon(\bar{t}, \bar{y}) \Big) \\
&\quad + \Big(b_\alpha^T(\bar{t}, x) D_x \phi_\beta^\lambda(\bar{t}, \bar{x}, \bar{y}) - b_\alpha^T(\bar{t}, y) \big(- D_y \phi_\beta^\lambda(\bar{t}, \bar{x}, \bar{y}) \big) \Big) \\
&\quad + \Big(\operatorname{tr}\big(\sigma_\alpha(\bar{t}, x) \sigma_\alpha^T(\bar{t}, x) X \big) - \operatorname{tr}\big(\sigma_\alpha(\bar{t}, y) \sigma_\alpha^T(\bar{t}, y) Y \big) \Big),
\end{aligned} \tag{2.35}$$

which allows us to treat the killing, drift and diffusion term separately: By assumption

$$M = M^{(\beta, \lambda, \varepsilon)} := \sup \{ u^\varepsilon(t, x) - v_\varepsilon(t, y) - \phi_\beta^\lambda(t, x, y) \mid (t, x, y) \in (0, T) \times \mathbb{R}^d \times \mathbb{R}^d \}$$
$$= u^\varepsilon(\bar{t}, \bar{x}) - v_\varepsilon(\bar{t}, \bar{y}) - \phi_\beta^\lambda(\bar{t}, \bar{x}, \bar{y}) \geq 0,$$

which shows $\phi_\beta^\lambda(\bar{t}, \bar{x}, \bar{y}) + M \geq 0$. Therefore, the monotonicity assumption (B5), the continuity assumption (B3), and $\limsup_{\varepsilon \to \infty} \|v_\varepsilon\|_p < \infty$ from Lemma 2.8 imply

$$\begin{aligned}
&c_\alpha(\bar{t}, x) u^\varepsilon(\bar{t}, \bar{x}) - c_\alpha(\bar{t}, y) v_\varepsilon(\bar{t}, \bar{y}) \\
&= c_\alpha(\bar{t}, x) \Big(v_\varepsilon(\bar{t}, \bar{y}) + \phi_\beta^\lambda(\bar{t}, \bar{x}, \bar{y}) + M \Big) - c_\alpha(\bar{t}, y) v_\varepsilon(\bar{t}, \bar{y}) \\
&= \big(c_\alpha(\bar{t}, x) - c_\alpha(\bar{t}, y) \big) v_\varepsilon(\bar{t}, \bar{y}) + c_\alpha(\bar{t}, x) \Big(\phi_\beta^\lambda(\bar{t}, \bar{x}, \bar{y}) + M \Big) \leq \omega(x - y) C \big(1 + |\bar{y}|^p \big)
\end{aligned}$$

for $(\alpha, x, y) \in \mathcal{A}_c$. Since $\lim_{\varepsilon \to 0}(x, y) = (\bar{x}, \bar{y})$ from $\varphi(x - \bar{x}) \vee \varphi(y - \bar{y}) \leq \varepsilon(\delta(\bar{x}) \vee \delta(\bar{y}))$ for $(\alpha, x, y) \in \mathcal{A}_\varepsilon$, this leads to

$$\begin{aligned}
&\limsup_{\varepsilon \to 0} \limsup_{\kappa \to 0} \sup_{(\alpha, x, y) \in \mathcal{A}_\varepsilon} \Big(c_\alpha(\bar{t}, x) u^\varepsilon(\bar{t}, \bar{x}) - c_\alpha(\bar{t}, y) v_\varepsilon(\bar{t}, \bar{y}) \Big) \\
&\leq \omega\Big(|\bar{x} - \bar{y}| \big(1 + |\bar{x}|^p + |\bar{y}|^p \big) \Big),
\end{aligned} \tag{2.36}$$

as required. The construction of $\phi_\beta^\lambda : (0, T) \times \mathbb{R}^d \times \mathbb{R}^d \to \mathbb{R}$ in Equation (2.32) shows

$$\begin{aligned}
D_x \phi_\beta^\lambda(\bar{t}, \bar{x}, \bar{y}) &= 2\lambda(\bar{x} - \bar{y}) + q\beta e^{\mu t} |\bar{x}|^{q-2} \bar{x} \\
D_y \phi_\beta^\lambda(\bar{t}, \bar{x}, \bar{y}) &= 2\lambda(\bar{y} - \bar{x}) + q\beta e^{\mu t} |\bar{y}|^{q-2} \bar{y}.
\end{aligned} \tag{2.37}$$

Consequently, the boundedness (B1) and continuity assumption (B3) imply

$$\begin{aligned}
&b_\alpha^T(\bar{t}, x) D_x \phi_\beta^\lambda(\bar{t}, \bar{x}, \bar{y}) - b_\alpha^T(\bar{t}, y) \big(- D_y \phi_\beta^\lambda(\bar{t}, \bar{x}, \bar{y}) \big) \\
&= 2\lambda \big(b_\alpha(\bar{t}, x) - b_\alpha(\bar{t}, y) \big)^T (\bar{x} - \bar{y}) + C\beta e^{\mu t} \Big(|\bar{x}|^{q-2} b_\alpha^T(\bar{t}, x) \bar{x} + |\bar{y}|^{q-2} b_\alpha^T(\bar{t}, y) \bar{y} \Big) \\
&\leq C\lambda |x - y| |\bar{x} - \bar{y}| + C\beta e^{\mu t} \Big(|\bar{x}|^{q-1} C(1 + |x|) + |\bar{y}|^{q-1} C(1 + |y|) \Big)
\end{aligned}$$

for $(\alpha, x, y) \in \mathcal{A}_\varepsilon$. Similar as before, $\lim_{\varepsilon \to 0}(x, y) = (\bar{x}, \bar{y})$ leads to

$$\begin{aligned}
&\limsup_{\varepsilon \to 0} \limsup_{\kappa \to 0} \sup_{(\alpha, x, y) \in \mathcal{A}_\varepsilon} \Big(b_\alpha^T(\bar{t}, x) D_x \phi_\beta^\lambda(\bar{t}, \bar{x}, \bar{y}) - b_\alpha^T(\bar{t}, y) \big(- D_y \phi_\beta^\lambda(\bar{t}, \bar{x}, \bar{y}) \big) \Big) \\
&\leq C\lambda |\bar{x} - \bar{y}|^2 + C\beta e^{\mu t} \Big(1 + |\bar{x}|^q + |\bar{y}|^q \Big).
\end{aligned} \tag{2.38}$$

2.3. Hamilton–Jacobi–Bellman Equations

The assumptions on $X, Y \in \mathbb{S}^{d\times d}$ from Equation (2.34) imply

$$\xi^T X \xi - \eta^T Y \eta \leq 4\lambda |\xi - \eta|^2 + q(q-1)\beta e^{\mu t}\left(|\overline{x}|^{q-2}|\xi|^2 + |\overline{y}|^{q-2}|\eta|^2\right)$$

for all $\xi, \eta \in \mathbb{R}^d$. Therefore, the continuity assumption (B3) and the properties of the trace operator from [89, Section 4.1.2, p. 43] (such as $\operatorname{tr}(ABC) = \operatorname{tr}(CAB) = \operatorname{tr}(BCA)$ for $A, B, C \in \mathbb{S}^{d\times d}$ and $\operatorname{tr}(A) = \sum_{j=1}^d e_j^T A e_j$ for the standard basis $(e_j)_{1\leq j\leq d}$) imply

$$\operatorname{tr}\left(\sigma_\alpha(\bar{t},x)\sigma_\alpha^T(\bar{t},x)X\right) - \operatorname{tr}\left(\sigma_\alpha(\bar{t},y)\sigma_\alpha^T(\bar{t},y)Y\right)$$
$$= \operatorname{tr}\left(\sigma_\alpha^T(\bar{t},x)X\sigma_\alpha(\bar{t},x) - \sigma_\alpha^T(\bar{t},y)Y\sigma_\alpha(\bar{t},y)\right)$$
$$= \sum_{j=1}^d \left((\sigma_\alpha(\bar{t},x)e_j)^T X (\sigma_\alpha(\bar{t},x)e_j) - (\sigma_\alpha(\bar{t},y)e_j)^T Y (\sigma_\alpha(\bar{t},y)e_j)\right)$$
$$\leq 4d\lambda |\sigma_\alpha(\bar{t},x) - \sigma_\alpha(\bar{t},y)|^2 + C\beta e^{\mu t}\left(|\overline{x}|^{q-2}|\sigma_\alpha(\bar{t},x)|^2 + |\overline{y}|^{q-2}|\sigma_\alpha(\bar{t},y)|^2\right)$$
$$\leq C\lambda |x-y|^2 + C\beta e^{\mu t}\left(|\overline{x}|^{q-2}(1+|x|)^2 + |\overline{y}|^{q-2}(1+|y|)^2\right),$$

for $(\alpha, x, y) \in \mathcal{A}_\varepsilon$. The convergence $\lim_{\varepsilon\to 0}(x,y) = (\overline{x}, \overline{y})$ yields

$$\limsup_{\varepsilon\to 0}\limsup_{\kappa\to 0}\sup_{(\alpha,x,y)\in\mathcal{A}_\varepsilon}\left(\operatorname{tr}\left(\sigma_\alpha(\bar{t},x)\sigma_\alpha^T(\bar{t},x)X\right) - \operatorname{tr}\left(\sigma_\alpha(\bar{t},y)\sigma_\alpha^T(\bar{t},y)Y\right)\right) \quad (2.39)$$
$$\leq C\lambda |\overline{x}-\overline{y}|^2 + C\beta e^{\mu t}\left(1 + |\overline{x}|^q + |\overline{y}|^q\right).$$

Finally, a combination of Equation (2.35) together with Equation (2.36), Equation (2.38) and Equation (2.39) leads to

$$\limsup_{\varepsilon\to 0}\limsup_{\kappa\to 0}\sup_{(\alpha,x,y)\in\mathcal{A}_\varepsilon}\Big(\mathcal{L}_\alpha(\bar{t}, x, u^\varepsilon(\bar{t},\overline{x}), D_x\phi_\beta^\lambda(\bar{t},\overline{x},\overline{y}), X)$$
$$- \mathcal{L}_\alpha(\bar{t}, y, v_\varepsilon(\bar{t},\overline{y}), -D_y\phi_\beta^\lambda(\bar{t},\overline{x},\overline{y}), Y)\Big)$$
$$\leq C\lambda |\overline{x}-\overline{y}|^2 + C\beta e^{\mu t}\left(1 + |\overline{x}|^q + |\overline{y}|^q\right),$$

as required.

For the nonlocal part $\mathcal{I}_\alpha^\kappa$, we find for $(\alpha, x, y) \in \mathcal{A}_\varepsilon$ that

$$\mathcal{I}_\alpha^\kappa(\bar{t}, x, u^\varepsilon(\bar{t},\cdot)\circ\tau_{\overline{x}-x}, \phi_\beta^\lambda(\bar{t},\cdot,\overline{y})\circ\tau_{\overline{x}-x}) - \mathcal{I}_\alpha^\kappa(\bar{t}, y, v_\varepsilon(\bar{t},\cdot)\circ\tau_{\overline{y}-y}, \phi_\beta^\lambda(\bar{t},\overline{x},\cdot)\circ\tau_{\overline{y}-y})$$
$$\leq \left(\check{\mathcal{I}}_\alpha^\kappa(\bar{t}, x, \phi_\beta^\lambda(\bar{t},\cdot,\overline{y})\circ\tau_{\overline{x}-x}) - \check{\mathcal{I}}_\alpha^\kappa(\bar{t}, y, \phi_\beta^\lambda(\bar{t},\overline{x},\cdot)\circ\tau_{\overline{y}-y})\right)$$
$$+ \Big(\overline{\mathcal{I}}_\alpha^\kappa(\bar{t}, x, u^\varepsilon(\bar{t},\cdot)\circ\tau_{\overline{x}-x}, \phi_\beta^\lambda(\bar{t},\cdot,\overline{y})\circ\tau_{\overline{x}-x})$$
$$- \overline{\mathcal{I}}_\alpha^\kappa(\bar{t}, y, v_\varepsilon(\bar{t},\cdot)\circ\tau_{\overline{y}-y}, \phi_\beta^\lambda(\bar{t},\overline{x},\cdot)\circ\tau_{\overline{y}-y})\Big)$$
$$+ \left(\hat{\mathcal{I}}_\alpha^\kappa(\bar{t}, x, u^\varepsilon(\bar{t},\cdot)\circ\tau_{\overline{x}-x}) - \hat{\mathcal{I}}_\alpha^\kappa(\bar{t}, y, v_\varepsilon(\bar{t},\cdot)\circ\tau_{\overline{y}-y})\right),$$

which allows us to treat the small jump, medium jump and large jump term separately:

2. Integro-Differential Equations

For the small jump term $\check{\mathcal{I}}_\alpha^\kappa$ of the nonlocal part $\mathcal{I}_\alpha^\kappa$, we find

$$\check{\mathcal{I}}_\alpha^\kappa(t,x,\psi) = \int_{|z|\leq \kappa} \int_0^1 (1-\xi) j_\alpha^T(t,x,z) D^2\psi\big(t, x+\xi j_\alpha(t,x,z)\big) j_\alpha(t,x,z)\, d\xi\, m_\alpha(dz)$$

for all $(t,x,\psi) \in (0,T)\times \mathbb{R}^d \times C^2(\mathbb{R}^d)$, using the multivariate Taylor's theorem from [2, Theorem 7.5, p. 278]. Hence, the continuity assumption (B3) implies

$$\check{\mathcal{I}}_\alpha^\kappa(\bar{t},x,\phi_\beta^\lambda(\bar{t},\cdot,\bar{y})\circ \tau_{\bar{x}-x}) - \check{\mathcal{I}}_\alpha^\kappa(\bar{t},y,\phi_\beta^\lambda(\bar{t},\bar{x},\cdot)\circ \tau_{\bar{y}-y})$$

$$\leq \left|\check{\mathcal{I}}_\alpha^\kappa(\bar{t},x,\phi_\beta^\lambda(\bar{t},\cdot,\bar{y})\circ \tau_{\bar{x}-x})\right| + \left|\check{\mathcal{I}}_\alpha^\kappa(\bar{t},y,\phi_\beta^\lambda(\bar{t},\bar{x},\cdot)\circ \tau_{\bar{y}-y})\right|$$

$$\leq \int_{|z|\leq \kappa} \int_0^1 (1-\xi)\left|D^2\phi_\beta^\lambda(\bar{t},\bar{x}+\xi j_\alpha(\bar{t},x,z),\bar{y})\right|\cdot \left|j_\alpha(\bar{t},x,z)\right|^2 d\xi\, m_\alpha(dz)$$

$$+ \int_{|z|\leq \kappa} \int_0^1 (1-\xi)\left|D^2\phi_\beta^\lambda(\bar{t},\bar{x},\bar{y}+\xi j_\alpha(\bar{t},y,z))\right|\cdot \left|j_\alpha(\bar{t},y,z)\right|^2 d\xi\, m_\alpha(dz)$$

$$\leq \sup_{|\zeta|\leq 1} \left|D^2\phi_\beta^\lambda\big(\bar{t},\bar{x}+C(1+|x|)\zeta,\bar{y}\big)\right| C\big(1+|x|^2\big) \sup_{\alpha\in \mathcal{A}} \int_{|z|\leq \kappa} |z|^2\, m_\alpha(dz)$$

$$+ \sup_{|\eta|\leq 1} \left|D^2\phi_\beta^\lambda\big(\bar{t},\bar{x},\bar{y}+C(1+|y|)\eta\big)\right| C\big(1+|y|^2\big) \sup_{\alpha\in \mathcal{A}} \int_{|z|\leq \kappa} |z|^2\, m_\alpha(dz).$$

for $(\alpha,x,y) \in \mathcal{A}_\varepsilon$. Finally, the tightness assumption (B2) leads to

$$\limsup_{\kappa\to 0} \sup_{(\alpha,x,y)\in \mathcal{A}_\varepsilon} \big(\check{\mathcal{I}}_\alpha^\kappa(\bar{t},x,\phi_\beta^\lambda(\bar{t},\cdot,\bar{y})\circ \tau_{\bar{x}-x}) - \check{\mathcal{I}}_\alpha^\kappa(\bar{t},y,\phi_\beta^\lambda(\bar{t},\bar{x},\cdot)\circ \tau_{\bar{y}-y})\big) \leq 0,$$

as required.

For the medium jump term $\overline{\mathcal{I}}_\alpha^\kappa$ of the nonlocal part $\mathcal{I}_\alpha^\kappa$, we find

$$\big(u^\varepsilon(\bar{t},\bar{x}+j_\alpha(\bar{t},x,z)) - u^\varepsilon(\bar{t},\bar{x}) - D_x\phi_\beta^\lambda(\bar{t},\bar{x},\bar{y}) j_\alpha(\bar{t},x,z)\big)$$
$$- \big(v_\varepsilon(\bar{t},\bar{y}+j_\alpha(\bar{t},y,z)) - v_\varepsilon(\bar{t},\bar{y}) + D_y\phi_\beta^\lambda(\bar{t},\bar{x},\bar{y}) j_\alpha(\bar{t},y,z)\big)$$
$$= \big(u^\varepsilon(\bar{t},\bar{x}+j_\alpha(\bar{t},x,z)) - v_\varepsilon(\bar{t},\bar{y}+j_\alpha(\bar{t},y,z)) - \phi_\beta^\lambda(\bar{t},\bar{x}+j_\alpha(\bar{t},x,z),\bar{y}+j_\alpha(\bar{t},y,z))\big)$$
$$- \big(u^\varepsilon(\bar{t},\bar{x}) - v_\varepsilon(\bar{t},\bar{y}) - \phi_\beta^\lambda(\bar{t},\bar{x},\bar{y})\big)$$
$$+ \big(\phi_\beta^\lambda(\bar{t},\bar{x}+j_\alpha(\bar{t},x,z),\bar{y}+j_\alpha(\bar{t},y,z)) - \phi_\beta^\lambda(\bar{t},\bar{x},\bar{y})\big)$$
$$- \big(D_y\phi_\beta^\lambda(\bar{t},\bar{x},\bar{y}) j_\alpha(\bar{t},y,z) + D_x\phi_\beta^\lambda(\bar{t},\bar{x},\bar{y}) j_\alpha(\bar{t},x,z)\big)$$
$$\leq \lambda\big(|\bar{x}+j_\alpha(\bar{t},x,z) - (\bar{y}+j_\alpha(\bar{t},y,z))|^2 - |\bar{x}-\bar{y}|^2\big)$$
$$+ \beta e^{\mu\bar{t}}\big(|\bar{x}+j_\alpha(\bar{t},x,z)|^q + |\bar{y}+j_\alpha(\bar{t},y,z)|^q - (|\bar{x}|^q + |\bar{y}|^q)\big)$$
$$- \big(2\lambda(\bar{y}-\bar{x}) + q\beta e^{\mu\bar{t}}|\bar{y}|^{q-2}\bar{y}\big) j_\alpha(\bar{t},y,z)$$
$$- \big(2\lambda(\bar{x}-\bar{y}) + q\beta e^{\mu\bar{t}}|\bar{x}|^{q-2}\bar{x}\big) j_\alpha(\bar{t},x,z)$$

$$= \lambda |j_\alpha(\bar{t},x,z) - j_\alpha(\bar{t},y,z)|^2$$
$$+ \beta e^{\mu \bar{t}}\Big(|\bar{x} + j_\alpha(\bar{t},x,z)|^q - |\bar{x}|^q - qj_\alpha(\bar{t},x,z)|\bar{x}|^{q-2}x\Big)$$
$$+ \beta e^{\mu \bar{t}}\Big(|\bar{y} + j_\alpha(\bar{t},y,z)|^q - |\bar{y}|^q - qj_\alpha(\bar{t},y,z)|\bar{y}|^{q-2}y\Big)$$

for $(\alpha,x,y) \in \mathcal{A}_\varepsilon$ and $|z| \leq 1$, using $(\bar{t},\bar{x},\bar{y}) \in \arg\max(u^\varepsilon - v_\varepsilon - \phi_\beta^\lambda)$ by assumption, the definition of $\phi_\beta^\lambda(\cdot)$ from Equation (2.32), and the form of $D\phi_\beta^\lambda(\cdot)$ from Equation (2.37). Furthermore, Taylor's theorem and the growth assumption (B4) imply

$$\big||\bar{x} + j_\alpha(\bar{t},x,z)|^q - |\bar{x}|^q - qj_\alpha(\bar{t},x,z)|\bar{x}|^{q-2}x\big|$$
$$\leq \frac{1}{2}q(q-1)\sup_{\xi \in [0,1]} |\bar{x} + \xi j_\alpha(\bar{t},x,z)|^{q-2} \cdot |j_\alpha(\bar{t},x,z)|^2$$
$$\leq q(q-1)2^{q-2}\Big(|\bar{x}|^{q-2} + |j_\alpha(\bar{t},x,z)|^{q-2}\Big)|j_\alpha(\bar{t},x,z)|^2 \leq C\Big(1 + |\bar{x}|^{q-2} \cdot |x|^2 + |\bar{x}|^q\Big)|z|^2$$

for all $(\alpha,x,y) \in \mathcal{A}_\varepsilon$ and $|z| \leq 1$. Replacing $(x,\bar{x}) \in \mathbb{R}^d \times \mathbb{R}^d$ by $(y,\bar{y}) \in \mathbb{R}^d \times \mathbb{R}^d$ shows

$$\big||\bar{y} + j_\alpha(\bar{t},y,z)|^q - |\bar{y}|^q - qj_\alpha(\bar{t},y,z)|\bar{y}|^{q-2}y\big| \leq C\Big(1 + |\bar{y}|^{q-2} \cdot |y|^2 + |\bar{y}|^{q-2}\Big)|z|^2$$

for $(\alpha,x,y) \in \mathcal{A}_\varepsilon$ and $|z| \leq 1$. A combination the preceding three inequalities with the continuity assumption (B3) leads to

$$\Big(u^\varepsilon(\bar{t},\bar{x}+j_\alpha(\bar{t},x,z)) - u^\varepsilon(\bar{t},\bar{x}) - D_x\phi_\beta^\lambda(\bar{t},\bar{x},\bar{y})j_\alpha(\bar{t},x,z)\Big)$$
$$- \Big(v_\varepsilon(\bar{t},\bar{y}+j_\alpha(\bar{t},y,z)) - v_\varepsilon(\bar{t},\bar{y}) + D_y\phi_\beta^\lambda(\bar{t},\bar{x},\bar{y})j_\alpha(\bar{t},y,z)\Big)$$
$$\leq \lambda|x-y|^2|z|^2 + \beta e^{\mu \bar{t}} C\Big(1 + |\bar{x}|^{q-2} \cdot |x|^2 + |\bar{x}|^q + |\bar{y}|^{q-2} \cdot |y|^2 + |\bar{y}|^q\Big)|z|^2$$

for $(\alpha,x,y) \in \mathcal{A}_\varepsilon$ and $|z| \leq 1$. In particular, we obtain

$$\overline{\mathcal{I}}_\alpha^\kappa(\bar{t},x,u^\varepsilon(\bar{t},\cdot) \circ \tau_{\bar{x}-x}, \phi_\beta^\lambda(\bar{t},\cdot,\bar{y}) \circ \tau_{\bar{x}-x}) - \overline{\mathcal{I}}_\alpha^\kappa(\bar{t},y,v_\varepsilon(\bar{t},\cdot) \circ \tau_{\bar{y}-y}, \phi_\beta^\lambda(\bar{t},\bar{x},\cdot) \circ \tau_{\bar{y}-y})$$
$$\leq \beta e^{\mu \bar{t}} C\Big(|\bar{x}|^{q-2}(|x|^2 - |\bar{x}|^2) + |\bar{y}|^{q-2}(|y|^2 - |\bar{y}|^2)\Big)\int_{|z|\leq 1}|z|^2 m_\alpha(dz)$$
$$+ \Big(\lambda|x-y|^2 + \beta e^{\mu \bar{t}} C(1 + |\bar{x}|^q + |\bar{y}|^q)\Big)\int_{|z|\leq 1}|z|^2 m_\alpha(dz)$$

for $(\alpha,x,y) \in \mathcal{A}_\varepsilon$. Finally, the convergence $\lim_{\varepsilon \to 0}(x,y) = (\bar{x},\bar{y})$ and the boundedness assumption (B1) yield

$$\limsup_{\varepsilon \to 0} \limsup_{\kappa \to 0} \sup_{(\alpha,x,y)\in\mathcal{A}_\varepsilon} \Big(\overline{\mathcal{I}}_\alpha^\kappa(\bar{t},x,u^\varepsilon(\bar{t},\cdot) \circ \tau_{\bar{x}-x}, \phi_\beta^\lambda(\bar{t},\cdot,\bar{y}) \circ \tau_{\bar{x}-x})$$
$$- \overline{\mathcal{I}}_\alpha^\kappa(\bar{t},y,v_\varepsilon(\bar{t},\cdot) \circ \tau_{\bar{y}-y}, \phi_\beta^\lambda(\bar{t},\bar{x},\cdot) \circ \tau_{\bar{y}-y})\Big)$$
$$\leq C\lambda|\bar{x}-\bar{y}|^2 + C\beta e^{\mu \bar{t}}\Big(1 + |\bar{x}|^q + |\bar{y}|^q\Big),$$

as required.

2. Integro-Differential Equations

For the large jump term $\hat{\mathcal{I}}_\alpha^\kappa$ of the nonlocal part $\mathcal{I}_\alpha^\kappa$ with $p > 0$, we find

$$\left(u^\varepsilon\left(\bar{t},\bar{x}+j_\alpha(\bar{t},x,z)\right) - u^\varepsilon\left(\bar{t},\bar{x}\right)\right) - \left(v_\varepsilon\left(\bar{t},\bar{y}+j_\alpha(\bar{t},y,z)\right) - v_\varepsilon\left(\bar{t},\bar{y}\right)\right)$$
$$= \left(u^\varepsilon\left(\bar{t},\bar{x}+j_\alpha(\bar{t},x,z)\right) - v_\varepsilon\left(\bar{t},\bar{y}+j_\alpha(\bar{t},y,z)\right)\right) - \left(u^\varepsilon\left(\bar{t},\bar{x}\right) - v_\varepsilon\left(\bar{t},\bar{y}\right)\right)$$
$$\leq \phi_\beta^\lambda\left(\bar{t},\bar{x}+j_\alpha(\bar{t},x,z),\bar{y}+j_\alpha(\bar{t},y,z)\right) - \phi_\beta^\lambda(\bar{t},\bar{x},\bar{y})$$
$$\leq \lambda\left|\left(\bar{x}+j_\alpha(\bar{t},x,z)\right) - \left(\bar{y}+j_\alpha(\bar{t},y,z)\right)\right|^2 + \beta e^{\mu\bar{t}}\left(|\bar{x}+j_\alpha(\bar{t},x,z)|^q + |\bar{y}+j_\alpha(\bar{t},y,z)|^q\right)$$

for $(\alpha,x,y) \in \mathcal{A}_\varepsilon$ and $|z| > 1$, using $(\bar{t},\bar{x},\bar{y}) \in \arg\max(u^\varepsilon - v_\varepsilon - \phi_\beta^\lambda)$ by assumption, the definition of $\phi_\beta^\lambda(\cdot)$ from Equation (2.32), and $\phi_\beta^\lambda(\bar{t},\bar{x},\bar{y}) - \delta(T-\bar{t})^{-1} \geq 0$. Hence, the continuity (B3) and growth assumption (B4) imply

$$\hat{\mathcal{I}}_\alpha^\kappa(\bar{t},x,u^\varepsilon(\bar{t},\cdot)\circ\tau_{\bar{x}-x}) - \hat{\mathcal{I}}_\alpha^\kappa(\bar{t},y,v_\varepsilon(\bar{t},\cdot)\circ\tau_{\bar{y}-y})$$
$$= \int_{1<|z|}\left(u^\varepsilon\left(\bar{t},\bar{x}+j_\alpha(\bar{t},x,z)\right) - u^\varepsilon\left(\bar{t},\bar{x}\right)\right) - \left(v_\varepsilon\left(\bar{t},\bar{y}+j_\alpha(\bar{t},y,z)\right) - v_\varepsilon(\bar{t},\bar{y})\right) m_\alpha(dz)$$
$$\leq C\left(\lambda(|\bar{x}-\bar{y}|^2 + |x-y|^2) + \beta e^{\mu\bar{t}}\left(1 + |\bar{x}|^q + |x|^q + |\bar{y}|^q + |y|^q\right)\right)\int_{|z|>1}|z|^q m_\alpha(dz)$$

for $(\alpha,x,y) \in \mathcal{A}_\varepsilon$. Finally, since $\lim_{\varepsilon\to 0}(x,y) = (\bar{x},\bar{y})$ and

$$\sup_{\alpha\in\mathcal{A}}\int_{|z|>1}|z|^q m_\alpha(dz) < \infty$$

by assumption, this leads to

$$\limsup_{\varepsilon\to 0}\limsup_{\kappa\to 0}\sup_{(\alpha,x,y)\in\mathcal{A}_\varepsilon}\left(\hat{\mathcal{I}}_\alpha^\kappa(\bar{t},x,u^\varepsilon(\bar{t},\cdot)\circ\tau_{\bar{x}-x}) - \hat{\mathcal{I}}_\alpha^\kappa(\bar{t},y,v_\varepsilon(\bar{t},\cdot)\circ\tau_{\bar{y}-y})\right)$$
$$\leq C\lambda|\bar{x}-\bar{y}|^2 + C\beta e^{\mu\bar{t}}\left(1+|\bar{x}|^q + |\bar{y}|^q\right),$$

as required.

For the large jump term $\hat{\mathcal{I}}_\alpha^\kappa$ of the nonlocal part $\mathcal{I}_\alpha^\kappa$ with $p = 0$, we find

$$\hat{\mathcal{I}}_\alpha^\kappa(\bar{t},x,u^\varepsilon(\bar{t},\cdot)\circ\tau_{\bar{x}-x}) - \hat{\mathcal{I}}_\alpha^\kappa(\bar{t},y,v_\varepsilon(\bar{t},\cdot)\circ\tau_{\bar{y}-y})$$
$$= \int_{1<|z|}\left(u^\varepsilon\left(\bar{t},\bar{x}+j_\alpha(\bar{t},x,z)\right) - u^\varepsilon\left(\bar{t},\bar{x}\right)\right) - \left(v_\varepsilon\left(\bar{t},\bar{y}+j_\alpha(\bar{t},y,z)\right) - v_\varepsilon(\bar{t},\bar{y})\right) m_\alpha(dz)$$
$$\leq C\left(\lambda(|\bar{x}-\bar{y}|^2 + |x-y|^2) + \beta e^{\mu\bar{t}}\left(1 + |\bar{x}|^q + |x|^q + |\bar{y}|^q + |y|^q\right)\right)\int_{1<|z|\leq R}|z|^q m_\alpha(dz)$$
$$+ C\left(\|u\|_\infty + \|v\|_\infty\right)\int_{|z|>R}1\,m_\alpha(dz)$$

for $(\alpha,x,y) \in \mathcal{A}_\varepsilon$ and $R > 1$, using $\|u^\varepsilon\|_\infty + \|v_\varepsilon\|_\infty \leq C(\|u\|_\infty + \|v\|_\infty)$ from Lemma 2.8.

Since $\lim_{\varepsilon \to 0}(x, y) = (\overline{x}, \overline{y})$ by construction, this implies

$$\limsup_{\varepsilon \to 0} \sup_{(\alpha,x,y) \in \mathcal{A}_\varepsilon} \left(\hat{\mathcal{I}}_\alpha^\kappa(\overline{t}, x, u^\varepsilon(\overline{t}, \cdot) \circ \tau_{\overline{x}-x}) - \hat{\mathcal{I}}_\alpha^\kappa(\overline{t}, y, v_\varepsilon(\overline{t}, \cdot) \circ \tau_{\overline{y}-y}) \right)$$
$$\leq C\left(\lambda(|\overline{x} - \overline{y}|^2) + \beta e^{\mu \overline{t}}(1 + |\overline{x}|^q + |\overline{y}|^q) \right) \int_{1 < |z| \leq R} |z|^q \, m_\alpha(dz) \quad (2.40)$$
$$+ C\left(\|u\|_\infty + \|v\|_\infty \right) \int_{|z| > R} 1 \, m_\alpha(dz)$$

for all $R > 1$. Note that (similar to the second part of the proof of Theorem 2.22) it is possible to show that $\limsup_{\beta \to 0} \limsup_{\lambda \to \infty} \beta e^{\mu \overline{t}}(1 + |\overline{x}|^q + |\overline{y}|^q) = 0$ as well as $\limsup_{\lambda \to \infty} \lambda|\overline{x} - \overline{y}|^2 = 0$, using the ideas of the proof of Lemma 2.16: First, we obtain

$$M_{\beta/2}^{\lambda/2} := \sup_{(t,x,y) \in (0,T) \times \mathbb{R}^d \times \mathbb{R}^d} \left(u(t,x) - v(t,y) - \phi_{\beta/2}^{\lambda/2}(t,x,y) \right)$$
$$\geq \sup_{(t,x,y) \in (0,T) \times \mathbb{R}^d \times \mathbb{R}^d} \left(u(t,x) - v(t,y) - \phi_\beta^\lambda(t,x,y) \right) + \frac{\lambda}{2}|\overline{x} - \overline{y}|^2 + \frac{\beta}{2}\left(|\overline{x}|^q + |\overline{y}|^q \right)$$
$$= M_\beta^\lambda + \frac{\lambda}{2}|\overline{x} - \overline{y}|^2 + \frac{\beta}{2}\left(|\overline{x}|^q + |\overline{y}|^q \right)$$

for all $\beta, \lambda > 0$ from the definition of $\phi_\beta^\lambda(\cdot)$ in Equation (2.32). Moreover, the boundedness of $u : \mathbb{R}^d \to \mathbb{R}$ and $v : \mathbb{R}^d \to \mathbb{R}$ implies that

$$\lim_{\beta \to 0} \lim_{\lambda \to \infty} M_\beta^\lambda$$

exists in $\mathbb{R}$, which shows that

$$\limsup_{\lambda \to \infty} \lambda|\overline{x} - \overline{y}|^2 = 0$$
$$\limsup_{\beta \to 0} \limsup_{\lambda \to \infty} \beta \left(|\overline{x}|^q + |\overline{y}|^q \right) = 0.$$

Finally, the boundedness assumption (B1) and Equation (2.40) yield

$$\limsup_{\beta \to 0} \limsup_{\lambda \to \infty} \limsup_{\varepsilon \to 0} \sup_{(\alpha,x,y) \in \mathcal{A}_\varepsilon} \left(\hat{\mathcal{I}}_\alpha^\kappa(\overline{t}, x, u^\varepsilon(\overline{t}, \cdot) \circ \tau_{\overline{x}-x}) - \hat{\mathcal{I}}_\alpha^\kappa(\overline{t}, y, v_\varepsilon(\overline{t}, \cdot) \circ \tau_{\overline{y}-y}) \right)$$
$$\leq C(\|u\|_\infty + \|v\|_\infty) \int_{|z| > R} 1 \, m_\alpha(dz)$$

for all $R > 1$. Sending $R \to \infty$ together with the tightness assumption (B2) thus shows

$$\limsup_{\beta \to 0} \limsup_{\lambda \to \infty} \limsup_{\varepsilon \to 0} \sup_{(\alpha,x,y) \in \mathcal{A}_\varepsilon} \left(\hat{\mathcal{I}}_\alpha^\kappa(\overline{t}, x, u^\varepsilon(\overline{t}, \cdot) \circ \tau_{\overline{x}-x}) - \hat{\mathcal{I}}_\alpha^\kappa(\overline{t}, y, v_\varepsilon(\overline{t}, \cdot) \circ \tau_{\overline{y}-y}) \right) \leq 0,$$

as required. $\square$

2. Integro-Differential Equations

It is interesting to note that the preceding proof generalizes the approach pursued in [108] even in the bounded case (i.e. $p = 0$): In this article, the author employs an additional integrability condition (similar to our unbounded case), even though it is somewhat hidden in [108, Chapter 2] and [108, Chapter 4]. However, the author incorrectly claims in [108, Example 2] to cover the α-stable case for any $\alpha \in (0, 2)$, although the assumptions only cover $\alpha \in (0, \frac{1}{2})$. In the unbounded case (i.e. $p > 0$), our arguments also depend on such an additional integrability condition, in order to show at the end of the preceding proof that the large jump term satisfies a regularity condition: This is mainly due to the fact that

$$\limsup_{\beta \to 0} \limsup_{\lambda \to \infty} \beta \left(|\overline{x}|^q + |\overline{y}|^q \right) = 0$$

does not hold in general. It would be interesting to study if it is possible to relax the additional integrability condition (in the unbounded case) by using an alternative approach to bound the large jump term.

Corollary 2.34 (Comparison Principle). *Suppose that*

$$u, v \in SC_p([0, T] \times \mathbb{R}^d)$$

with $p \geq 0$ and $T > 0$ are viscosity solutions in $(0, T) \times \mathbb{R}^d$ of

$$\partial_t u(t, x) + G(t, x, u(t, x), Du(t, x), D^2 u(t, x), u(t, \cdot)) \leq 0$$
$$\partial_t v(t, x) + G(t, x, v(t, x), Dv(t, x), D^2 v(t, x), v(t, \cdot)) \geq 0$$

for some nonlinear Hamilton-Jacobi-Bellman operator G (as defined in Definition 2.27), which satisfies all assumptions from Remark 2.28. In the unbounded case (i.e. $p > 0$), suppose further that the additional integrability assumption

$$\sup_{\alpha \in \mathcal{A}} \int_{|z| > 1} |z|^q \, m_\alpha(dz) < \infty$$

holds for some $q > p$ with $q \geq 2$. If the initial values $u(0, \cdot), v(0, \cdot)$ are continuous and

$$u(0, x) \leq v(0, x)$$

for all $x \in \mathbb{R}^d$, then $u(t, x) \leq v(t, x)$ for all $(t, x) \in (0, T) \times \mathbb{R}^d$.

Proof. According to Lemma 2.32 and Proposition 2.33, the general comparison theorem from Corollary 2.23 can be applied to obtain the desired result. □

A detailed inspection of the main line of argument shows that (under appropriate integrability assumptions on the related jump measures) it seems feasible to extend our comparison principle from Corollary 2.34 to unbounded solutions with more general growth assumptions (as long as they are bounded by a smooth quasidistance that equals the squared norm in a neighborhood of zero, so that Lemma 1.12 can be applied). An example in [9, Remark 3.6] shows, however, that it is not possible to obtain a general comparison principle for unbounded viscosity solutions with arbitrary growth at infinity.

2.3. Hamilton–Jacobi–Bellman Equations

Corollary 2.35 (Subadditivity & Convexity of Solutions). *Suppose that*

$$u^\varphi \in C_p([0,T] \times \mathbb{R}^d)$$

with $p \geq 0$, $T > 0$ and $u^\varphi(0, \cdot) = \varphi \in C_p(\mathbb{R}^d)$ are viscosity solutions in $(0,T) \times \mathbb{R}^d$ of

$$\partial_t u^\varphi(t,x) + G(t, x, u^\varphi(t,x), Du^\varphi(t,x), D^2 u^\varphi(t,x), u^\varphi(t, \cdot)) = 0$$

for some nonlinear Hamilton-Jacobi-Bellman operator G (as defined in Definition 2.27), which satisfies all assumptions from Remark 2.28. In the unbounded case (i.e. $p > 0$), suppose further that the additional integrability assumption

$$\sup_{\alpha \in \mathcal{A}} \int_{|z|>1} |z|^q \, m_\alpha(dz) < \infty$$

holds for some $q > p$ with $q \geq 2$. For all initial values $\varphi, \psi \in C_p(\mathbb{R}^d)$, we have

$$u^{\varphi+\psi}(t,x) \leq u^\varphi(t,x) + u^\psi(t,x)$$
$$u^{\lambda\varphi+(1-\lambda)\psi}(t,x) \leq \lambda u^\varphi(t,x) + (1-\lambda) u^\psi(t,x)$$

for all $(t,x) \in (0,T) \times \mathbb{R}^d$ and $\lambda \in (0,1)$.

Proof. It is easy to check that the operators

$$G(t,x,\cdot) : \mathbb{R} \times \mathbb{R}^d \times \mathbb{S}^{d \times d} \times C_p^2(\mathbb{R}^d) \longrightarrow \mathbb{R}$$

are superlinear and concave for all $(t,x) \in (0,T) \times \mathbb{R}^d$, since they are the infimum of a family of linear operators

$$G_\alpha(t,x,\cdot) : \mathbb{R} \times \mathbb{R}^d \times \mathbb{S}^{d \times d} \times C_p^2(\mathbb{R}^d) \longrightarrow \mathbb{R}$$

with $\alpha \in \mathcal{A}$. According to Lemma 2.32 and Proposition 2.33, the general results from Corollary 2.25 and Corollary 2.26 can be applied to obtain the desired result. □

3. Sublinear Expectations

As the name suggests, sublinear expectations generalize classical linear expectations (i.e. typically Lebesgue integrals with respect to probability measures) by relaxing its linearity property. They are a special case of nonlinear expectations, which are used in a variety of different fields (see e.g. [128, Chapter 15, p. 303] or [43] for an overview), and are often interpreted as worst-case or best-case bounds for risk evaluations under the presence of (Knightian) uncertainty. In the pioneering work [5], general sublinear expectation were first characterized and studied axiomatically under the name coherent risk measures by Philippe Artzner, Freddy Delbaen, Jean-Marc Eber and David Heath, as a unified framework to model financial risks. Shige Peng later introduced a corresponding notion for the distribution, independence and convergence of random variables for sublinear expectations in [102], in order to derive classical results from probability theory (such as the central limit theorem and the weak law of large numbers) in this generalized context of sublinear expectation spaces.

The purpose of this chapter is to summarize the intrinsic approach towards sublinear expectations from Shige Peng in [102] and to fix important, related (partly refined) definitions required for Chapter 4 – intrinsic in the sense, that the approach is a-priorily not based on a classical probability or measurable space. In the first half, in Section 3.1, we introduce the notion of sublinear expectation spaces and remind the reader of a well-known characterization, which relates sublinear expectations to classical linear expectations and in turn motivates its common interpretation as worst-case or best-case bounds in applications. In the second half, in Section 3.2, we establish refined notions (compared to the ones from Shige Peng in [102]) for the distribution, independence and convergence of random variables on sublinear expectation spaces, recall the central limit theorem and law of large numbers for sublinear expectations from [104], and introduce the corresponding limiting distributions.

3.1. Expectation Spaces

Sublinear expectations have been studied extensively in the context of decision theory and related topics in mathematical finance under the name coherent risk measures. For a thorough introduction into monetary risk measures and their applications, we refer the interested reader to the monograph [55, Chapter 4, p. 157].

As we will see in Chapter 4 of this thesis, one important question for sublinear expectations is the extent of their domains. For that reason, we introduce sublinear expectations on rather general spaces, instead of working with classical measurable spaces directly.

3. Sublinear Expectations

Definition 3.1 (Space of Integrands). Let Ω be a given set, the so-called *ground set*, and $\mathcal{H}$ a linear space of real-valued functions on Ω. The space $\mathcal{H}$ is called a *space of integrands* on Ω if it contains all constant functions and $X \in \mathcal{H}$ implies $|X| \in \mathcal{H}$.

Remark 3.2 ($\mathcal{H}$ are Vector Lattices). A linear space $\mathcal{H}$ of real-valued functions defined on a common set Ω is called a *vector lattice* if it is stable under the minimum and maximum operators, i.e. for every $f, g \in \mathcal{H}$ we have $f \wedge g = \min\{f, g\} \in \mathcal{H}$ and $f \vee g = \max\{f, g\} \in \mathcal{H}$. Since the representations

$$f^+ = \max\{f, 0\} = \tfrac{1}{2}(|f| + f)$$
$$f^- = \min\{f, 0\} = \tfrac{1}{2}(|f| - f)$$
$$f \vee g = \max\{f, g\} = (f - g)^+ + g$$
$$f \wedge g = \min\{f, g\} = (f - g)^- + g$$

holds, every space of integrands is also a vector lattice. Moreover, a vector lattice that contains all constant functions is a space of integrands due to $|f| = \max\{f, -f\}$.

Definition 3.3 (Expectation Spaces). Let $\mathcal{H}$ be a space of integrands on the set Ω. A *nonlinear expectation* is a functional $E : \mathcal{H} \to \mathbb{R}$ such that

(i) $\forall X \leq Y : E(X) \leq E(Y)$ \hfill (Monotonicity)

(ii) $\forall c \in \mathbb{R} : E(c) = c$ \hfill (Preservation of Constants)

and the triplet $(\Omega, \mathcal{H}, E)$ is called a *nonlinear expectation space*. If we additionally have

(iii) $\forall X, Y \in \mathcal{H} : E(X + Y) \leq E(X) + E(Y)$ \hfill (Subadditivity)

(iv) $\forall \lambda \geq 0 : E(\lambda X) = \lambda E(X)$ \hfill (Positive Homogeneity)

the functional $E : \mathcal{H} \to \mathbb{R}$ is called *sublinear expectation* and the triplet $(\Omega, \mathcal{H}, E)$ the corresponding *sublinear expectation space*.

It is interesting to note that, although we restrict our attention in this thesis to sublinear expectations, many interesting results can be extended to nonlinear expectations using a standard procedure (see [105, Chapter 8] for an exemplary account), as long as they are (in a certain sense) dominated by sublinear expectations.

Let us quickly motivate, why we generalize the linearity of expectations instead of the additivity of probability measures: First of all, many interesting nonlinear expectations are not constructed on (and cannot be extended uniquely to) a set of measurable functions, so that it is a-priorily unclear how to relate nonlinear expectations to measurable spaces. An interesting example related to this problem for sublinear expectations can be found in [98, Example 5.1]. On top of that, nonlinear expectations (even if defined on measurable spaces) contain in general more (for many applications very vital) information compared to its underlying generalization of probability measures, which can be seen in the following example: Suppose that $\mathbb{E}(\cdot)$ is a classical linear expectation, then for every increasing, measurable bijection $f : \mathbb{R} \to \mathbb{R}$ with $f(x) = x$ for all $x \in [0,1]$,

$$E(X) := f^{-1}(\mathbb{E}(f(X)))$$

defines a nonlinear expectation such that $E(\mathbb{1}_A) = \mathbb{E}(\mathbb{1}_A)$ holds for measurable $A \subset \Omega$. In contrast to this general statement, we will now show that in case of sublinear expectations on measurable spaces, there exists a one-to-one correspondence to the underlying sublinear generalizations of probability measures (often referred to as capacities) under some additional conditions.

Lemma 3.4 (Properties of Sublinear Expectations). *A sublinear expectation space $(\Omega, \mathcal{H}, E)$ has the following basic properties:*

(v) $\forall \alpha \in [0,1] : E(\alpha X + (1-\alpha)Y) \leq \alpha E(X) + (1-\alpha) E(Y)$ \hfill (Convexity)

(vi) $\forall c \in \mathbb{R} : E(X + c) = E(X) + c$ \hfill (Simple Translation Invariance)

(vii) $E(Y) = -E(-Y) \implies E(X+Y) = E(X) + E(Y)$ \hfill (Translation Invariance)

(viii) $\forall \lambda \in \mathbb{R} : E(\lambda X) = \lambda^+ E(X) + \lambda^- E(-X)$ \hfill (Homogeneity)

In fact, the homogeneity property (viii) is equivalent to the positive homogeneity property (iv) from Definition 3.3. Moreover, the convexity (v) and homogeneity (viii) are equivalent to the sublinearity, i.e. property (iii) and (iv).

Proof. A simple application of the subadditivity (iii) and positive homogeneity (iv) implies the convexity (v). For the opposite direction note that

$$\begin{aligned} E(X+Y) &= E\left(\alpha \left(\tfrac{1}{\alpha}X\right) + (1-\alpha)\left(\tfrac{1}{1-\alpha}Y\right)\right) \\ &\leq \alpha E\left(\tfrac{1}{\alpha}X\right) + (1-\alpha) E\left(\tfrac{1}{1-\alpha}Y\right) \\ &= E(X) + E(Y), \end{aligned}$$

where we used the convexity (v) in the second and the homogeneity (iv) in the last step. For the translation invariance properties consider

$$\begin{aligned} E(X) + E(Y) &= E(X) - E(-Y) \\ &\leq E(X - (-Y)) \\ &= E(X+Y) \\ &\leq E(X) + E(Y), \end{aligned}$$

where we applied the subadditivity (iii) in the third and last step. The equivalence of the homogeneity properties follows directly from the definition of the positive part $\lambda^+ = \max\{\lambda, 0\}$ and the negative part $\lambda^- = \max\{-\lambda, 0\}$. $\square$

3. Sublinear Expectations

Theorem 3.5 (Representation of Sublinear Functionals). *Let $E : \mathcal{H} \to \mathbb{R}$ be a sublinear functional on a linear space $\mathcal{H}$, i.e. it satisfies subadditivity (iii) and positive homogeneity property (iv) from Definition 3.3. There exists a family $\{E_\theta : \theta \in \Theta\}$ of linear functionals on $\mathcal{H}$ such that*

$$E(X) = \sup_{\theta \in \Theta} E_\theta(X)$$

for all $X \in \mathcal{H}$. In fact, for every $X \in \mathcal{H}$ there exists $\theta_X \in \Theta$ such that $E(X) = E_{\theta_X}(X)$. Moreover, if E is a sublinear expectation then all E_θ are linear expectations, i.e. they also satisfy monotonicity (i) and constant preservation property (ii) from Definition 3.3.

Proof. Let $\Theta := \{\theta : \mathcal{H} \to \mathbb{R} : \theta \text{ is linear and } \theta \leq E\}$ and denote $E_\theta := \theta$ for $\theta \in \Theta$. Since $E \geq \sup_{\theta \in \Theta} E_\theta$ by construction, it suffices to prove (for the first part of the theorem) that for every $X \in \mathcal{H}$ there exists $\theta_X \in \Theta$ such that $E(X) = E_{\theta_X}(X)$:

Suppose that $X \in \mathcal{H}$ and $\mathcal{L}_X := \{aX : a \in \mathbb{R}\} \subset \mathcal{H}$. Since for each $a \in \mathbb{R}$

$$aE(X) = a^+ E(X) - a^- E(X) = E(a^+ X) - E(a^- X) \leq E(a^+ X - a^- X) = E(aX)$$

follows from the subadditivity property (iii) in Definition 3.3, the linear functional

$$f_X : \mathcal{L}_X \longrightarrow \mathbb{R}$$
$$(aX) \longmapsto f_X(aX) := aE(X)$$

is dominated by $E(\cdot)$, i.e. $f_X \leq E$ on $\mathcal{L}_X$. Accordingly, the Hahn-Banach theorem (cf. Theorem A.9) implies that there exists $\theta_X \in \Theta$ such that

$$E(X) = f_X(X) = \theta_X(X) = E_{\theta_X}(X).$$

For the second part of the theorem, assume that $E : \mathcal{H} \to \mathbb{R}$ is a sublinear expectation, i.e. it also satisfies monotonicity (i) and preservation of constants (ii) from Definition 3.3. For $X \geq Y$ and $\theta \in \Theta$, we find

$$E_\theta(X) - E_\theta(Y) = E_\theta(X - Y) = -E_\theta(Y - X)$$
$$\geq -\sup_{\theta \in \Theta} E_\theta(Y - X) \geq 0,$$

using the linearity of E_θ and monotonicity (i) of $E = \sup_{\theta \in \Theta} E_\theta$. Finally, we obtain for each $c \in \mathbb{R}$ and $\theta \in \Theta$

$$c = -\sup_{\theta \in \Theta} E(-c) \leq -E_\theta(-c) = E_\theta(c) \leq \sup_{\theta \in \Theta} E_\theta(c) = c,$$

using the preservation of constants (ii) of $E = \sup_{\theta \in \Theta} E_\theta$ and the linearity of E_θ. □

The preceding characterization of sublinear expectations in Theorem 3.5 was already obtained in the original work [5] on coherent risk measures (with finite ground sets Ω). It suggests to interpret sublinear expectations as best-case or worst-case bounds for models with incomplete knowledge on all related parameters. In particular, it seems reasonable to model risk bounds under the presence of (Knightian) uncertainty by sublinear expectations.

3.1. Expectation Spaces

Remark 3.6 (Properties of Parameter Sets Θ). The previous theorem states that every sublinear functional E is of the form $E = \sup_{\theta \in \Theta} E_\theta$ for a family $\{E_\theta : \theta \in \Theta\}$ of linear functionals. Moreover, its proof entails that we can assume (without loss of generality) that the parameter set Θ is of the form

$$\Theta = \{\theta : \mathcal{H} \to \mathbb{R} : \theta \text{ is linear and } \theta \leq E\},$$

which is convex and closed (with respect to the operator norm induced by $\|\cdot\| := E[|\cdot|]$), and $E_\theta = \theta$ for $\theta \in \Theta$. In particular, if this representation of Θ is characterized (in an isomorphic way) by a subset of a topological space, it can be replaced by the closure without changing the resulting sublinear functional.

Theorem 3.7 (Daniell-Stone). *Suppose $\mathcal{H}$ is a space of integrands on a ground set Ω and $I : \mathcal{H} \to \mathbb{R}$ is a monotone, linear functional such that*

$$\lim_{n \to \infty} I(f_n) = 0$$

for each $(f_n)_n \subset \mathcal{H}$ with $f_n \downarrow 0$ as $n \to \infty$, i.e. $f_1 \geq f_2 \geq \ldots \geq 0$ and $\lim_{n \to \infty} f_n(x) = 0$ for all $x \in \Omega$. There exists a unique measure μ on $\sigma(\mathcal{H})$ (i.e. the smallest σ-algebra on Ω such that all functions in $\mathcal{H}$ are Borel-measurable) such that

$$I(f) = \int_\Omega f(x)\,\mu(dx)$$

for all $f \in \mathcal{H}$. Furthermore, if $I : \mathcal{H} \to \mathbb{R}$ is even a linear expectation (i.e. it also constant-preserving), then $\mu : \sigma(\mathcal{H}) \to [0, \infty)$ is a probability measure.

Proof. $\hookrightarrow$ [46, Theorem 4.5.2, p. 144] □

Linear functionals as in Theorem 3.7 were studied extensively by Percy Daniell in [37] (as an alternative way towards integration) and are therefore often referred to as *Daniell integrals*. The relation between Daniell and Lebesgue integrals from Theorem 3.7 was first obtained by Marshall Stone in [122].

Corollary 3.8 (Representation of Sublinear Expectations). *Let $E : \mathcal{H} \to \mathbb{R}$ be a sublinear expectation on Ω such that*

$$\lim_{n \to \infty} E(X_n) = 0$$

for $(X_n)_n \subset \mathcal{H}$ with $X_n \downarrow 0$ as $n \to \infty$, i.e. $X_1 \geq X_2 \geq \ldots \geq 0$ and $\lim_{n \to \infty} X_n(\omega) = 0$ for all $\omega \in \Omega$. There exists a family $\{\mu_\theta : \theta \in \Theta\}$ of probability measures on $\sigma(\mathcal{H})$ (i.e. the smallest σ-algebra on Ω such that all functions in $\mathcal{H}$ are Borel measurable) with

$$E(X) = \sup_{\theta \in \Theta} \int_\Omega X(\omega)\,\mu_\theta(d\omega)$$

for all $X \in \mathcal{H}$. Moreover, for each $X \in \mathcal{H}$ there exists $\theta_X \in \Theta$ such that $E(X) = \int_\Omega X(\omega)\,\mu_{\theta_X}(d\omega)$ holds. A family $(\mu_\theta)_{\theta \in \Theta}$ of probability measures related to such a representation is often referred to as uncertainty subset *of the sublinear expectation.*

Proof. The statement is a combination of Theorem 3.5 and Theorem 3.7 with $I = E_\theta$. □

3. Sublinear Expectations

It is interesting to note that, if a sublinear expectation $E : \mathcal{H} \to \mathbb{R}$ is defined on a space $\mathcal{H}$ of bounded, measurable functions, then [55, Theorem A.33, p. 394] shows that we can always obtain a representation of the form

$$E(X) = \sup_{\theta \in \Theta} \int_\Omega X(\omega)\, \mu_\theta(d\omega)$$

for a family $(\mu_\theta)_{\theta \in \Theta}$ of finitely additive measures – even without the additional continuity assumption from Corollary 3.8. Moreover, if $\mathcal{H}$ consists of bounded, Borel-measurable functions on a Polish space Ω, the continuity condition in Corollary 3.8 to obtain σ-additive (and not only finitely additive) measures can be weakened to sequences $(X_n)_{n \in \mathbb{N}} \subset C_b(\Omega)$ of continuous functions, according to [55, Theorem 4.22, p. 171].

Remark 3.9 (Dominated Convergence). The classical monotone convergence theorem and the interchangeability of suprema show that sublinear expectations, which can be represented as Lebesgue integrals with respect to (σ-additive) probability measures, are continuous from below, in the sense that

$$E\left(\sup_{n \in \mathbb{N}} X_n\right) = \sup_{n \in \mathbb{N}} E(X_n)$$

holds for every increasing sequence $(X_n)_{n \in \mathbb{N}} \subset \mathcal{H}$ with $\sup_{n \in \mathbb{N}} X_n \in \mathcal{H}$. Together with Corollary 3.8, this implies that sublinear expectations, which are continuous from above, are also continuous from below. Moreover, a simple calculation (as in [55, Lemma 4.16, p. 168]) shows that, in this case, sublinear expectations even satisfy a dominated convergence theorem, in the sense that

$$E\left(\lim_{n \to \infty} X_n\right) = \lim_{n \to \infty} E(X_n)$$

for every sequence $(X_n)_{n \in \mathbb{N}} \subset \mathcal{H}$ such that $\sup_{n \in \mathbb{N}} |X_n| \in \mathcal{H}$ and $\lim_{n \to \infty} X_n \in \mathcal{H}$.

Another popular notion of nonlinear integrals – the so-called *Choquet integrals* – was introduced by Gustave Choquet in [26]. A detailed introduction into the theory of Choquet integrals can be found in the monograph [42]. One important property all Choquet integrals share is their comonotonic additivity, i.e. they are additive for comonotone functions. Recall that two functions $f, g : \Omega \to \mathbb{R}$ are referred to as comonotone if

$$\big(f(\omega_1) - f(\omega_2)\big)\big(g(\omega_1) - g(\omega_2)\big) \geq 0$$

for all $\omega_1, \omega_2 \in \Omega$. According to Greco's representation theorem (as in [42, Theorem 13.2, p. 158]), which was originally obtained by Gabriele Greco in [59], sublinear expectations from Definition 3.3 are Choquet integrals if they are comonotonic additive and continuous from below, i.e. $E(\sup_{n \in \mathbb{N}} X_n) = \sup_{n \in \mathbb{N}} E(X_n)$ holds for every increasing sequence $(X_n)_{n \in \mathbb{N}} \subset \mathcal{H}$ with $\sup_{n \in \mathbb{N}} X_n \in \mathcal{H}$. Note, however, that not every sublinear expectation is comonotonic additive, cf. e.g. [109, Section 2.2.3.2, p. 60].

3.2. Important Distributions

One of the main contribution by Shige Peng in [102] was to obtain an analog of the central limit theorem for sublinear expectations using intrinsic notions for the distribution, independence and convergence of random variables. Interestingly, the proof of this novel central limit theorem did not rely on any results from classical probability theory. In this section, we present partly revised versions of these notions for sublinear expectations and recall the exact statement of the resulting limit theorem (in a slightly generalized version from [103]), including a sketch for the main idea of its proof.

In order to understand the main challenge of the following definitions, note that it is an important issue for introducing a notion of ($\mathcal{S}$-valued) random variables

$$X : \Omega \longrightarrow \mathcal{S}$$

and their distributions, for which functions $\varphi : \mathcal{S} \to \mathbb{R}$ expressions of the form $E(\varphi(X))$ are well-defined. This is due to the fact that sublinear expectations $(\Omega, \mathcal{H}, E)$ are (in general) only defined on a space of integrands $\mathcal{H}$ in Definition 3.3.

Definition 3.10 (Random Variables & Test Functions). Suppose $\mathcal{S}$ is an arbitrary set, which is usually referred to as the *state space*. An ($\mathcal{S}$-valued) *random variable* on a sublinear expectation space $(\Omega, \mathcal{H}, E)$ is a map

$$X : \Omega \longrightarrow \mathcal{S}$$

with its associated (maximal) set of test functions

$$\mathcal{T}_X := \{\varphi : \mathcal{S} \to \mathbb{R} \mid \varphi \circ X \in \mathcal{H}\}.$$

It is called *adapted* to a family of test functions $\mathcal{T}$ (i.e. a space of integrands on the state space $\mathcal{S}$ as in Definition 3.1) if $\mathcal{T} \subset \mathcal{T}_X$, or equivalently $\varphi \circ X \in \mathcal{H}$ for all $\varphi \in \mathcal{T}$. Given a family of test functions $\mathcal{T}$ on the state space $\mathcal{S}$,

$$L(\Omega, \mathcal{H}; \mathcal{S}, \mathcal{T}) := \{X : \Omega \to \mathcal{S} \mid \mathcal{T} \subset \mathcal{T}_X\} = \{X : \Omega \to \mathcal{S} \mid \forall \varphi \in \mathcal{T} : \varphi \circ X \in \mathcal{H}\}$$

denotes the space of all random variables that are adapted to $\mathcal{T}$.

Definition 3.11 (Distributions). The *distribution* of $X \in L(\Omega, \mathcal{H}; \mathcal{S}, \mathcal{T})$ (i.e. a random variable $X : \Omega \to \mathcal{S}$ on a sublinear expectation space $(\Omega, \mathcal{H}, E)$ adapted to a family of test functions $\mathcal{T}$) is the sublinear expectation space $(\mathcal{S}, \mathcal{T}, E_X)$ with

$$E_X(\varphi) := E(\varphi(X))$$

for all test functions $\varphi \in \mathcal{T}$.

The intrinsic notion of distributions in Definition 3.11 implies in a trivial way that for every (sublinear) distribution, there exists a sublinear expectation and a random variable with that specific distribution: Simply use the distribution itself as the sublinear expectation space and the identity map as the random variable. Similar to classical probability theory, it therefore often suffices to study distributions irregardless of their underlying sublinear expectation spaces.

3. Sublinear Expectations

Remark 3.12 (Equivalence in Distribution). Note that Definition 3.11 (similar to classical probability theory) already determines what it means for two random variables (not necessarily defined on the same sublinear expectation space) to have the same distribution with respect to a fixed family of test functions $\mathcal{T}$ on $\mathcal{S}$:

If $X : \Omega \to \mathcal{S}$ and $\hat{X} : \hat{\Omega} \to \mathcal{S}$ are two $\mathcal{S}$-valued random variables on sublinear expectation spaces $(\Omega, \mathcal{H}, E)$ and $(\hat{\Omega}, \hat{\mathcal{H}}, \hat{E})$ respectively, then they have the same distribution, or short $X \sim \hat{X}$, if their distributions (w.r.t. the family of test functions $\mathcal{T}$) coincide, i.e.

$$E(\varphi(X)) = E_X(\varphi) = \hat{E}_{\hat{X}}(\varphi) = \hat{E}(\varphi(\hat{X}))$$

for all $\varphi \in \mathcal{T}$. It is important to realize that this notion significantly depends on the choice of the family of test functions: If the family $\mathcal{T}$ is too small (e.g. only contains continuous functions), the corresponding distributions sometimes do not contain enough information on the random variables. In fact, many tightness arguments, that are frequently used in classical probability theory, to show that the distribution with respect to $\mathcal{T}$ is already uniquely determined by a smaller subfamily of test functions $\hat{\mathcal{T}} \subset \mathcal{T}$, do not work in many interesting examples, cf. [40, Section 2] and [98, Example 5.1].

Definition 3.13 (Independence). A random variable $X : \Omega \to \mathcal{S}_X$ is said to be *independent* of another random variable $Y : \Omega \to \mathcal{S}_Y$ (or short $X \perp\!\!\!\perp Y$) on a sublinear expectation space $(\Omega, \mathcal{H}, E)$ with respect to a family of test functions $\mathcal{T}$ on $\mathcal{S}_X \times \mathcal{S}_Y$ if

$$E(\varphi(X,Y)) = E(E(\varphi(X,y))|_{y=Y})$$

for all $\varphi \in \mathcal{T}$. In order for all terms to be well-defined, this implicitly assumes that $\mathcal{T} \subset \mathcal{T}_{(X,Y)}$, $\varphi(\,\cdot\,,y) \in \mathcal{T}_X$ and $E(\varphi(X,\cdot\,)) \in \mathcal{T}_Y$ for all $\varphi \in \mathcal{T}$ and $y \in \mathcal{S}_Y$.

Note that the construction of independent random variables for sublinear expectations can be obtained by a similar product construction as in classical probability theory, using the right-hand side of the defining equation for independence (cf. [105, Section 1.3, p. 6]).

Remark 3.14 (Symmetry of Independence). In contrast to classical probability theory, the independence of random variables for sublinear expectations is (in general) not symmetric anymore: Suppose that $X, Y : \Omega \to \mathbb{R}$ are random variables on a sublinear expectation space $(\Omega, \mathcal{H}, E)$ such that $X, X^2, XY^2 \in \mathcal{H}$ with

$$E(|X|) > 0 = E(X) = E(-X)$$
$$E(X^2) > -E(-X^2)$$

and $X \sim Y$, i.e. X and Y have the same distribution. If $X \perp\!\!\!\perp Y$, then

$$\begin{aligned}E(YX^2) &= E\left(E(yX^2)|_{y=Y}\right) = E\left(Y^+E(X^2) + Y^-E(-X^2)\right)\\ &= E\left(Y^+\left(E(X^2) - E(-X^2)\right) - YE(-X^2)\right)\\ &= E(Y^+)\left(E(X^2) - E(-X^2)\right)\\ &= \tfrac{1}{2}E(|Y|)\left(E(X^2) - E(-X^2)\right) > 0\end{aligned}$$

using $Y \sim X$, the identity $Y^+ = \tfrac{1}{2}(|Y| + Y)$, and the translation invariance (vii) and

3.2. Important Distributions

homogeneity property (viii) for sublinear expectations from Lemma 3.4. If, on the other hand, $Y \perp\!\!\!\perp X$, a similar calculation shows that

$$E(YX^2) = E\left(E(Yx^2)|_{x=X^2}\right) = E\left(E(Y)X^2\right) = 0,$$

which implies that we can have either $X \perp\!\!\!\perp Y$ or $Y \perp\!\!\!\perp X$, but not both.

As a next step, we introduce the family of stable distributions and their connection to certain nonlinear partial differential equations. Similar to classical probability theory, stable distributions are the limiting distributions of the generalized central limit theorem for sublinear expectations (as we will see later in Theorem 3.18).

Definition 3.15 (Stable Distributions). A d-dimensional real-valued random variable $X : \Omega \to \mathbb{R}^d$ on a sublinear expectation space $(\Omega, \mathcal{H}, E)$ is said to have a *stable distribution* with *stability index* $\alpha > 0$ if

$$aX + bX' \sim (a^\alpha + b^\alpha)^{1/\alpha} X$$

holds for all $a, b \geq 0$ and independent copies $X' : \Omega \to \mathbb{R}^d$, i.e. $X' \sim X$ and $X' \perp\!\!\!\perp X$. Furthermore, stable distributions with $\alpha = 1$ are also known as *maximal distributions*, and stable distributions with $\alpha = 2$ as (sublinear) *normal distributions*.

Remark 3.16 (Characterization of Stable Distributions). Suppose that a random variable $(X, Y) : \Omega \to \mathbb{R}^d \times \mathbb{R}^d$ on a sublinear expectation space $(\Omega, \mathcal{H}, E)$ is adapted to the family $\mathcal{T} \subset C(\mathbb{R}^d \times \mathbb{R}^d)$, that contains all continuous functions $\varphi \in C(\mathbb{R}^d \times \mathbb{R}^d)$ for which there exists $p \geq 0$ and $C \geq 0$ such that

$$|\varphi(x) - \varphi(y)| \leq C(1 + |x|^p + |y|^p)|x - y|$$

holds for all $x, y \in \mathbb{R}^d \times \mathbb{R}^d$. Moreover, assume that the stability

$$a(X, Y) + b(X', Y') \sim (\sqrt{a^2 + b^2} X, (a + b)Y)$$

holds for all $a, b \geq 0$ and independent copies (X', Y') of (X, Y) (with respect to $\mathcal{T}$). In particular, X is normally distributed and Y maximally distributed. One can show (cf. [104, Proposition 4.8]) that the functions $u^\varphi : \mathbb{R}_+ \times \mathbb{R}^d \times \mathbb{R}^d \to \mathbb{R}$ with

$$u^\varphi(t, x, y) := E(\varphi(x + \sqrt{t}X, y + tY))$$

for $(t, x, y) \in \mathbb{R}_+ \times \mathbb{R}^d \times \mathbb{R}^d$ and $\varphi \in \mathcal{T}$ are the unique (viscosity) solutions of

$$u^\varphi(t, x, y) - G\left(D_y u^\varphi(t, x, y), D_x^2 u^\varphi(t, x, y)\right) = 0$$

with $u^\varphi(0, x, y) = \varphi(x, y)$ for $x, y \in \mathbb{R}^d$, where $G : \mathbb{R}^d \times \mathbb{S}^{d \times d} \to \mathbb{R}$ is defined by

$$G(p, A) := E\left(\tfrac{1}{2} X^T A X + p^T Y\right) = E\left(\tfrac{1}{2} \operatorname{tr}(XX^T A) + p^T Y\right)$$

for $p \in \mathbb{R}^d$ and $A \in \mathbb{S}^{d \times d}$. The central idea is to establish and utilize a dynamic programming principle (which easily follows from the stability of (X,Y))

$$u^\varphi(t+s,x,y) = E(u^\varphi(t, x+\sqrt{s}X, y+sY))$$

for $t, s \geq 0$ and $x, y \in \mathbb{R}^d$, and then exploit the regularity of $u^\varphi : \mathbb{R}_+ \times \mathbb{R}^d \times \mathbb{R}^d \to \mathbb{R}$ with Taylor's theorem. Since the functional $G : \mathbb{R}^d \times \mathbb{S}^{d \times d} \to \mathbb{R}$ is sublinear, Theorem 3.5 implies the existence of a family $(b_\alpha, \sigma_\alpha) \subset \mathbb{R}^d \times \mathbb{S}^{d \times d}$ such that

$$G(p, A) = \sup_{\alpha \in \mathcal{A}} \left(b_\alpha^T p + \tfrac{1}{2} \operatorname{tr}(\sigma_\alpha \sigma_\alpha^T A) \right)$$

holds for all $p \in \mathbb{R}^d$ and $A \in \mathbb{S}^{d \times d}$. In particular, maximally distributed Y are uniquely characterized by a bounded subset of $(b_\alpha)_{\alpha \in \mathcal{A}} \subset \mathbb{R}^d$ and normally distributed X by a bounded subset of $(\sigma_\alpha)_{\alpha \in \mathcal{A}} \subset \mathbb{S}^{d \times d}$ via the unique solutions

$$u^\varphi(t, x) := E(\varphi(x + \sqrt{t}X))$$
$$v^\varphi(t, y) := E(\varphi(y + tY))$$

for $t \geq 0$ and $x, y \in \mathbb{R}^d$ of the Hamilton-Jacobi-Bellman equations (as in Definition 2.27)

$$u^\varphi(t, x) - \sup_{\alpha \in \mathcal{A}} \left(\tfrac{1}{2} \operatorname{tr}(\sigma_\alpha \sigma_\alpha^T D_x^2 u^\varphi(t, x)) \right) = 0$$
$$v^\varphi(t, y) - \sup_{\alpha \in \mathcal{A}} \left(b_\alpha^T D_y v^\varphi(t, y) \right) = 0$$

in the viscosity sense, where $u^\varphi(0, \cdot) = v^\varphi(0, \cdot) = \varphi \in C(\mathbb{R}^d)$ such that

$$|\varphi(x) - \varphi(y)| \leq C(1 + |x|^p + |y|^p)|x - y|$$

holds for all $x, y \in \mathbb{R}^d$ with some $C > 0$ and $p \geq 0$.

The preceding remark shows, why it is reasonable to refer to stable distributions with $\alpha = 2$ as (sublinear) normal distributions: For linear expectations, they correspond to the classical heat equation, and therefore to classical normal distributions (with mean zero). Similarly, linear maximal distributions (i.e. stable distributions with $\alpha = 1$) correspond to deterministic distributions (i.e. almost surely constant random variables).

Definition 3.17 (Convergence in Distribution). A sequence $(X^n)_{n \in \mathbb{N}}$ of random variable $X_n : \Omega^n \to \mathcal{S}$ on sublinear expectation spaces $(\Omega^n, \mathcal{H}^n, E^n)$ (adapted to a family $\mathcal{T}$ of test functions) is said to *converge in distribution* (with respect to $\mathcal{T}$) if

$$E(\varphi(X)) = \lim_{n \to \infty} E^n(\varphi(X^n))$$

for all test functions $\varphi \in \mathcal{T}$ and some random variable $X : \Omega \to \mathcal{S}$ (adapted to $\mathcal{T}$) on a sublinear expectation space $(\Omega, \mathcal{H}, E)$.

Although the classical tool of characteristic functions for proving a central limit theorem for linear expectations is not readily available, Shige Peng showed in [102] that it is possible to use regularity results for the characterizing equations from Remark 3.16 instead, in order to obtain the following central limit theorem for sublinear expectations:

3.2. Important Distributions

Theorem 3.18 (Central Limit Theorem & Law of Large Numbers). *Suppose that $(X_n, Y_n)_{n \in \mathbb{N}}$ is a sequence of random variables*

$$(X_n, Y_n) : \Omega \longrightarrow \mathbb{R}^d \times \mathbb{R}^d$$

on a sublinear expectation space $(\Omega, \mathcal{H}, E)$ adapted to the family

$$\mathcal{T} := \left\{ \varphi \in C(\mathbb{R}^d \times \mathbb{R}^d) \;\middle|\; \exists p \geq 0\, \exists C > 0\, \forall x, y \in \mathbb{R}^d \times \mathbb{R}^d : \right.$$
$$\left. |\varphi(x) - \varphi(y)| \leq C(1 + |x|^p + |y|^p)|x - y| \right\}$$

of locally Lipschitz-continuous functions with polynomially growing Lipschitz constant. If the sequence $(X_n, Y_n)_{n \in \mathbb{N}}$ is independent and identically distributed, i.e.

$$(X_{n+1}, Y_{n+1}) \perp\!\!\!\perp (X_n, Y_n), \ldots, (X_1, Y_1)$$

with respect to $\mathcal{T} \times \mathcal{T}^n$ and $(X_{n+1}, Y_{n+1}) \sim (X_n, Y_n)$ with respect to $\mathcal{T}$ for all $n \in \mathbb{N}$, then

$$\lim_{n \to \infty} \left(\frac{1}{\sqrt{n}} \sum_{k=1}^{n} X_k, \frac{1}{n} \sum_{k=1}^{n} Y_k \right) = (X, Y)$$

in distribution with respect to $\mathcal{T}$, where $X : \hat{\Omega} \to \mathbb{R}^d$ is a normally distributed random variable and $Y : \hat{\Omega} \to \mathbb{R}^d$ a maximally distributed random variable (adapted to $\mathcal{T}$) with

$$\hat{E}\left(\tfrac{1}{2} X^T A X + p^T Y\right) = E\left(\tfrac{1}{2} X_1^T A X_1 + p^T Y_1\right)$$

for all $p \in \mathbb{R}^d$ and $A \in \mathbb{S}^{d \times d}$ on a sublinear expectation space $(\hat{\Omega}, \hat{\mathcal{H}}, \hat{E})$.

Proof. ↪ [104, Theorem 5.1] □

It is easy to check that the connection between stable random variables and nonlinear equations from Remark 3.16 can be extended to $\alpha \in (1, 2)$ with basically the same arguments. The resulting characterizing equations, however, are (as one can already see in classical probability theory) not local but nonlocal, i.e. they are partial integro-differential instead of differential equations as in Chapter 2. Employing the appropriate regularity theory for nonlocal equations, Erhan Bayraktar and Alexander Munk showed in [13] that the proof of the preceding central limit theorem can be lifted to this generalized context. The weak law of large numbers, which is contained in the proof of Theorem 3.18, can be viewed as a special instance of this more general central limit theorem for stable distributions. Moreover, it is interesting for applications, that Zengjing Chen recently showed in [25], that it is also possible to prove a strong law of large numbers for sublinear expectations using Choquet integral methods.

3. Sublinear Expectations

Triggered by the original work [102] from Shige Peng on the generalization of the central limit theorem to sublinear expectation spaces, numerous successful attempts were made to obtain additional results from classical probability theory in this new context. An overview over the recent developments in this area can be found in the draft monograph [105] from Shige Peng. In the rest of this thesis, we restrict our attention on the study of stochastic processes for sublinear expectations: More precisely, we will generalize the relation between stable distributions and nonlinear nonlocal equations from Remark 3.16 in Chapter 4, in order to construct and characterize a broad class of spatially inhomogeneous jump processes on sublinear expectation spaces.

4. Stochastic Processes

Stochastic processes related to nonlinear expectations were studied and applied in the past using a variety of different approaches, including stochastic control and game theory. In the pioneering work [102], Shige Peng introduced an intrinsic approach to stochastic processes for sublinear expectations (in the sense that it does a-priorily not depend on a classical probability space) and constructed stochastic processes with continuous paths, called G-Brownian motions, which possess independent, stationary increments and whose marginals are the limiting distributions of the generalized central limit theorem for sublinear expectations from Section 3.2. His main idea was to extend the strong connection from Remark 3.16, between normal distributions for sublinear expectations and certain nonlinear partial differential equations, from single distributions to stochastic processes. This approach was later lifted to jump processes (for sublinear expectations) with independent, stationary increments, called G-Lévy processes, by Mingshang Hu and Shige Penge in [67], and substantially advanced by Ariel Neufeld and Marcel Nutz in [93]. In this chapter, we construct and characterize an even broader class of spatially inhomogeneous jump processes, called Lévy-type processes for sublinear expectations, as a special case of Markov processes for sublinear expectations, and study related nonlinear nonlocal equations. Such processes could be interpreted as Lévy-type processes under uncertainty in their characteristics and might also be referred to as G-Lévy-type processes for historical reasons. The related results seem to be interesting not only from a theoretical standpoint, but also for applications in fields such as mathematical finance, where classical Lévy-type processes (without uncertainty) are frequently used to model asset and bond prices: For a thorough introduction into financial modeling with jump processes, we refer the interested reader to [31].

This chapter is organized as follows: In Section 4.1, we introduce the concept of sublinear Markov semigroups and Markov processes for sublinear expectations, and show how the characterization via generator equations from classical probability theory can be extended to this generalized set-up. After that, we construct a broad class of sublinear Markov semigroups in Section 4.2 and prove the explicit form of their generators. In particular, we recover the construction of Lévy processes for sublinear expectations from [93] as a special case. In Section 4.3 and Section 4.4, we demonstrate how stochastic integration and stochastic differential equation theory for classical probability spaces can be lifted to sublinear expectations. Moreover, we combine the results from the first three sections at the end of Section 4.4, in order to construct and characterize a large class of Lévy-type processes for sublinear expectations (with bounded coefficients).

Since we study stochastic processes with jumps, let us quickly recall the common choice for the path space of jump processes, that we will adopt throughout this chapter, and its associated topology, which makes it a Polish space (and henceforth separable):

4. Stochastic Processes

Definition 4.1 (Skorokhod Space). The (d-dimensional) *Skorokhod space* is the family of right-continuous functions with finite left limits

$$D(\mathbb{R}_+) = D(\mathbb{R}_+; \mathbb{R}^d)$$
$$= \{\omega : \mathbb{R}_+ \to \mathbb{R}^d \mid \forall s \in \mathbb{R}_+ : \lim_{t \to s+} \omega(t) = \omega(s) \text{ and } \lim_{t \to s-} \omega(t) \in \mathbb{R}^d \text{ exists}\},$$

where $\lim_{t \to 0-} \omega(t) := \omega(0)$ for $\omega \in D(\mathbb{R}_+; \mathbb{R}^d)$, and whose elements are referred to as *càdlàg functions* (French "continue à droite, limite à gauche"). Furthermore, let

$$D_x(\mathbb{R}_+) = D_x(\mathbb{R}_+; \mathbb{R}^d) = \{\omega \in D(\mathbb{R}_+; \mathbb{R}^d) \mid \omega(0) = x\}$$

denote the subspace of all càdlàg functions starting at $x \in \mathbb{R}^d$.

Lemma 4.2 (Skorokhod Topology). *There exists a metrizable topology on $D(\mathbb{R}_+)$, called the* Skorokhod topology, *which makes it a Polish space (i.e. it is complete and separable). A sequence $(\omega_n)_{n \in \mathbb{N}} \subset D(\mathbb{R}_+)$ converges to $\omega \in D(\mathbb{R}_+)$ with respect to this topology if and only if there exists a sequence $(\lambda_n)_{n \in \mathbb{N}}$ of strictly increasing functions*

$$\lambda_n : \mathbb{R}_+ \longrightarrow \mathbb{R}_+$$

with $\lim_{t \to \infty} \lambda_n(t) = \infty$ such that $\lim_{n \to \infty} \sup_{t \geq 0} |\lambda_n(t) - t| = 0$ and

$$\lim_{n \to \infty} \sup_{t \in [0,T]} |\omega_n(\lambda_n(t)) - \omega(t)| = 0$$

for all $T > 0$. Moreover, the corresponding Borel σ-algebra $\mathcal{B}(D(\mathbb{R}_+))$ is generated by the family $(\pi_t)_{t \geq 0}$ of projections $\pi_t : D(\mathbb{R}_+) \to \mathbb{R}^d$ with $\phi_t(\omega) = \omega(t)$.

Proof. ↪ [73, Chapter 6, Section 1b, p. 327], [18, Chaper 3, Section 14, p. 109] □

In order to clarify the notation, let us fix what we mean, when we talk about a stochastic process for sublinear expectations in this thesis. In accordance with our notion of random variables from Chapter 3, we introduce stochastic processes on sublinear expectation spaces as path mappings into the Skorokhod space:

Definition 4.3 (Stochastic Processes). A random variable $X : \Omega \to D(\mathbb{R}_+; \mathbb{R}^d)$ on a sublinear expectation space $(\Omega, \mathcal{H}, E)$ is called a (d-dimensional) *stochastic process for sublinear expectations* (with càdlàg paths). The index set $\mathbb{R}_+$ is referred to as *time*.

Remember that for sublinear expectations (as discussed in Section 3.2), we have to pay special attention to the family of adapted test functions for a stochastic process

$$X : \Omega \longrightarrow D(\mathbb{R}_+; \mathbb{R}^d)$$

on $(\Omega, \mathcal{H}, E)$, i.e. for which $\varphi : D(\mathbb{R}_+; \mathbb{R}^d) \to \mathbb{R}$ expressions of the form $E(\varphi(X))$ are well-defined. As we will see in Section 4.2, this issue often causes additional nontrivial problems for sublinear expectations compared to classical probability theory.

4.1. Markov Processes

Markov processes are stochastic processes for which, broadly speaking, the future evolution at any given time only depends on its current position and not on its past evolution. These memoryless processes were first introduced in classical probability theory (for discrete times) by Andrey Markov in [90]. During the following century, Markov processes and related concepts were studied in detail and applied in a large variety of different contexts. For a modern introduction into Markov processes in classical probability theory and a concise overview of a range of different applications, we refer the interested reader to [83] or [96, Chapter 5, p. 170] respectively.

Over the years, numerous advances were made to generalize the underlying linearity of Markov processes and their related semigroups with various different objectives in mind. In this section, we present an intrinsic generalization of (time-homogeneous) Markov processes for sublinear expectations, and obtain a connection to sublinear Markov semigroups and related generator equations similar to the classical linear theory. It is interesting to note that the resulting sublinear Markov semigroups in this section are closely related to Nisio semigroups, which were originally introduced in the context of stochastic control theory by Makiko Nisio in [95]. A modern account on stochastic control theory (for processes with continuous paths) and its connection to Nisio semigroups (of local operators) can be found in [54].

Definition 4.4 (Markov Semigroups). A family $(T_t)_{t\geq 0}$ of sublinear operators on $\mathcal{H}$ (i.e. $T_t(u+v) \leq T_t(u) + T_t(v)$ and $T_t(\alpha u) = \alpha T_t(u)$ for all $\alpha, t \in \mathbb{R}_+$ and $u, v \in \mathcal{H}$)

$$T_t : \mathcal{H} \longrightarrow \mathcal{H}$$

with a convex cone $\mathcal{H}$ of real-valued functions on $\mathbb{R}^d$ that contains all constant functions (i.e. $(\alpha u + \beta v) \in \mathcal{H}$ for all $\alpha, \beta \in \mathbb{R}_+$ and $u, v \in \mathcal{H}$, as well as $c \in \mathcal{H}$ for all $c \in \mathbb{R}$), which satisfies the following three properties

(i) $\forall t, s \geq 0 : T_{t+s} = T_t T_s$ and $T_0 = \mathrm{id}$ \hfill (Semigroup Property)

(ii) $\forall u, v \in \mathcal{H}\ \forall t \geq 0 : u \leq v \Longrightarrow T_t(u) \leq T_t(v)$ \hfill (Monotonicity)

(iii) $\forall c \in \mathbb{R}\ \forall t \geq 0 : T_t(c) = c,$ \hfill (Preservation of Constants)

is called a *sublinear Markov semigroup* on $\mathcal{H}$.

Lemma 4.5 (Properties of Markov Semigroups). *A Markov semigroup $(T_t)_{t\geq 0}$ on a convex cone $\mathcal{H}$ of real-valued functions on $\mathbb{R}^d$ (that contains all constant functions) satisfies the following basic properties:*

(iv) $\forall u \in \mathcal{H}\ \forall t \geq 0 : 0 \leq u \leq 1 \Longrightarrow 0 \leq T_t(u) \leq 1$ \hfill (Sub-Markov Property)

(v) $\forall u \in \mathcal{H} : \sup_{x \in \mathbb{R}^d} |T_t(u)(x)| \leq \sup_{x \in \mathbb{R}^d} |u(x)|$ \hfill (Contractivity)

(vi) $\forall u \in \mathcal{H}\ \forall c \in \mathbb{R}\ \forall t \geq 0 : T_t(u + c) = T_t(u) + c$ \hfill (Translation Invariance)

4. Stochastic Processes

Moreover, if $\mathcal{H}$ is a space of integrands on $\mathbb{R}^d$ (as in Definition 3.1), then the functional

$$\mathcal{H} \ni u \longmapsto T_t(u)(x) \in \mathbb{R}$$

is a sublinear expectation (as in Definition 3.3) for every $t \geq 0$ and $x \in \mathbb{R}^d$.

Proof. The sub-Markov property (iv) and contractivity (v) are an obvious consequence of the monotonicity (ii) and preservation of constants (iii). The translation invariance (vi) can be proved exactly as for sublinear expectations in Lemma 3.4, i.e. for $\varphi \in \mathcal{H}$, $c \in \mathbb{R}$ and $t \geq 0$ the sublinearity of $T_t : \mathcal{H} \to \mathcal{H}$ and the constant preservation (iii) show

$$T_t(\varphi) + c = T_t(\varphi) - T_t(-c) \leq T_t(\varphi - (-c)) = T_t(\varphi + c) \leq T_t(\varphi) + T_t(c) = T_t(\varphi) + c.$$

Finally, it is easy to see from Definition 3.3 that $u \longmapsto T_t(u)(x)$ for fixed $t \geq 0$ and $x \in \mathbb{R}^d$ is (as a monotone and constant-preserving sublinear functional) a sublinear expectation, if $\mathcal{H}$ is a space of integrands on $\mathbb{R}^d$. □

Note that the translation invariance assumption (vi) for sublinear Markov semigroups is completely unrelated to its spatial homogeneity. The term "translation invariance" is common in the literature on sublinear expectations (which explains why we want to use it here) and describes a certain kind of additivity of sublinear operators.

Although our monotonicity and preservation-of-constants assumptions for sublinear Markov semigroups in Definition 4.4 are equivalent to the sub-Markov property and conservativeness assumption for linear operators (which are typically postulated for linear Markov semigroups, as in [21, Section 1.1, p. 14] for example), those assumptions are strictly stronger in case of sublinear operators. Moreover, since our Markov semigroups are families of sublinear instead of linear operators, it is reasonable to assume that the underlying spaces are only convex cones instead of linear spaces. In fact, it is impossible in some important cases (as demonstrated in [98, Section 5]) to construct sublinear Markov semigroups on a rich linear space. Nevertheless, we introduced sublinear expectations in Chapter 3 as functionals on linear spaces, in accordance with the established literature. In view of this discrepancy, it would be interesting to examine, if it is reasonable to introduce sublinear expectations as functionals on convex cones instead of linear spaces.

As a next step, we propose a novel definition of Markov processes for sublinear expectations and discuss its defining assumptions. Despite the fact, that its definition might seem abstract at first, we provide a construction and characterization for a broad class of such Markov processes in Section 4.4, which can be interpreted as Lévy-type processes under uncertainty in their characteristics and are therefore potentially valuable for applications in mathematical finance.

Definition 4.6 (Markov Processes). A (d-dimensional) *Markov process for sublinear expectations* is a stochastic process $X : \Omega \to D(\mathbb{R}_+; \mathbb{R}^d)$ on a family of sublinear expectation spaces $(\Omega, \mathcal{H}, E^x)_{x \in \mathbb{R}^d}$ together with a family of test functions $\mathcal{T}$ on $\mathbb{R}^d$ (as in Definition 3.10) such that $X_t : \Omega \to \mathbb{R}^d$ are adapted to $\mathcal{T}$ for each $t \geq 0$, and such that

$$T_t\varphi(x) := E^x(\varphi(X_t))$$

with $t \geq 0$, $\varphi \in \mathcal{T}$ and $x \in \mathbb{R}$ defines a sublinear Markov semigroup on $\mathcal{T}$.

4.1. Markov Processes

Remark 4.7 (Notion of Markov Processes). Note that in Definition 4.6 of Markov processes $(X_t)_{t\geq 0}$ for sublinear expectations, the corresponding family $(T_t)_{t\geq 0}$ with

$$T_t\varphi(x) = E^x(\varphi(X_t))$$

consists of sublinear operators such that the monotonicity (ii) and preservation of constants (iii) from Definition 4.4 are obviously satisfied, due to the defining properties of the sublinear expectation spaces $(\Omega, \mathcal{H}, E^x)_{x\in\mathbb{R}}$. The crucial assumptions thus stem from the semigroup property (i) of Definition 4.4: First of all, $T_0 = \mathrm{id}$ is equivalent to

$$E^x\varphi(X_0) = T_0\varphi(x) = \mathrm{id}\,\varphi(x) = \varphi(x)$$

for all $x \in \mathbb{R}$ and $\varphi \in \mathcal{T}$, which means that $X_0 \sim x$ under E^x or, in other words, the process $(X_t)_{t\geq 0}$ starts (without uncertainty) at $X_0 = x$ under E^x. Secondly, $T_{t+s} = T_t T_s$ for $t,s \geq 0$ is equivalent to the so-called Markov property

$$E^x\big(\varphi(X_{t+s})\big) = T_{t+s}\varphi(x) = T_t T_s\varphi(x) = T_t\big(y \mapsto E^y\big(\varphi(X_s)\big)\big)(x) = E^x\big(E^{X_t}\big(\varphi(X_s)\big)\big)$$

for all $x \in \mathbb{R}$ and $\varphi \in \mathcal{T}$, which means that the behavior of the process remains unchanged if it is stopped at a fixed time $t \geq 0$ and restarted at the stopping point X_t. Note that it is important that $T_s\varphi \in \mathcal{T}$ for all $s \geq 0$ and $\varphi \in \mathcal{T}$, in order to be able to make such calculations rigorous.

Corollary 4.8 (Characterization of Markov Processes). *Suppose that*

$$X : \Omega \longrightarrow D(\mathbb{R}_+; \mathbb{R}^d)$$

is a (d-dimensional) stochastic process on a family of sublinear expectation spaces

$$(\Omega, \mathcal{H}, E^x)_{x \in \mathbb{R}}$$

and $\mathcal{T}$ is a family of test functions on $\mathbb{R}^d$, such that the marginals $X_t : \Omega \to \mathbb{R}^d$ are adapted to $\mathcal{T}$ for all $t \geq 0$. If $E^x(\varphi(X_0)) = \varphi(x)$ for all $x \in \mathbb{R}$ and $\varphi \in \mathcal{T}$,

$$\big(x \mapsto E^x\big(\varphi(X_t)\big)\big) \in \mathcal{T}$$

for all $t \geq 0$ and $\varphi \in \mathcal{T}$, as well as

$$E^x\big(\varphi(X_{t+s})\big) = E^x\big(E^{X_t}\big(\varphi(X_s)\big)\big)$$

for all $x \in \mathbb{R}$, $t,s \geq 0$ and $\varphi \in \mathcal{T}$, then $(X_t)_{t\geq 0}$ is a Markov process for sublinear expectations (as in Definition 4.6).

In the rest of this section, we generalize the connection between linear Markov semigroups and their generators (see [21, Section 1.4, p. 28] for example) to our sublinear setting. We want to remind the reader that this connection is one of the important reasons, why Markov processes are interesting for applications, since it allows us to evaluate related quantities by solving the generator equation analytically or numerically.

4. Stochastic Processes

Definition 4.9 (Generators). Suppose that $(T_t)_{t\geq 0}$ is a sublinear Markov semigroup on a convex cone $\mathcal{H}$ of real-valued functions on $\mathbb{R}^d$ that contains all constant functions. The *sublinear generator* $A : \mathcal{D}(A) \to \mathcal{H}$ of $(T_t)_{t\geq 0}$ is defined as

$$Au(x) := \lim_{\delta \downarrow 0} \frac{T_\delta u(x) - T_0 u(x)}{\delta} = \lim_{\delta \downarrow 0} \frac{T_\delta u(x) - u(x)}{\delta}$$

for all $x \in \mathbb{R}^d$ and $u \in \mathcal{D}(A) \subset \mathcal{H}$, where

$$\mathcal{D}(A) := \left\{ u \in \mathcal{H} \,\middle|\, \exists \varphi \in \mathcal{H} \,\forall x \in \mathbb{R}^d : \varphi(x) = \lim_{\delta \downarrow 0} \frac{T_\delta u(x) - u(x)}{\delta} \right\}$$

is the domain of the sublinear operator A.

Since sublinear Markov semigroups $(T_t)_{t\geq 0}$ are defined on arbitrary convex cones $\mathcal{H}$, it is difficult to compare our definition of sublinear generators to classical Markov generators in general. However, if $(T_t)_{t\geq 0}$ is a classical Feller semigroup of linear operators (i.e. $\mathcal{H} = C_\infty(\mathbb{R}^d)$ is the linear space of continuous functions vanishing at infinity and

$$\lim_{t \downarrow 0} \|T_t \varphi - \varphi\|_\infty = 0$$

for all $\varphi \in \mathcal{H} = C_\infty(\mathbb{R}^d)$), then the sublinear generator $A : \mathcal{D}(A) \to \mathcal{H}$ coincides with the weak or pointwise generator from the literature (cf. e.g. [21, Definition 1.32, p. 33]). It is well-known (cf. [21, Theorem 1.33, p. 34]) that the weak generator (i.e. defined with respect to pointwise convergence) of linear Feller semigroups coincides with the strong, classical Markov generator (i.e. defined with respect to uniform convergence). Moreover, [21, Theorem 2.21, p. 56] shows that $C_c^2(\mathbb{R}^d) \subset \mathcal{D}(A)$ implies that the generator of a linear Feller semigroup is a Hamilton-Jacobi-Bellman operator with measurable coefficients, as introduced in Section 2.3 (cf. also the discussion after Definition 2.27). It seems interesting to study, if it is possible to relate the uniform and pointwise convergence for sublinear generators similarly, and maybe also obtain a general representation for sublinear generators as Hamilton-Jacobi-Bellman operators. However, since those two ideas have no direct impact on the applicability of our current results, we leave it for future work to study if they can be generalized to our sublinear setting.

Proposition 4.10 (Generator Equations). *Assume that $(T_t)_{t\geq 0}$ is a Markov semigroup on a convex cone $\mathcal{H}$ of real-valued functions on $\mathbb{R}^d$ containing all constant functions and $A : \mathcal{D}(A) \to \mathcal{H}$ its sublinear generator. If $C_b^\infty(\mathbb{R}^d) \subset \mathcal{D}(A)$ and $\varphi \in \mathcal{H}$ such that*

$$u^\varphi : \mathbb{R}_+ \times \mathbb{R}^d \longrightarrow \mathbb{R}$$
$$(t, x) \longmapsto u^\varphi(t, x) := T_t \varphi(x)$$

is continuous, then u^φ is a viscosity solution in $(0, \infty) \times \mathbb{R}^d$ of

$$\partial_t u^\varphi(t, x) - A(u^\varphi(t, \cdot))(x) = 0$$

with $u(0, x) = \varphi(x)$ for all $x \in \mathbb{R}^d$, in the following sense: For all $\psi \in C_b^\infty(\mathbb{R}_+ \times \mathbb{R}^d)$ such that $u^\varphi - \psi$ has a global maximum (global minimum) in $(t, x) \in (0, \infty) \times \mathbb{R}^d$ with $u^\varphi(t, x) = \psi(t, x)$, the inequality $\partial_t \psi(t, x) - A(\psi(t, \cdot))(x) \leq 0$ (≥ 0) holds.

4.1. Markov Processes

Proof. First of all, since $T_0 = \text{id}$, we find $u^\varphi(0, x) = T_0 \varphi(x) = \varphi(x)$ for all $x \in \mathbb{R}^d$. Secondly, suppose that $\psi \in C_b^\infty(\mathbb{R}_+ \times \mathbb{R}^d)$ such that $\psi(t, x) = u^\varphi(t, x)$ and $u^\varphi - \psi$ has global maximum in $(t, x) \in (0, \infty) \times \mathbb{R}^d$. According to the semigroup property

$$\psi(t, x) = u^\varphi(t, x) = T_t \varphi(x) = T_\delta T_{t-\delta} \varphi(x) = T_\delta(u^\varphi(t - \delta, \cdot))(x) \leq T_\delta(\psi(t - \delta, \cdot))(x)$$

holds for all $\delta > 0$. Moreover, Taylor's theorem implies that

$$\psi(t - \delta, \cdot) - \psi(t, x) = \psi(t, \cdot) - \psi(t, x) - \delta \int_0^1 \partial_t \psi(t - \xi \delta, \cdot) \, d\xi$$
$$= \psi(t, \cdot) - \psi(t, x) - \delta \partial_t \psi(t, x) + \delta(R_1(\cdot) + R_{2,\delta}(\cdot)),$$

where the remainder terms are of the form $R_1(\cdot) := \partial_t \psi(t, x) - \partial_t \psi(t, \cdot)$ and

$$R_{2,\delta}(\cdot) := \int_0^1 (\partial_t \psi(t, \cdot) - \partial_t \psi(t - \xi \delta, \cdot)) \, d\xi$$

for $\delta > 0$. Combining those two facts leads to

$$0 \leq T_\delta(\psi(t - \delta, \cdot) - \psi(t, x))(x)$$
$$\leq T_\delta(\psi(t, \cdot) - \psi(t, x))(x) - \delta \partial_t \psi(t, x) + \delta(T_\delta(R_1)(x) + T_\delta(R_{2,\delta})(x)),$$

using the subadditivity and translation invariance of $T_\delta : \mathcal{H} \to \mathcal{H}$ for $\delta > 0$. Hence,

$$\partial_t \psi(t, x) \leq \frac{T_\delta(\psi(t, \cdot) - \psi(t, x))(x)}{\delta} + T_\delta(R_1)(x) + T_\delta(R_{2,\delta})(x),$$

which would imply the desired inequality as $\delta \downarrow 0$ using the definition of the generator

$$A(\psi(t, \cdot))(x) = \lim_{\delta \downarrow 0} \frac{T_\delta(\psi(t, \cdot) - \psi(t, x))(x)}{\delta},$$

if we had $T_\delta(R_1)(x) + T_\delta(R_{2,\delta})(x) \to 0$ as $\delta \downarrow 0$. According to the mean value theorem

$$\|T_\delta(\pm R_{2,\delta})\|_\infty \leq \|R_{2,\delta}\|_\infty \leq \frac{1}{2} \delta \|\partial_t^2 \psi\|_\infty$$

for all $\delta > 0$. Moreover, since $\partial_t \psi(t, \cdot) \in C_b^\infty(\mathbb{R}^d) \subset \mathcal{D}(A)$, the definition shows

$$\lim_{\delta \downarrow 0} \delta^{-1} T_\delta(\pm R_1)(x) = \lim_{\delta \downarrow 0} \frac{T_\delta(\mp \partial_t \psi(t, \cdot))(x) - (\mp \partial_t \psi(t, x))}{\delta} = A(\mp \partial_t \psi(t, \cdot))(x),$$

which implies $T_\delta(\pm R_1)(x) \to 0$ as $\delta \downarrow 0$. Thus, $\partial_t \psi(t, x) - A(\psi(t, \cdot))(x) \leq 0$ holds.

In order to show the opposite inequality, suppose that $\psi \in C_b^\infty(\mathbb{R}_+ \times \mathbb{R}^d)$ such that $u^\varphi - \psi$ has a global minimum in $(t, x) \in (0, \infty) \times \mathbb{R}^d$ with $u^\varphi(t, x) = \psi(t, x)$. The same arguments as before lead to

$$0 \geq T_\delta(\psi(t - \delta, \cdot) - \psi(t, x))(x)$$
$$\geq T_\delta(\psi(t, \cdot) - \psi(t, x))(x) - \delta \partial_t \psi(t, x) - \delta(T_\delta(-R_1)(x) + T_\delta(-R_{2,\delta})(x)),$$

which implies $\partial_t \psi(t, x) - A(\psi(t, \cdot))(x) \geq 0$, by dividing by $\delta > 0$ and sending $\delta \downarrow 0$. □

4. Stochastic Processes

Let us compare and discuss the assumptions in the preceding statement to the results for linear Markov semigroups: The assumption $C_b^\infty(\mathbb{R}^d) \subset \mathcal{D}(A)$ and the continuity of

$$(t, x) \longmapsto T_t\varphi(x)$$

for fixed $\varphi \in \mathcal{H}$ can be understood as a generalization of the classical C_b-Feller property (which assumes that $T_t C_b(\mathbb{R}^d) \subset C_b(\mathbb{R}^d)$ for all $t \geq 0$, cf. [21, Definition 1.6, p. 19]) together with a richness condition for the domain of the extended generator (cf. [21, Theorem 2.37, p. 68]). Moreover, if we had a comparison principle for the viscosity solutions of the generator equation as in Corollary 2.23, we could relax the continuity assumption to functions $\varphi \in C_b^\infty(\mathbb{R}^d)$ with the following idea: If

$$(t, x) \longmapsto T_t\varphi(x)$$

was continuous for all $\varphi \in C_b^\infty(\mathbb{R}^d)$, the current proof would actually show that the upper and lower semicontinuous envelopes of

$$u^\psi(t, x) = T_t\psi(x)$$

for $\psi \in C_b(\mathbb{R}^d) \cap \mathcal{H}$ (i.e. the smallest upper or the largest lower semicontinuous function above or below $u^\psi : \mathbb{R}_+ \times \mathbb{R}^d \to \mathbb{R}$) are viscosity sub- or supersolutions respectively. An application of the comparison principle would therefore imply that the upper and lower envelope coincide, which would prove that $u^\psi : [0, \infty) \times \mathbb{R}^d \to \mathbb{R}$ is continuous. Since we typically have no uniqueness of viscosity solutions without a comparison principle, this idea leads to an interesting generalization of Proposition 4.10.

As soon as the related initial value problem has at most one viscosity solution for each initial value (e.g. due to a comparison principle), Proposition 4.10 shows that there is a one-to-one correspondence between certain sublinear Markov semigroups and their nonlinear nonlocal generator equations. This connection can not only be used to evaluate related quantities for the application of Markov processes for sublinear expectations, as advocated earlier. It also allows to derive new results for nonlinear nonlocal equations, such as the regularity of viscosity solutions for example. We therefore deem it to be useful on its own, to study if the connection from Proposition 4.10 can be generalized even further, e.g. to unbounded viscosity solutions.

It is worth mentioning that similar results to Proposition 4.10 exist in the closely related field of stochastic control theory (cf. e.g. [54, Chapter 2, Theorem 5.1, p. 72]), but they usually assume (instead of proving) that the convergence

$$\lim_{\delta \to 0} \frac{T_\delta(\psi(t - \delta, \cdot) - \psi(t, x))(x)}{\delta} = \partial_t \psi(t, x) - A(\psi(t, \cdot))(x)$$

holds for all smooth $\psi \in \mathcal{D}(A)$, and check this convergence on a case-by-case basis for a limited set of applications. Not only does this assumption simplify the proof substantially, but it is also not suitable for the general connection between sublinear Markov semigroups and their generator equations, that we want to establish here.

4.2. Construction of Processes

One of the first important, nontrivial Markov processes for sublinear expectations related to our setting from Section 4.1 was introduced in [102] by Shige Peng: The so-called G-Brownian motion $(B_t)_{t\geq 0}$ is a continuous stochastic process with independent and stationary increments, whose generator equation (as in Proposition 4.10) is of the form

$$\partial_t u(t,x) - \sup_{\alpha \in \mathcal{A}} \left(\frac{1}{2} \operatorname{tr}\left(c_\alpha D^2 u(t,x)\right) \right) = 0 \qquad (4.1)$$

for a family $(c_\alpha)_{\alpha \in \mathcal{A}} \subset \mathbb{S}_+^{d\times d}$ of positive semi-definite, symmetric matrices and is referred to as G-heat equation. Shige Peng's original approach in [102] used existence results (based on Perron's method, cf. Remark 2.24) for the G-heat equation to construct

$$[0,\infty) \times \mathbb{R}^d \ni (t,x) \longmapsto E\left(\varphi(x+B_t)\right)$$

as the unique viscosity solutions $u^\varphi : [0,\infty) \times \mathbb{R}^d \to \mathbb{R}$ of the related initial value problem with $u^\varphi(0,\cdot) = \varphi(\cdot) \in C_b^{\mathrm{Lip}}(\mathbb{R}^d)$. This approach was later generalized by Mingshang Hu and Shige Peng in [67] to nonlocal generator equations of the form

$$\partial_t u(t,x) - \sup_{\alpha \in \mathcal{A}} \left(b_\alpha Du(t,x) + \frac{1}{2} \operatorname{tr}\left(c_\alpha D^2 u(t,x)\right) \right.$$
$$\left. + \int_{\mathbb{R}^d} \left(u(t,x+z) - u(t,x) - Du(t,x) z \mathbb{1}_{|z|\leq 1} \right) F_\alpha(dz) \right) = 0$$

for a family $(b_\alpha, c_\alpha, F_\alpha)_{\alpha \in \mathcal{A}} \subset \mathbb{R}^d \times \mathbb{S}_+^{d\times d} \times \mathfrak{L}(\mathbb{R}^d)$ of Lévy triplets (cf. Remark 4.17), in order to construct stochastic jump processes with independent and stationary increments, which they refer to as G-Lévy processes. Unfortunately, it is a-priorily unclear how to extend this analytical construction of the sublinear Markov semigroup (via the generator equation) to merely measurable functions $\varphi : \mathbb{R}^d \to \mathbb{R}$. However, Laurent Denis, Mingshang Hu and Shige Peng managed to show in [40] that G-Brownian motions $(B_t)_{t\geq 0}$ can be represented by a standard Brownian motion $(W_t)_{t\geq 0}$ on a classical probability space $(\Omega, \mathcal{A}, \mathbb{P})$ (and its corresponding linear expectation $\mathbb{E}$), in the sense that

$$E\left(\varphi(x+B_t)\right) = \sup_{\theta \in \Theta} \mathbb{E}\left(\varphi\left(x + \int_0^t \theta_s \, dW_s \right) \right)$$

for $\varphi \in C_b^{\mathrm{Lip}}(\mathbb{R}^d)$, where the supremum is taken over the family Θ of all $\mathbb{S}^{d\times d}$-valued predictable integrands $\theta = (\theta_t)_{t\geq 0}$ such that

$$\theta_t(\omega) \theta_t(\omega)^T \in \bigcup_{\alpha \in \mathcal{A}} c_\alpha := \{c_\alpha : \alpha \in \mathcal{A}\} \subset \mathbb{S}_+^{d\times d}$$

holds $\lambda(dt) \times \mathbb{P}(d\omega)$-almost everywhere for the uncertainty coefficients $(c_\alpha)_{\alpha \in \mathcal{A}}$ of the corresponding G-heat equation as in Equation (4.1). This representation not only justifies the interpretation of G-Brownian motions as a generalization of classical Brownian

4. Stochastic Processes

motions under volatility uncertainty, but it allows to extend $E\left(\varphi(x + B_t)\right)$ to measurable $\varphi : \mathbb{R}^d \to \mathbb{R}$ in a canonical way. Marcel Nutz and Ramon van Handel extended this purely measure-theoretic idea in [98] to construct a broad class of continuous stochastic processes for sublinear expectations. Moreover, a similar construction and representation for G-Lévy processes was developed in [93] by Ariel Neufeld and Marcel Nutz.

In this section, we combine and generalize the results from [93] and [98] to obtain a measure-theoretic construction of semimartingales with jumps for sublinear expectations. Based on this construction, we prove the existence of a large class of (spatially inhomogeneous) sublinear Markov semigroups, and determine the explicit form of their associated generators in terms of integro-differential operators. Moreover, we prove a compactness result for the corresponding uncertainty subsets, and discuss how this result relates to the C_b-Feller property of our newly constructed sublinear Markov semigroups. In order to fix the notation and provide a nearly self-contained presentation of our results, we include some of the important definitions and statements for the semigroup construction from [92] and [93] by Ariel Neufeld and Marcel Nutz. Furthermore, we recall relevant parts of the classical semimartingale theory from the monograph [73] by Jean Jacod and Albert Shiryaev, and adopt their notions for classical probability theory throughout this chapter.

Definition 4.11 (Semimartingales). A stochastic process $(X_t)_{t \geq 0}$ with càdlàg paths on a classical probability space $(\Omega, \mathcal{A}, \mathbb{P})$ adapted to a filtration $\mathcal{F} = (\mathcal{F}_t)_{t \geq 0}$ is called a $\mathbb{P}$-$\mathcal{F}$-*semimartingale* if there exists a càdlàg $\mathbb{P}$-$\mathcal{F}$-local martingale $(M_t)_{t \geq 0}$ and a càdlàg $\mathcal{F}$-adapted bounded variation process $(A_t)_{t \geq 0}$ such that

$$X = X_0 + M + A$$

and $M_0 = A_0 = 0$ $\mathbb{P}$-almost surely.

For the following well-known characterization of semimartingales, recall that a random measure on $\mathbb{R}_+ \times \mathbb{R}^d$ is a family of Borel measures $(\nu(\omega, dt, dx))_{\omega \in \Omega}$ on $\mathbb{R}_+ \times \mathbb{R}^d$ with $\nu(\omega, \{0\} \times \mathbb{R}^d) = 0$ for all $\omega \in \Omega$ (cf. [73, Chapter 2, Section 1, p. 64]).

Remark 4.12 (Integral Processes). Suppose that ν is a random measure on $\mathbb{R}_+ \times \mathbb{R}^d$ and X is a bounded variation process. If $V : \Omega \times \mathbb{R}_+ \times \mathbb{R}^d \to \mathbb{R}$ and $W : \Omega \times \mathbb{R}_+ \to \mathbb{R}$ are measurable, then $V * \nu$ and $W * X$ denote the integral processes defined by

$$(V * \nu)(\omega, t) = \int_{[0,t] \times \mathbb{R}^d} V(\omega, s, x)\, \nu(\omega, ds, dx)$$
$$(W * X)(\omega, t) = \int_{[0,t]} W(\omega, s)\, dX(\omega, s)$$

for $(\omega, t) \in \Omega \times \mathbb{R}_+$, where the second integral is an (ω-wise defined) Lebesgue-Stieltjes integral. Recall that $V : \Omega \times \mathbb{R}_+ \times \mathbb{R}^d \to \mathbb{R}$ is called optional or predictable, if it is measurable with respect to $\mathcal{O} \otimes \mathcal{B}(\mathbb{R}^d)$ or $\mathcal{P} \otimes \mathcal{B}(\mathbb{R}^d)$ respectively, where $\mathcal{O}$ is the optional and $\mathcal{P}$ the predictable σ-algebra on $\Omega \times \mathbb{R}_+$. Consequently, a random measure on $\mathbb{R}_+ \times \mathbb{R}^d$ is referred to as optional (or predictable), if $V * \nu$ is optional (or predictable) for every

4.2. Construction of Processes

optional (or predictable) $V : \Omega \times \mathbb{R}_+ \times \mathbb{R}^d \to \mathbb{R}$. For a thorough introduction into the topic of integration with respect to random measures and bounded variation processes, see [73, Chapter 2, Section 1, p. 64] and [73, Chapter 1, Section 3, p. 27].

Definition 4.13 (Semimartingale Characteristics). Suppose $(X_t)_{t\geq 0}$ is a d-dimensional $\mathbb{P}$-$\mathcal{F}$-semimartingale and $h : \mathbb{R}^d \to \mathbb{R}^d$ is a *truncation function* (i.e. it is bounded and Borel-measurable with $h(x) \equiv x$ in a neighborhood of $x = 0$). A triplet (B, C, ν) consisting of (cf. [73, Chapter 2, Theorem 2.42, p. 86])

- a predictable, càdlàg, bounded variation process $B = (B^i)_{i \leq d}$
- a continuous, bounded variation process $C = (C^{i,j})_{i,j \leq d}$
- a predictable random measure ν on $\mathbb{R}_+ \times \mathbb{R}^d$

is called (predictable) *semimartingale characteristics* of $(X_t)_{t\geq 0}$ (associated to the truncation function $h = (h^i)_{i \leq d}$), if for every bounded $\varphi \in C^2(\mathbb{R}^d)$ the process

$$\varphi(X) - \varphi(X_0) - \sum_{j \leq d} D_j \varphi(X_-) * B^j - \frac{1}{2} \sum_{j,k \leq d} D_{jk} \varphi(X_-) * C^{jk}$$
$$- \left(\varphi(X_- + \cdot) - \varphi(X_-) - \sum_{j \leq d} D_j \varphi(X_-) h^j \right) * \nu$$

is a $\mathbb{P}$-$\mathcal{F}$-local martingale. Note that there exist slightly different notions of semimartingale characteristics in the literature (cf. [73, Chaper 2, Remark 2.8, p. 76]), especially in terms of the predictability of the related processes.

As a next step, we discuss the existence and uniqueness of semimartingale characteristics using a well-known explicit construction. This construction also motivates, why it is reasonable to refer to the three components of the characteristics (B, C, ν) as the bounded variation part B, the diffusion part C and the jump part ν.

Remark 4.14 (Construction of Characteristics). If $(X_t)_{t\geq 0}$ is a d-dimensional $\mathbb{P}$-$\mathcal{F}$-semimartingale and $h : \mathbb{R}^d \to \mathbb{R}^d$ is a truncation function, then

$$\hat{X}_t := X_t - \sum_{s \leq t} (\Delta X_s - h(\Delta X_s))$$

for $t \geq 0$ defines a (d-dimensional) $\mathbb{P}$-$\mathcal{F}$-semimartingale, because the process we subtract from $(X_t)_{t\geq 0}$ is a $\mathcal{F}$-adapted, càdlàg, bounded variation process. Since $\Delta(\hat{X}) = h(\Delta X)$, the resulting semimartingale $\hat{X}$ has bounded jumps. Therefore, it admits a unique (up to indistinguishability) Doob-Meyer decomposition (cf. [73, Chapter 1, Section 4c, p. 43])

$$\hat{X} = \hat{X}_0 + \hat{M} + \hat{A}$$

for some càdlàg $\mathbb{P}$-$\mathcal{F}$-local martingale $\hat{M}$ and some predictable, càdlàg, bounded variation process $\hat{A}$ with $\hat{M}_0 = \hat{A}_0 = 0$. If $B := \hat{A}$, C is the (predictable) covariation process of the continuous part of $\hat{M}$ (cf. [73, Chapter 1, Theorem 4.2, p. 38] or Remark 4.34) and ν is the compensator of the random measure (cf. [73, Chapter 2, Theorem 1.8, p. 66])

$$\mu^X(w;dt,dx) := \sum_{s \geq 0} \mathbb{1}_{\{\Delta X_s(\omega) \neq 0\}} \delta_{(s, \Delta X(s))}(dt, dx)$$

on $\mathbb{R}_+ \times \mathbb{R}^d$, then (cf. [73, Chapter 2, Section 2d, p. 43]) (B, C, ν) are the unique (up to indistinguishability) characteristics (associated to the truncation function h) of $(X_t)_{t \geq 0}$.

The previous remark entails that only the bounded variation part $B = (B^i)_{i \leq d}$ of the semimartingale characteristics (B, C, ν) depends on the truncation function $h : \mathbb{R}^d \to \mathbb{R}^d$. In fact, it is easy to show (cf. [73, Chapter 2, Proposition 2.24, p. 81]) that if $\hat{h} : \mathbb{R}^d \to \mathbb{R}^d$ is another truncation function, then

$$\hat{B} = B + (\hat{h} - h) * \nu$$

for the semimartingale characteristics $(\hat{B}, C, \nu)$ associated to $\hat{h} : \mathbb{R}^d \to \mathbb{R}^d$. In the rest of this section, we fix one truncation function $h : \mathbb{R}^d \to \mathbb{R}^d$, if not stated otherwise.

Many publications on semimartingales only work with filtrations $\mathcal{F} = (\mathcal{F}_t)_{t \geq 0}$ that satisfy the usual conditions (i.e. they contain all subsets of null sets and are right continuous in the sense that $\mathcal{F}_{t+} := \cap_{s \geq t} \mathcal{F}_s = \mathcal{F}_t$ holds for all $t \geq 0$). The following lemma shows how we can work with the original filtration (instead of the augmentation, cf. Remark 4.44) and still apply results from the literature that require the usual conditions.

Lemma 4.15 (Filtration of Semimartingales). *Suppose that $X : \Omega \to \mathbb{R}^d$ is a càdlàg stochastic process on a probability space $(\Omega, \mathcal{A}, \mathbb{P})$ adapted to a filtration $\mathcal{F} = (\mathcal{F}_t)_{t \geq 0}$ of sub-σ-algebras of $\mathcal{A}$. Let $\mathcal{F}_+^{\mathbb{P}}$ be the augmentation of $\mathcal{F}$ under $\mathbb{P}$ as in Remark 4.44. Then X is a $\mathbb{P}$-$\mathcal{F}$-semimartingale if and only if X is a $\overline{\mathbb{P}}$-$\mathcal{F}_+^{\mathbb{P}}$-semimartingale, and the associated semimartingale characteristics coincide $\mathbb{P}$-almost surely.*

Proof. $\hookrightarrow$ [92, Proposition 2.2] □

In order to construct an analogue of semimartingales with jumps for sublinear expectations, we have to study the measurability of semimartingale characteristics (B, C, ν) with respect to the underlying probability measure $\mathbb{P}$. Before we can formalize the related results from [92], however, we need to introduce topological structures on the set of probability measures $\mathfrak{P}(\Omega)$ and the family of semimartingale characteristics.

Remark 4.16 (Topology on Measures). It is well-known (cf. [20, Theorem 8.9.4, p. 213] and [20, Theorem 7.2.2, p. 74]) that if (X, d) is a separable metric space, so is the family of bounded (non-negative) Borel measures $\mathfrak{M}_b^+(X)$ with respect to the topology associated to $C_b(X)$-weak convergence. More precisely, the *Kantorovich-Rubinstein metric* (cf. [20, Section 8.3, p. 191])

$$d_{\mathfrak{M}_b^+(X)}(\mu, \nu) := \sup \left\{ \int_X f(x)\,\mu(dx) - \int_X f(x)\,\nu(dx) \; \middle| \right.$$
$$\left. f : X \to \mathbb{R} \text{ with } \sup_{x \in X} |f(x)| \leq 1 \text{ and } \sup_{x \neq y} \frac{|f(x) - f(y)|}{d(x,y)} \leq 1 \right\}$$

for $\mu, \nu \in \mathfrak{M}_b^+(X)$ generates the topology associated to the $C_b(X)$-weak convergence.

In particular, the family of (Borel) probability measures on a separable metric space X

$$\mathfrak{P}(X) := \{\mathbb{P} \in \mathfrak{M}_b^+(X) \mid \mathbb{P}(X) = 1\} = \{\mathbb{P} \mid \mathbb{P} \text{ is a (Borel) probability measure on } X\}$$

is a separable metric space, since it is a subset of a separable metric space (cf. [62, Theorem 2.4.1, p. 65]). Moreover, the Kantorovich-Rubinshtein metric generalizes the *Lévy–Prokhorov metric* (cf. [20, Theorem 8.3.2, p. 193]) in that

$$d_{\mathfrak{M}_b^+(X)}(\mathbb{P}, \mathbb{Q}) = \inf\{\varepsilon > 0 \mid \forall B \in \mathcal{B}(X) : \mathbb{P}(B) \leq \mathbb{Q}(B^\varepsilon) + \varepsilon \text{ and } \mathbb{Q}(B) \leq \mathbb{P}(B^\varepsilon) + \varepsilon\}$$

for all $\mathbb{P}, \mathbb{Q} \in \mathfrak{P}(X)$, where $B^\varepsilon := \{x \in X : d(x, B) < \varepsilon\}$ for $B \subset X$. On top of that, if (X, d) is also complete, then so are $\mathfrak{M}_b^+(X)$ and $\mathfrak{P}(X)$.

Remark 4.17 (Lévy Measures). The Kantorovich-Rubinshtein metric on bounded Borel measures from Remark 4.16 can be used to introduce a metric (and hence a Borel-σ-algebra) on the families of Lévy measures on $\mathbb{R}^d$ and $\mathbb{R}_+ \times \mathbb{R}^d$

$$\mathfrak{L}(\mathbb{R}^d) := \{\nu \in \mathfrak{M}^+(\mathbb{R}^d) \mid \nu(\{0\}) = 0 \text{ and } \int_{\mathbb{R}^d} (|x|^2 \wedge 1)\, \nu(dx) < \infty\}$$
$$\mathfrak{L}(\mathbb{R}_+ \times \mathbb{R}^d) := \{\nu \in \mathfrak{M}^+(\mathbb{R}_+ \times \mathbb{R}^d) \mid \nu(\{0\} \times \mathbb{R}^d) = 0$$
$$\text{and } \forall T > 0 : \nu([0,T), dx) \in \mathfrak{L}(\mathbb{R}^d)\}$$
$$= \{\nu \in \mathfrak{M}^+(\mathbb{R}_+ \times \mathbb{R}^d) \mid \nu(\{0\} \times \mathbb{R}^d) = \nu(\mathbb{R}_+ \times \{0\}) = 0$$
$$\text{and } \forall T > 0 : \int_{[0,T)} \int_{\mathbb{R}^d} (|x|^2 \wedge 1)\, \nu(dt, dx) < \infty\}$$

in the following way: For every $\nu \in \mathfrak{L}(\mathbb{R}^d)$ and $B \in \mathcal{B}(\mathbb{R}^d)$

$$\hat{\nu}(B) := \int_B (|x|^2 \wedge 1)\, \nu(dx)$$

defines a bounded Borel measure on $\mathbb{R}^d$. It is easy to see (cf. [92, Lemma 2.3]) that

$$d_{\mathfrak{L}(\mathbb{R}^d)}(\nu, \mu) := d_{\mathfrak{M}_b^+(\mathbb{R}^d)}(\hat{\nu}, \hat{\mu})$$

for $\nu, \mu \in \mathfrak{L}(\mathbb{R}^d)$ defines a separable metric on $\mathfrak{L}(\mathbb{R}^d)$. In a similar manner, we assign to every $\nu \in \mathfrak{L}(\mathbb{R}_+ \times \mathbb{R}^d)$ a sequence of bounded Borel measures by

$$\hat{\nu}_n(B) := \int_{[0,n)} \int_{\mathbb{R}^d} (|x|^2 \wedge 1)\, \mathbb{1}_B(t, x) \nu(dt, dx)$$

for $B \in \mathcal{B}(\mathbb{R}_+ \times \mathbb{R}^d)$ and $n \in \mathbb{N}$, and construct with

$$d_{\mathfrak{L}(\mathbb{R}_+ \times \mathbb{R}^d)}(\nu, \mu) := \sum_{n \in \mathbb{N}} 2^{-n} d_{\mathfrak{M}_b^+(\mathbb{R}_+ \times \mathbb{R}^d)}(\hat{\nu}_n, \hat{\mu}_n)$$

for $\nu, \mu \in \mathfrak{L}(\mathbb{R}_+ \times \mathbb{R}^d)$ a separable metric on $\mathfrak{L}(\mathbb{R}_+ \times \mathbb{R}^d)$.

4. Stochastic Processes

Theorem 4.18 (Measurability of Characteristics). *If $(X_t)_{t\geq 0}$ is a càdlàg, $\mathbb{R}^d$-valued process on a separable metric space $(\Omega, \mathcal{B}(\Omega))$ adapted to a filtration $(\mathcal{F}_t)_{t\geq 0}$ of sub-σ-algebras of $\mathcal{B}(\Omega)$, then the family of all probability measures $\mathbb{P} \in \mathfrak{P}(\Omega)$ under which $(X_t)_{t\geq 0}$ is a $\mathbb{P}$-$\mathcal{F}$-semimartingale*

$$\mathfrak{P}_{sem}(\Omega) = \{\mathbb{P} \in \mathfrak{P}(\Omega) \mid (X_t)_{t\geq 0} \text{ is a } \mathbb{P}\text{-}\mathcal{F}\text{-semimartingale}\}$$

is Borel-measurable. Moreover, there exists a Borel-measurable map

$$\mathfrak{P}_{sem}(\Omega) \times \Omega \times \mathbb{R}_+ \longrightarrow \mathbb{R}^d \times \mathbb{S}_+^{d\times d} \times \mathfrak{L}(\mathbb{R}_+ \times \mathbb{R}^d)$$
$$(\mathbb{P}, \omega, t) \longmapsto (B_t^{\mathbb{P}}(\omega), C_t^{\mathbb{P}}(\omega), \nu_t^{\mathbb{P}}(\omega))$$

such that $(B^{\mathbb{P}}, C^{\mathbb{P}}, \nu^{\mathbb{P}})$ are (predictable) semimartingale characteristics of $(X_t)_{t\geq 0}$ under each $\mathbb{P} \in \mathfrak{P}_{sem}(\Omega)$.

Proof. $\hookrightarrow$ [92, Theorem 2.5] $\square$

Remark 4.19 (Differential Characteristics). One can show (cf. [73, Chapter 2, Proposition 2.9, p. 77]) that for every $\mathbb{P}$-$\mathcal{F}$-semimartingale $(X_t)_{t\geq 0}$ there exists a predictable, increasing, locally integrable process $(A_t)_{t\geq 0}$ (which is continuous if and only if X is quasi-left-continuous, i.e. $\mathbb{P}(\Delta X_\tau \neq 0, \tau < \infty) = 0$ for every predictable stopping time τ, cf. [73, Chapter 1, Definition 2.7, p. 17]) such that

$$(B, C, \nu) \ll dA_t$$

holds $\mathbb{P}$-almost surely. In other words, there exist predictable processes $b = (b^i)_{i\leq d}$ on $\mathbb{R}^d$ and $c = (c^{i,j})_{i,j\leq d}$ on $\mathbb{S}^{d\times d}$, as well as a predictable kernel $F = (F_t(dx))_{t\geq 0}$ on $(\mathbb{R}^d, \mathcal{B}(\mathbb{R}^d))$ given $(\Omega \times \mathbb{R}_+)$ (i.e. for every $t \geq 0$ the projection $F_t(dx)$ is a random Borel measure on $\mathbb{R}^d$, and for every $B \in \mathcal{B}(\mathbb{R}^d)$ the process $(F_t(B))_{t\geq 0}$ is predictable) such that

$$(B_t, C_t, \nu(dt, dx))(\omega) = (b_t\, dA_t, c_t\, dA_t, F_t(dx)\, dA_t)(\omega)$$

holds $\lambda(dt) \times \mathbb{P}(d\omega)$-almost everywhere. In fact, a canonical choice for $(A_t)_{t\geq 0}$ is

$$A = \sum_{i\leq d} \text{var}(B^i) + \sum_{i,j\leq d} \text{var}(C^{i,j}) + \int_{[0,\cdot]} \int_{\mathbb{R}^d} (1 \wedge |z|^2)\, \nu(ds, dz),$$

where $\text{var}(\cdot)$ is the bounded variation operator. For a large class of stochastic processes, e.g. most Feller processes (cf. [21, Theorem 2.44, p. 73]), we even have

$$(B, C, \nu) \ll dt$$

$\mathbb{P}$-almost surely, i.e. the semimartingale characteristics are absolutely continuous with respect to the Lebesgue measure. In this case, triplets (b, c, F) such that

$$(B_t, C_t, \nu(dt, dx))(\omega) = (b_t\, dt, c_t\, dt, F_t(dx)\, dt)(\omega)$$

holds $\lambda(dt) \times \mathbb{P}(d\omega)$-almost everywhere are called *differential characteristics* of $(X_t)_{t\geq 0}$. Note that differential characteristics are unique up to $\mathbb{P}$-indistinguishability due to the uniqueness of (predictable) semimartingale characteristics (cf. Remark 4.14).

4.2. Construction of Processes

The standard literature (such as [73, Chapter 2, Proposition 2.9, p. 77]) introduces the identity $\nu(dt, dx) = F_t(dx)\, dt$ for differential characteristics in the weak sense that

$$\int_{[0,t]\times \mathbb{R}^d} V(\omega, s, x)\, \nu(\omega, ds, dx) = \int_{[0,t]} \int_{\mathbb{R}^d} V(\omega, s, x)\, F(\omega, s, dx)\, ds$$

holds $\lambda(dt) \times \mathbb{P}(d\omega)$-almost surely for all measurable $V : \Omega \times \mathbb{R}_+ \times \mathbb{R}^d \to \mathbb{R}$. It is interesting to note that this identity can equivalently be understood as Bochner integrals for each fixed $\omega \in \Omega$, as discussed after Remark A.11.

Theorem 4.20 (Measurability of Differential Characteristics). *If $(X_t)_{t\geq 0}$ is a càdlàg, $\mathbb{R}^d$-valued process on a separable metric space $(\Omega, \mathcal{B}(\Omega))$ adapted to a filtration $(\mathcal{F}_t)_{t\geq 0}$ of sub-σ-algebras of $\mathcal{B}(\Omega)$, then the family of all probability measures $\mathbb{P} \in \mathfrak{P}(\Omega)$ under which $(X_t)_{t\geq 0}$ is a $\mathbb{P}$-$\mathcal{F}$-semimartingale with absolutely continuous characteristics with respect to the Lebesgue measure*

$$\mathfrak{P}^{ac}_{sem}(\Omega) = \{\mathbb{P} \in \mathfrak{P}_{sem}(\Omega) \mid (B^\mathbb{P}, C^\mathbb{P}, \nu^\mathbb{P}) \ll dt\ \mathbb{P}\text{-}a.s.\}$$

is Borel-measurable. Moreover, there exists a Borel-measurable map

$$\mathfrak{P}^{ac}_{sem}(\Omega) \times \Omega \times \mathbb{R}_+ \longrightarrow \mathbb{R}^d \times \mathbb{S}^{d\times d}_+ \times \mathfrak{L}(\mathbb{R}^d)$$
$$(\mathbb{P}, \omega, t) \longmapsto (b^\mathbb{P}_t(\omega), c^\mathbb{P}_t(\omega), F^\mathbb{P}_t(\omega))$$

such that $(b^\mathbb{P}, c^\mathbb{P}, F^\mathbb{P})$ are differential characteristics of $(X_t)_{t\geq 0}$ under each $\mathbb{P} \in \mathfrak{P}^{ac}_{sem}(\Omega)$.

Proof. ↪ [92, Theorem 2.6] □

As announced earlier, the next step towards the existence of a large class of sublinear Markov semigroups is the construction of an analogue for the notion of semimartingales for sublinear expectations. However, instead of constructing a stochastic process on a given sublinear expectation space, we are going to construct sublinear generalizations of conditional expectations $(E^x_t)_{t\geq 0}$, such that the *coordinate mapping process*

$$X : \Omega \times [0, \infty) \longrightarrow \mathbb{R}^d$$
$$(\omega, t) \longmapsto \omega_t$$

on $\Omega = D_x(\mathbb{R}_+; \mathbb{R}^d)$ with $x \in \mathbb{R}^d$ can be interpreted as a semimartingale under uncertainty in its characteristics. More precisely, given a measurable, set-valued function

$$\Theta : \mathbb{R}_+ \times D(\mathbb{R}_+; \mathbb{R}^d) \longrightarrow 2^{\mathbb{R}^d \times \mathbb{S}^{d\times d}_+ \times \mathfrak{L}(\mathbb{R}^d)},$$

we construct sublinear operators $E^x_t : \mathcal{H} \to \mathcal{H}$ for $t \geq 0$ and $x \in \mathbb{R}^d$ on a convex cone $\mathcal{H}$ of measurable functions on $D_x(\mathbb{R}_+; \mathbb{R}^d)$, which behave like a conditional generalization of the sublinear expectations $E^x_0(\varphi(X)) = \sup_{\mathbb{P}\in\mathcal{P}^\Theta_x} \mathbb{E}_\mathbb{P}(\varphi(X))$ for $\varphi \in \mathcal{H}$, where the uncertainty subsets are defined by

$$\mathcal{P}^\Theta_x = \left\{\mathbb{P} \in \mathfrak{P}^{ac}_{sem}(D_x(\mathbb{R}_+)) \mid (b^\mathbb{P}_s, c^\mathbb{P}_s, F^\mathbb{P}_s)(\omega) \in \Theta_s(\omega) \quad \lambda(ds) \times \mathbb{P}(d\omega)\text{-a.e.}\right\}.$$

161

4. Stochastic Processes

This construction suggests to interpret the coordinate mapping process $(X_t)_{t\geq 0}$ together with the conditional sublinear expectations $(E_t^x)_{t\geq 0}$ as a generalization of semimartingales (starting at $x \in \mathbb{R}^d$) under uncertainty in their semimartingale characteristics. Note that the preceding results on the measurability of the related (differential) characteristics are applicable in this context, since the Skorokhod space $D(\mathbb{R}_+; \mathbb{R}^d)$ is a Polish space according to Lemma 4.2.

We are left with the problem of how to construct sublinear generalizations of classical conditional expectations. A straightforward approach similar to the construction of sublinear expectations from Section 3.1 – as the supremum of the classical linear expectations $(\mathbb{E}_\mathbb{P})_{\mathbb{P}\in\mathcal{P}}$ for a given uncertainty subset $\mathcal{P} \subset \mathfrak{P}(\Omega)$ – unfortunately leads to the so-called aggregation problem (cf. [28, Theorem 3.16]): A-priori, it is unclear how to construct a single measurable random variable that coincides with the supremum

$$\text{“}\sup_{\mathbb{P}\in\mathcal{P}} \mathbb{E}_\mathbb{P}\left(\varphi(X)\,|\,\mathcal{F}_t\right)\text{”}$$

$\mathbb{P}$-almost surely under all $\mathbb{P} \in \mathcal{P}$ for an uncertainty subset $\mathcal{P} \subset \mathfrak{P}(\Omega)$, since classical conditional expectations $\mathbb{E}_\mathbb{P}\left(\varphi(X)\,|\,\mathcal{F}_t\right)$ are only defined $\mathbb{P}$-almost surely. Shige Peng circumvented the aggregation problem for G-Brownian motions in [102] by constructing the related sublinear conditional expectations only for very regular functions $\varphi : \Omega \to \mathbb{R}$ via the associated G-heat equations. However, since the evaluation of sublinear expectations for merely measurable $\varphi : \Omega \to \mathbb{R}$ is important in many applications, we do not follow this approach here. Instead, we develop a straightforward extension of the construction (for continuous processes starting at the origin) from Marcel Nutz and Ramon van Häendel in [98], which is based on the following concept of regular conditional distributions from classical probability theory:

Remark 4.21 (Non-Integrable Functions). In order to avoid cumbersome notation, we adopt the convention that $\infty - \infty := -\infty$ from [98]. In particular, we will work with measurable functions $\xi : \Omega \to \overline{\mathbb{R}}$ that take values in the two-point compactification $\overline{\mathbb{R}} = \mathbb{R} \cup \{\pm\infty\}$ and suppose that all (classical) conditional expectations are defined as

$$\mathbb{E}(\xi|\mathcal{F}) := \mathbb{E}(\xi^+|\mathcal{F}) - \mathbb{E}(\xi^-|\mathcal{F}).$$

This notation proves useful in the rest of this chapter, in order to state the (almost sure) equivalence of representations without a distinction with respect to the integrability of the related terms.

Definition 4.22 (Operations on Skorokhod Space). The concatenation $\omega \otimes_\tau \bar{\omega}$ of càdlàg paths $\omega, \bar{\omega} \in D(\mathbb{R}_+)$ at a random time $\tau : D(\mathbb{R}_+) \to [0, \infty]$ is defined as

$$\left(\omega \otimes_\tau \bar{\omega}\right)(t) := \omega(t)\,\mathbb{1}_{[0,\tau(\omega))}(t) + \left(\omega\bigl(\tau(\omega)\bigr) + \bar{\omega}\bigl(t - \tau(\omega)\bigr) - \bar{\omega}(0)\right)\mathbb{1}_{[\tau(\omega),\infty)}(t)$$

for $t \geq 0$. Moreover, given $\xi : D(\mathbb{R}_+) \to \overline{\mathbb{R}}$, define the conditioned $\xi^{\tau,\omega} : D(\mathbb{R}_+) \to \overline{\mathbb{R}}$ by

$$\xi^{\tau,\omega}(\bar{\omega}) := \xi(\omega \otimes_\tau \bar{\omega}) \tag{4.2}$$

for $\bar{\omega} \in D(\mathbb{R}_+)$, where $\tau : D(\mathbb{R}_+) \to [0, \infty]$ and $\omega \in D(\mathbb{R}_+)$.

4.2. Construction of Processes

Remark 4.23 (Regular Conditional Distributions). Suppose that $\Omega = D_x(\mathbb{R}_+)$ for $x \in \mathbb{R}^d$ is equipped with the Borel-σ-algebra $\mathcal{B}(\Omega)$ and the filtration $(\mathcal{F}_t)_{t\geq 0}$ generated by the coordinate-mapping process $(X_t)_{t\geq 0}$. One can show (cf. [19, Theorem 33.3, p. 439]) that for any probability measure $\mathbb{P} \in \mathfrak{P}(\Omega)$ and any finite stopping time $\tau : \Omega \to [0, \infty)$, there exists a *regular conditional distribution* $(\mathbb{P}_{\tau,\omega})_{\omega \in \Omega}$ – which means that $\mathbb{P}_{\tau,\omega} \in \mathfrak{P}(\Omega)$ for all $\omega \in \Omega$, the maps $\omega \mapsto \mathbb{P}_{\tau,\omega}(B)$ are $\mathcal{F}_\tau$-measurable for all $B \in \mathcal{B}(\Omega)$ and

$$\mathbb{E}_{\mathbb{P}_{\tau,\omega}}(\xi) = \mathbb{E}(\xi \mid \mathcal{F}_\tau)(\omega)$$

holds $\mathbb{P}(d\omega)$-almost surely for all measurable $\xi : \Omega \to \overline{\mathbb{R}}$. According to [123, Theorem 1.3.4, p. 34], the probability measures $\mathbb{P}_{\tau,\omega}$ can chosen to be concentrated on the set of paths that coincide with $\omega \in \Omega$ up to time $\tau(\omega)$, i.e.

$$\mathbb{P}_{\tau,\omega}(\{\bar{\omega} \in \Omega \mid \forall s \leq \tau(\omega) : \bar{\omega}(s) = \omega(s)\}) = 1.$$

Hence, for every stopping time $\tau : \Omega \to [0, \infty]$ and $\omega \in \Omega$, the translation

$$\mathbb{P}^{\tau,\omega}(B) := \mathbb{P}_{\tau,\omega}(\omega \otimes_\tau B) = \mathbb{P}_{\tau,\omega}(\{\omega \otimes_\tau \bar{\omega} \in \Omega \mid \bar{\omega} \in B\})$$

with $B \in \mathcal{B}(\Omega)$ defines a probability measure on Ω such that

$$\mathbb{E}_{\mathbb{P}^{\tau,\omega}}(\xi^{\tau,\omega}) = \mathbb{E}_{\mathbb{P}_{\tau,\omega}}(\xi) = \mathbb{E}(\xi \mid \mathcal{F}_\tau)(\omega)$$

holds $\mathbb{P}(d\omega)$-almost surely for every measurable $\xi : \Omega \to \mathbb{R}^d$.

The main idea from Marcel Nutz and Ramon van Häendel in [98] is to use the representation of conditional expectations via regular conditional distributions from Remark 4.23, to define sublinear conditional expectations ω-wise. In fact, many of the important properties of conditional expectations (excluding the aggregation or tower property, as we will see in Theorem 4.29) can be verified ω-wise using the characterization of measurability with respect to the canonical filtration (of the coordinate-mapping process) as in [38, Theorem 97, p. 147]. However, although we can show that

$$\omega \longmapsto \mathbb{E}_{\mathbb{P}^{\tau,\omega}}(\xi^{\tau,\omega})$$

is Borel-measurable for Borel-measurable $\xi : \Omega \to \overline{\mathbb{R}}$ and fixed $\mathbb{P} \in \mathfrak{P}(\Omega)$, the ω-wise supremum over an uncountable family $\mathcal{P} \subset \mathfrak{P}(\Omega)$ is in general not Borel-measurable anymore. For that reason, we will work with the following extension of the family of Borel-measurable functions instead, which is stable under taking pointwise suprema.

Definition 4.24 (Analytic Sets). A subset $B \subset X$ of a Polish space (X, d) (i.e. a complete, separable metric space) is called *analytic* if it is the image

$$B = \Phi(B')$$

of a Borel set $B' \subset X'$ in another Polish space (X', d') under a continuous mapping $\Phi : X' \to X$. The family of all analytic subsets of (X, d) is denoted by

$$\mathfrak{A}(X) = \mathfrak{A}(X, d) := \{B \subset X \mid B \text{ is analytic}\}.$$

4. Stochastic Processes

Remark 4.25 (Properties of Analytics Sets). It is well-known (cf. [30, Proposition 8.2.2, p. 248]) that the family of analytic sets is stable under countable unions and intersections, but not under complementation, i.e. in general $B \in \mathfrak{A}(X)$ does not imply that $B^C = X \setminus B$ is analytic. Moreover, every Borel set $B \in \mathcal{B}(X)$ is obviously analytic, and according to [30, Corollary 8.3.3, p. 258], an analytic set $B \in \mathfrak{A}(X)$ is Borel if and only if its complement $B^C = X \setminus B$ is analytic as well. In addition, [30, Theorem 8.4.1, p. 262] shows that every analytic set is universally measurable and hence

$$\mathcal{B}(X) \subset \mathfrak{A}(X) \subset \sigma(\mathfrak{A}(X)) \subset \mathcal{B}(X)^* \subset \mathcal{B}(X)^\mathbb{P},$$

where $\mathcal{B}(X)^\mathbb{P}$ is the completion of $\mathcal{B}(X)$ under the probability measure $\mathbb{P} \in \mathfrak{P}(X)$ and

$$\mathcal{B}(X)^* = \cap_{\mathbb{P} \in \mathfrak{P}(X)} \mathcal{B}(X)^\mathbb{P}$$

is its universal completion. In particular, it is possible to integrate $\sigma(\mathfrak{A}(X))$-measurable functions with respect to any Borel measure $\mathbb{P} \in \mathfrak{P}(X)$, using its unique extension to the completion $\mathcal{B}(X)^\mathbb{P}$.

Definition 4.26 (Semi-Analytic Functions). A function $f : X \to \overline{\mathbb{R}}$ on a Polish space (X, d) is called *upper semi-analytic* if

$$\{f > c\} = \{x \in X \mid f(x) > c\} \in \mathfrak{A}(X)$$

is analytic for every $c \in \mathbb{R}$. A function $f : X \to \overline{\mathbb{R}}$ on a Polish space (X, d) is called *lower semi-analytic* if $(-f)$ is upper semi-analytic.

Remark 4.27 (Stability of Semi-Analytic Functions). Note that if a function $f : X \times Y \to \overline{\mathbb{R}}$ on Polish spaces X, Y is Borel-measurable, then

$$y \longmapsto \sup_{x \in X} f(x, y)$$

does not have to be Borel-measurable anymore. This is closely related to the fact that the projection $\pi_Y(B) \subset Y$ of a Borel set $B \subset X \times Y$ does not have to be Borel anymore, cf. [127, Chapter 18, p. 118]: Suppose that $B \subset X \times Y$ is one of those sets, then

$$f(x, y) := \mathbb{1}_B(x, y)$$

with $(x, y) \in X \times Y$ is Borel-measurable, whereas $y \mapsto \sup_{x \in X} f(x, y)$ is not, since

$$\{y \in Y \mid \sup_{x \in X} f(x, y) > c\} = \cup_{x \in X} \{y \in Y \mid f(x, y) > c\}$$
$$= \pi_Y(\{(x, y) \in X \times Y \mid f(x, y) > c\})$$

for $c \in \mathbb{R}$ and hence $\{\sup_{x \in X} f(x, \cdot) > 0\} = \pi_Y(\{f > 0\}) = \pi_Y(B) \notin \mathcal{B}(Y)$. The same arguments, together with the obvious fact that the projection of analytic sets is analytic, shows that if $f : X \times Y \to \overline{\mathbb{R}}$ on Polish spaces X, Y is upper semi-analytic, then $y \mapsto \sup_{x \in X} f(x, y)$ is upper semi-analytic as well.

Remark 4.28 (Assumptions on $\mathcal{P}$). Suppose that $\Omega = D_x(\mathbb{R}_+)$ for some $x \in \mathbb{R}^d$ and

$$\{\mathcal{P}(t,\omega) : (t,\omega) \in \mathbb{R}_+ \times \Omega\}$$

with $\mathcal{P}(t,\omega) \subset \mathfrak{P}(\Omega)$ for all $t \geq 0$ and $\omega \in \Omega$ is adapted, in that

$$\omega|_{[0,t]} = \bar{\omega}|_{[0,t]} \implies \mathcal{P}(t,\omega) = \mathcal{P}(t,\bar{\omega})$$

for all $t \geq 0$ and $\omega, \bar{\omega} \in \Omega$. In particular, the uncertainty subset $\mathcal{P} := \mathcal{P}(0,\omega)$ does not depend on $\omega \in \Omega$. In order to define a conditional sublinear expectation, we have to impose the following assumptions on the family $(\mathcal{P}(t,\omega))_{(t,\omega) \in \mathbb{R}_+ \times \Omega}$ of so-called *conditional uncertainty subsets*. All statements are supposed to hold for all $s \geq 0$, finite stopping times τ with $\tau \geq s$, paths $\bar{\omega} \in \Omega$, measures $\mathbb{P} \in \mathcal{P}(s,\bar{\omega})$ and resulting shifted stopping times $\sigma := \tau^{s,\bar{\omega}} - s$ (as defined in Equation (4.2) from Definition 4.22).

(C1) (Measurability) The graph of $\omega \mapsto \mathcal{P}(\tau(\omega), \omega)$, i.e. the set

$$\{(\omega, \mathbb{P}) \in \Omega \times \mathfrak{P}(\Omega) \mid \omega \in \Omega \text{ and } \mathbb{P} \in \mathcal{P}(\tau(\omega), \omega)\},$$

is analytic.

(C2) (Invariance under Conditioning) The regular conditional distribution (as introduced in Remark 4.23) $\mathbb{P}^{\sigma,\omega} \in \mathfrak{P}(\Omega)$ of $\mathbb{P} \in \mathcal{P}(s,\bar{\omega})$ satisfies

$$\mathbb{P}^{\sigma,\omega} \in \mathcal{P}(\tau(\bar{\omega} \otimes_s \omega), \bar{\omega} \otimes_s \omega)$$

for $\mathbb{P}$-almost every $\omega \in \Omega$.

(C3) (Stability under Pasting) For $\mathcal{F}_\sigma$-measurable kernels $\nu : \Omega \to \mathfrak{P}(\Omega)$ with

$$\nu(\omega) \in \mathcal{P}(\tau(\bar{\omega} \otimes_s \omega), \bar{\omega} \otimes_s \omega)$$

for $\mathbb{P}$-almost every $\omega \in \Omega$, the measure $\bar{\mathbb{P}} \in \mathfrak{P}(\Omega)$, defined by

$$\bar{\mathbb{P}}(B) = \int \int (\mathbb{1}_B)^{\sigma,\omega}(\tilde{\omega}) \, \nu(\omega)(d\tilde{\omega}) \, \mathbb{P}(d\omega)$$

for $B \in \mathcal{B}(\Omega)$ (with $(\mathbb{1}_B)^{\sigma,\omega}$ as in Definition 4.22), is an element of $\mathcal{P}(s,\bar{\omega})$.

Before we present the rigorous construction, let us quickly discuss the structural assumptions (C2) and (C3) for conditional uncertainty subsets $(\mathcal{P}(t,\omega))_{(t,w) \in \mathbb{R}_+ \times \Omega}$ from Remark 4.28: If $\mathcal{P}(t,\omega) = \{\mathbb{P}^{t,\omega}\}$ are singletons for $(t,\omega) \in \mathbb{R}_+ \times \Omega$, where $\mathbb{P}^{t,\omega}$ are regular conditional distributions (as in Remark 4.23) of one fixed $\mathbb{P} \in \mathfrak{P}(\Omega)$, then (C2) and (C3) follow from the aggregation (or tower) property of the corresponding classical conditional expectation. In general, assumptions (C2) and (C3) intuitively suggest that

$$\text{``}\mathcal{P}(t,\omega) = \{\mathbb{P}^{t,\omega} \mid \mathbb{P} \in \mathcal{P}\}\text{''}$$

holds for $(t,\omega) \in \mathbb{R}_+ \times \Omega$ – although this identity is not well-defined, since regular conditional distributions are only defined up to null sets. However, the following proof of Theorem 4.29 reveals that assumptions (C2) and (C3) for general conditional uncertainty subsets $(\mathcal{P}(t,\omega))_{(t,w) \in \mathbb{R}_+ \times \Omega}$ are the reason, why the corresponding sublinear generalization of conditional expectations satisfies the aggregation property.

4. Stochastic Processes

Theorem 4.29 (Sublinear Conditional Expectation). *Suppose that $\Omega_x := D_x(\mathbb{R}_+)$ for $x \in \mathbb{R}^d$ and $\{\mathcal{P}_x(t,\omega) \subset \mathfrak{P}(\Omega_x) : (t,\omega) \in \mathbb{R}_+ \times \Omega_x\}$ are conditional uncertainty subsets that satisfy all assumptions from Remark 4.28. For stopping times $\tau : \Omega_x \to [0,\infty)$ and upper semi-analytic functions $\xi : \Omega_x \to \overline{\mathbb{R}}$, the definition*

$$E_\tau^x(\xi)(\omega) := \sup_{\mathbb{P} \in \mathcal{P}_x(\tau(\omega),\omega)} \mathbb{E}_\mathbb{P}(\xi^{\tau(\omega),\omega})$$

with $\omega \in \Omega_x$ leads to upper semi-analytic $E_\tau^x(\xi) : \Omega_x \to \overline{\mathbb{R}}$, which are measurable with respect to $\mathcal{F}_\tau^*$ (i.e. the universal completion of $\mathcal{F}_\tau$ as in Remark 4.25). Moreover, for stopping times $\tau, \sigma : \Omega_x \to [0,\infty)$ with $\sigma \leq \tau$, the aggregation property

$$E_\sigma^x(E_\tau^x(\xi))(\omega) = E_\sigma^x(\xi)(\omega)$$

holds for all $\omega \in \Omega_x$ and upper semi-analytic functions $\xi : \Omega_x \to \overline{\mathbb{R}}$.

Proof. The result is a straightforward generalization of [98, Theorem 2.3] from the space of continuous paths starting at zero $\Omega = C_0(\mathbb{R}_+)$ to the space $\Omega = D_x(\mathbb{R}_+)$ for $x \in \mathbb{R}^d$. A careful inspection of the proof shows that all arguments also work for probability measures on the Skorokhod space, if we work with the path operations introduced in Definition 4.22. In order to clarify the role of the structural assumptions (C2) and (C3) in Remark 4.28, let us recall the arguments for the aggregation property here:

Suppose that $\tau, \sigma : \Omega_x \to [0,\infty)$ are stopping times with $\sigma \leq \tau$ and $\xi : \Omega_x \to \overline{\mathbb{R}}$ is an upper semi-analytic function. We have to show that

$$E_\sigma^x(E_\tau^x(\xi))(\bar{\omega}) = E_\sigma^x(\xi)(\bar{\omega})$$

holds for $\bar{\omega} \in \Omega_x$. Due to the $\bar{\omega}$-wise definition, this is equivalent to show that

$$E_s^x(E_\tau^x(\xi))(\bar{\omega}) = E_s^x(\xi)(\bar{\omega})$$

holds for $\bar{\omega} \in \Omega_x$ and $s \leq \tau(\bar{\omega})$. With the notation $\mathcal{P}_x(\tau,\omega) := \mathcal{P}_x(\tau(\omega),\omega)$, we obtain

$$E_s^x(\xi)(\bar{\omega}) = \sup_{\mathbb{P} \in \mathcal{P}_x(s,\bar{\omega})} \int \xi^{s,\bar{\omega}}(\omega)\, \mathbb{P}(d\omega) = \sup_{\mathbb{P} \in \mathcal{P}_x(s,\bar{\omega})} \int \xi(\bar{\omega} \otimes_s \omega)\, \mathbb{P}(d\omega)$$

$$E_s^x(E_\tau^x(\xi))(\bar{\omega}) = \sup_{\mathbb{P} \in \mathcal{P}_x(s,\bar{\omega})} \int \sup_{\tilde{\mathbb{P}} \in \mathcal{P}_x(\tau,\bar{\omega}\otimes_s\omega)} \int \xi((\bar{\omega} \otimes_s \omega) \otimes_\tau \tilde{\omega})\, \bar{\mathbb{P}}(d\tilde{\omega})\, \mathbb{P}(d\omega)$$

for $\bar{\omega} \in \Omega_x$ and $s \leq \tau(\bar{\omega})$. With $\sigma := \tau^{s,\bar{\omega}} - s$, the classical tower property for $\mathbb{P} \in \mathcal{P}_x(s,\bar{\omega})$ and the invariance under conditioning assumption (C2) lead to

$$\int \xi(\bar{\omega} \otimes_s \omega)\, \mathbb{P}(d\omega) = \int\int \xi(\bar{\omega} \otimes_s \tilde{\omega})\, \mathbb{P}_{\sigma,\omega}(d\tilde{\omega})\, \mathbb{P}(d\omega)$$

$$= \int\int \xi(\bar{\omega} \otimes_s (\omega \otimes_\sigma \tilde{\omega}))\, \mathbb{P}^{\sigma,\omega}(d\tilde{\omega})\, \mathbb{P}(d\omega)$$

$$= \int\int \xi((\bar{\omega} \otimes_s \omega) \otimes_\tau \tilde{\omega})\, \mathbb{P}^{\sigma,\omega}(d\tilde{\omega})\, \mathbb{P}(d\omega)$$

$$\leq \int \sup_{\tilde{\mathbb{P}} \in \mathcal{P}_x(\tau,\bar{\omega}\otimes_s\omega)} \int \xi((\bar{\omega} \otimes_s \omega) \otimes_\tau \tilde{\omega})\, \bar{\mathbb{P}}(d\tilde{\omega})\, \mathbb{P}(d\omega),$$

which shows $E_s^x(\xi)(\bar{\omega}) \leq E_s^x(E_\tau^x(\xi))(\bar{\omega})$ for $\bar{\omega} \in \Omega_x$ and $s \leq \tau(\bar{\omega})$.

Conversely, for fixed $\varepsilon > 0$ and $\bar{\omega} \in \Omega$, the definition of the supremum implies the existence of $\nu_\varepsilon(\omega) \in \mathcal{P}_x(\tau, \bar{\omega} \otimes_s \omega)$ for every $\omega \in \Omega$ such that

$$\sup_{\bar{\mathbb{P}} \in \mathcal{P}_x(\tau, \bar{\omega} \otimes_s \omega)} \int (\xi^{s,\bar{\omega}})^{\sigma,\omega}(\tilde{\omega}) \bar{\mathbb{P}}(d\tilde{\omega}) \leq \lim_{\varepsilon \to 0} \left(\int (\xi^{s,\bar{\omega}})^{\sigma,\omega}(\tilde{\omega}) \nu_\varepsilon(\omega)(d\tilde{\omega}) + \varepsilon \right)$$

holds. Hence, Fatou's lemma and the stability assumption (C3) for $\mathbb{P} \in \mathcal{P}_x(s, \bar{\omega})$ imply

$$\int \sup_{\bar{\mathbb{P}} \in \mathcal{P}_x(\tau, \bar{\omega} \otimes_s \omega)} \int (\xi^{s,\bar{\omega}})^{\sigma,\omega}(\tilde{\omega}) \bar{\mathbb{P}}(d\tilde{\omega}) \mathbb{P}(d\omega) \leq \liminf_{\varepsilon \to 0} \int \int (\xi^{s,\bar{\omega}})^{\sigma,\omega}(\tilde{\omega}) \nu_\varepsilon(\omega)(d\tilde{\omega}) \mathbb{P}(d\omega)$$

$$\leq \sup_{\mathbb{P} \in \mathcal{P}_x(s,\bar{\omega})} \int \xi^{s,\bar{\omega}}(\omega) \mathbb{P}(d\omega),$$

which shows $E_s^x(E_\tau^x(\xi))(\bar{\omega}) \leq E_s^x(\xi)(\bar{\omega})$ with $\bar{\omega} \in \Omega_x$ and $s \leq \tau(\bar{\omega})$, using the identity $(\xi^{s,\bar{\omega}})^{\sigma,\omega}(\tilde{\omega}) = \xi((\bar{\omega} \otimes_s \omega) \otimes_\tau \tilde{\omega})$ from Definition 4.22. $\square$

In the following few results, we show how the general construction for sublinear conditional expectations from Theorem 4.29 can be used to construct semimartingales for sublinear expectations and related sublinear Markov semigroups (as in Section 4.1).

Lemma 4.30 (Regularity of $x \mapsto E^x(\xi)$). *Suppose that for every $x \in \mathbb{R}^d$*

$$\{\mathcal{P}_x(t,\omega) \subset \mathfrak{P}(\Omega_x) : (t,\omega) \in \mathbb{R}_+ \times \Omega_x\}$$

are conditional uncertainty subsets on $\Omega_x := D_x(\mathbb{R}_+; \mathbb{R}^d)$ satisfying all assumptions from Remark 4.28. If the graph of $x \mapsto \mathcal{P}_x$ is analytic, i.e.

$$\{(x, \mathbb{P}) \in \mathbb{R}^d \times \mathfrak{P}(D(\mathbb{R}_+)) \mid x \in \mathbb{R}^d \text{ and } \mathbb{P}|_{\Omega_x} \in \mathcal{P}_x\} \in \mathfrak{A}\left(\mathbb{R}^d \times \mathfrak{P}(D(\mathbb{R}_+))\right),$$

then $x \mapsto E^x(\xi) := E^x(\xi|_{\Omega_x})$ is upper semi-analytic (with $E^x := E_0^x$ as in Theorem 4.29) for every upper semi-analytic function $\xi : D(\mathbb{R}_+) \to \overline{\mathbb{R}}$.

Proof. Using the convention from Remark 4.21, it is easy to see that for $x \in \mathbb{R}^d$

$$E^x(\xi) = \sup_{\mathbb{P} \in \mathcal{P}_x} \mathbb{E}_\mathbb{P}(\xi|_{D_x(\mathbb{R}_+)}) = \sup_{\mathbb{P} \in \mathcal{P}} \mathbb{E}_\mathbb{P}(\hat{\xi}(x, \cdot)),$$

where $\hat{\xi}(x, \omega) := \xi(\omega) \mathbb{1}_{D_x(\mathbb{R}_+)}(\omega) + (-\infty) \mathbb{1}_{D(\mathbb{R}_+) \setminus D_x(\mathbb{R}_+)}(\omega)$ for $(x, \omega) \in \mathbb{R}^d \times D(\mathbb{R}_+)$ and

$$\mathcal{P} := \{\mathbb{P} \in \mathfrak{P}(D(\mathbb{R}_+)) : \exists x \in \mathbb{R}^d : \mathbb{P}|_{D_x(\mathbb{R}_+)} \in \mathcal{P}_x\}.$$

According to [16, Proposition 7.47, p. 179], the analyticity of $\mathcal{P}$ implies that $x \mapsto E^x(\xi)$ is upper semi-analytic if $x \mapsto \mathbb{E}_\mathbb{P}(\hat{\xi}(x, \cdot))$ is upper semi-analytic for every $\mathbb{P} \in \mathcal{P}$. Moreover, [16, Proposition 7.48, p. 180] shows that $x \mapsto \mathbb{E}_\mathbb{P}(\hat{\xi}(x, \cdot))$ for fixed $\mathbb{P} \in \mathcal{P}$ is upper semi-analytic if $(x, \omega) \mapsto \hat{\xi}(x, \omega)$ is upper semi-analytic. Thus, it remains to show that

$$\mathbb{R}^d \times D(\mathbb{R}_+) \ni (x, \omega) \mapsto \hat{\xi}(x, \omega) \in \overline{\mathbb{R}}$$

is upper semi-analytic:

The construction of $\hat{\xi} : \mathbb{R}^d \times D(\mathbb{R}_+) \to \overline{\mathbb{R}}$ implies that

$$\{(x,\omega) \in \mathbb{R}^d \times D(\mathbb{R}_+) \mid \hat{\xi}(x,\omega) > c\}$$
$$= \{(x,\omega) \in \mathbb{R}^d \times D(\mathbb{R}_+) \mid \omega \in D_x(\mathbb{R}_+) \text{ and } \xi(\omega) > c\}$$
$$= \bigcup_{x \in \mathbb{R}^d} \left(\{x\} \times \left(D_x(\mathbb{R}_+) \cap \{\omega \in D(\mathbb{R}_+) \mid \xi(\omega) > c\} \right) \right)$$
$$= \left(\bigcup_{x \in \mathbb{R}^d} \left(\{x\} \times D_x(\mathbb{R}_+) \right) \right) \cap \left(\mathbb{R}^d \times \{\omega \in D(\mathbb{R}_+) \mid \xi(\omega) > c\} \right)$$

for all $c \in \mathbb{R}$, where $\mathbb{R}^d \times \{\xi > c\} \subset \mathbb{R}^d \times D(\mathbb{R}_+)$ is analytic, because $\xi : D(\mathbb{R}_+) \to \mathbb{R}$ is upper semi-analytic. But $\cup_{x \in \mathbb{R}^d} (\{x\} \times D_x(\mathbb{R}_+))$ is also analytic, since

$$\bigcup_{x \in \mathbb{R}^d} \left(\{x\} \times D_x(\mathbb{R}_+) \right) = \psi(\mathbb{R}^d \times D_0(\mathbb{R}_+))$$

for the continuous map $\psi : \mathbb{R}^d \times D_0(\mathbb{R}_+) \to \mathbb{R}^d \times D(\mathbb{R}_+)$ with $\psi(x,\omega) := (x, x+\omega)$ for all $(x,\omega) \in \mathbb{R}^d \times D_0(\mathbb{R}_+)$, which concludes the proof. $\square$

Proposition 4.31 (Admissibility of $\mathcal{P}^\Theta$). *Suppose that $\mathfrak{S} := \mathbb{R}^d \times \mathbb{S}_+^{d \times d} \times \mathfrak{L}(\mathbb{R}^d)$ and*

$$\Theta : D(\mathbb{R}_+; \mathbb{R}^d) \times \mathbb{R}_+ \longrightarrow 2^{\mathfrak{S}} = 2^{\mathbb{R}^d \times \mathbb{S}_+^{d \times d} \times \mathfrak{L}(\mathbb{R}^d)}$$

is a map that is adapted to the natural filtration $\mathcal{F} = (\mathcal{F}_t)_{t \geq 0}$ in that

$$\{(s,\omega,\theta) \in [0,t] \times D(\mathbb{R}_+) \times \mathfrak{S} \mid \theta \in \Theta_s(\omega)\} \in \mathcal{B}([0,t]) \otimes \mathcal{F}_t \otimes \mathcal{B}(\mathfrak{S})$$

is satisfied for all $t \geq 0$. For every $x \in \mathbb{R}^d$, the definition

$$\mathcal{P}_x^\Theta(t,\omega) := \{\mathbb{P} \in \mathfrak{P}_{\text{sem}}^{ac}(D_x(\mathbb{R}_+)) \mid (b_s^\mathbb{P}, c_s^\mathbb{P}, F_s^\mathbb{P})(\tilde{\omega}) \in \Theta_{t+s}(\omega \otimes_t \tilde{\omega}) \quad \lambda(ds) \times \mathbb{P}(d\tilde{\omega})\text{-a.e.}\}$$

for $t \geq 0$ and $\omega \in D_x(\mathbb{R}_+)$ leads to a family $\{\mathcal{P}_x^\Theta(t,\omega) : (t,\omega) \in \mathbb{R}_+ \times D_x(\mathbb{R}_+)\}$ that satisfies all assumptions from Remark 4.28 and Lemma 4.30.

Proof. Suppose that τ is a finite stopping time with $\tau \geq s \geq 0$, $\mathbb{P} \in \mathcal{P}_x^\Theta(s,\bar{\omega})$ for some $(x,\bar{\omega}) \in \mathbb{R}^d \times \Omega$, and set $\sigma := \tau^{s,\bar{\omega}} - s$ (cf. Definition 4.22). The measurability property (C1) can be proved as in the continuous paths setting in [98, Lemma 4.5]. The following arguments for the invariance under conditioning (C2) and stability under pasting (C3) are generalizations of [93, Corollary 3.2] and [93, Proposition 4.2] from path-independent to path-dependent uncertainty coefficients.

First of all, we derive the invariance under conditioning (C2): According to [93, Theorem 3.1], we have $\mathbb{P}^{\sigma,\omega} \in \mathfrak{P}_{\text{sem}}^{ac}(D_x(\mathbb{R}_+))$ for $\mathbb{P}$-almost every $\omega \in D_x(\mathbb{R}_+)$ with

$$(b_u^{\mathbb{P}^{\sigma,\omega}}, c_u^{\mathbb{P}^{\sigma,\omega}}, F_u^{\mathbb{P}^{\sigma,\omega}})(\tilde{\omega}) = (b_{\sigma(\omega)+u}^\mathbb{P}, c_{\sigma(\omega)+u}^\mathbb{P}, F_{\sigma(\omega)+u}^\mathbb{P})(\omega \otimes_{\sigma(\omega)} \tilde{\omega})$$

$\lambda(du) \times \mathbb{P}^{\sigma,\omega}(d\tilde{\omega})$-almost everywhere. Since we assume that $\mathbb{P} \in \mathcal{P}_x^\Theta(s,\bar{\omega})$, the inclusion

$$(b_u^\mathbb{P}, c_u^\mathbb{P}, F_u^\mathbb{P})(\tilde{\omega}) \in \Theta_{s+u}(\bar{\omega} \otimes_s \tilde{\omega})$$

holds $\lambda(du) \times \mathbb{P}(d\tilde{\omega})$-almost everywhere. Combining these two results with the construction

4.2. Construction of Processes

of $\mathbb{P}^{\sigma,\omega} \in \mathfrak{P}(D_x(\mathbb{R}_+))$ from $\mathbb{P} \in \mathfrak{P}(D_x(\mathbb{R}_+))$ (as introduced in Remark 4.23) implies that $\mathbb{P}^{\sigma,\omega} \in \mathfrak{P}^{ac}_{sem}(D_x(\mathbb{R}_+))$ for $\mathbb{P}$-almost every $\omega \in D_x(\mathbb{R}_+)$ with

$$(b_u^{\mathbb{P}^{\sigma,\omega}}, c_u^{\mathbb{P}^{\sigma,\omega}}, F_u^{\mathbb{P}^{\sigma,\omega}})(\tilde{\omega}) \in \Theta_{s+\sigma(\omega)+u}(\bar{\omega} \otimes_s (\omega \otimes_{\sigma(\omega)} \tilde{\omega}))$$
$$= \Theta_{\tau(\bar{\omega} \otimes_s \omega)+u}((\bar{\omega} \otimes_s \omega) \otimes_{\tau(\bar{\omega} \otimes_s \omega)} \tilde{\omega})$$

$\lambda(du) \times \mathbb{P}^{\sigma,\omega}(d\tilde{\omega})$-almost everywhere, i.e. $\mathbb{P}^{\sigma,\omega} \in \mathcal{P}_x^\Theta(\tau(\bar{\omega} \otimes_s \omega), \bar{\omega} \otimes_s \omega)$ for $\mathbb{P}$-almost every $\omega \in D_x(\mathbb{R}_+)$, as required.

As a next step, we prove the stability-under-pasting property (C3): Suppose that $\mathbb{P}_x^\Theta \in \mathcal{P}(s, \bar{\omega})$ and that $\bar{\mathbb{P}} \in \mathfrak{P}(D_x(\mathbb{R}_+))$ is defined by

$$\bar{\mathbb{P}}(B) = \int \int (\mathbb{1}_B)^{\sigma,\omega}(\tilde{\omega}) \, \nu(\omega)(d\tilde{\omega}) \, \mathbb{P}(d\omega)$$

for $B \in \mathcal{B}(D_x(\mathbb{R}_+))$, where $\nu : D_x(\mathbb{R}^d) \to \mathfrak{P}(D_x(\mathbb{R}^d))$ is a $\mathcal{F}_\sigma$-measurable kernel with

$$\nu(\omega) \in \mathcal{P}(\tau(\bar{\omega} \otimes_s \omega), \bar{\omega} \otimes_s \omega)$$

for $\mathbb{P}$-almost every $\omega \in D_x(\mathbb{R}^d)$. It suffices to show that $\bar{\mathbb{P}}$ is an element of $\mathcal{P}(s, \bar{\omega})$. According to [93, Proposition 4.1] and the first half of the proof of [93, Proposition 4.2], we have $\bar{\mathbb{P}} \in \mathfrak{P}^{ac}_{sem}(D_x(\mathbb{R}^d))$. Since $\bar{\mathbb{P}} = \mathbb{P}$ on $\mathcal{F}_\sigma$ by construction and $\mathbb{P} \in \mathcal{P}_x^\Theta(s, \bar{\omega})$ by assumption, we find that

$$\{(u, \tilde{\omega}) \in \mathbb{R}_+ \times D_x(\mathbb{R}_+) \mid (b_u^{\bar{\mathbb{P}}}, c_u^{\bar{\mathbb{P}}}, F_u^{\bar{\mathbb{P}}})(\tilde{\omega}) \notin \Theta_{s+u}(\bar{\omega} \otimes_s \tilde{\omega}) \text{ for } u < \sigma(\tilde{\omega})\}$$

is a $\lambda \times \bar{\mathbb{P}}$-null set. In order to show that $\bar{\mathbb{P}} \in \mathcal{P}_x^\Theta(s, \bar{\omega})$, it therefore remains to show that

$$B := \{(u, \tilde{\omega}) \in \mathbb{R}_+ \times D_x(\mathbb{R}_+) \mid (b_u^{\bar{\mathbb{P}}}, c_u^{\bar{\mathbb{P}}}, F_u^{\bar{\mathbb{P}}})(\tilde{\omega}) \notin \Theta_{s+u}(\bar{\omega} \otimes_s \tilde{\omega}) \text{ for } u \geq \sigma(\tilde{\omega})\}$$

is also a $\lambda \times \bar{\mathbb{P}}$-null set: According to [93, Theorem 3.1], $\bar{\mathbb{P}}^{\sigma,\omega} \in \mathfrak{P}^{ac}_{sem}(D_x(\mathbb{R}_+))$ for $\mathbb{P}$-almost every $\omega \in D_x(\mathbb{R}_+)$ with differential characteristics

$$(b_u^{\bar{\mathbb{P}}^{\sigma,\omega}}, c_u^{\bar{\mathbb{P}}^{\sigma,\omega}}, F_u^{\bar{\mathbb{P}}^{\sigma,\omega}})(\tilde{\omega}) = (b_{\sigma(\omega)+u}^{\bar{\mathbb{P}}}, c_{\sigma(\omega)+u}^{\bar{\mathbb{P}}}, F_{\sigma(\omega)+u}^{\bar{\mathbb{P}}})(\omega \otimes_{\sigma(\omega)} \tilde{\omega})$$

$\lambda(du) \times \bar{\mathbb{P}}^{\sigma,\omega}(d\tilde{\omega})$-almost everywhere. Hence, the identity

$$B^{\sigma,\omega} := \{(u, \tilde{\omega}) \in \mathbb{R}_+ \times D_x(\mathbb{R}_+) \mid (\mathbb{1}_B(u, \cdot))^{\sigma,\omega}(\tilde{\omega}) = 1\}$$
$$= \{(u, \tilde{\omega}) \in \mathbb{R}_+ \times D_x(\mathbb{R}_+) \mid (b_u^{\bar{\mathbb{P}}^{\sigma,\omega}}, c_u^{\bar{\mathbb{P}}^{\sigma,\omega}}, F_u^{\bar{\mathbb{P}}^{\sigma,\omega}})(\tilde{\omega}) \notin \Theta_{\tau+u}((\bar{\omega} \otimes_s \omega) \otimes_\tau \tilde{\omega})\}$$

holds $\lambda(du) \times \bar{\mathbb{P}}^{\sigma,\omega}(d\tilde{\omega})$-almost everywhere for $\mathbb{P}$-almost every $\omega \in D_x(\mathbb{R}_+)$, where we used the notation $\Theta_\tau(\omega) := \Theta_{\tau(\omega)}(\omega)$. Moreover, [98, Lemma 2.7] shows that $\bar{\mathbb{P}}^{\sigma,\omega} = \nu(\omega)$ for $\mathbb{P}$-almost every $\omega \in D_x(\mathbb{R}_+)$, which implies

$$(\lambda \times \nu(\omega))(B^{\sigma,\omega})$$
$$= (\lambda \times \nu(\omega))(\{(u, \tilde{\omega}) \in \mathbb{R}_+ \times D_x(\mathbb{R}_+) \mid (b_u^{\nu(\omega)}, c_u^{\nu(\omega)}, F_u^{\nu(\omega)})(\tilde{\omega}) \notin \Theta_{\tau+u}((\bar{\omega} \otimes_s \omega) \otimes_\tau \tilde{\omega})\})$$
$$= 0$$

$\mathbb{P}(d\omega)$-almost surely, using $\nu(\omega) \in \mathcal{P}_x^\Theta(\tau(\bar{\omega} \otimes_s \omega), \bar{\omega} \otimes_s \omega)$ for $\mathbb{P}$-almost every $\omega \in D_x(\mathbb{R}_+)$.

An application of Fubini's theorem leads to

$$(\lambda \times \bar{\mathbb{P}})(B) = \int_{\mathbb{R}_+} \int_{D_x(\mathbb{R}_+)} \int_{D_x(\mathbb{R}_+)} (\mathbb{1}_B(u, \cdot))^{\sigma, \omega}(\tilde{\omega})\, \nu(\omega)(d\tilde{\omega})\, \mathbb{P}(d\omega)\, du$$

$$= \int_{D_x(\mathbb{R}_+)} \int_{D_x(\mathbb{R}_+)} \int_{\mathbb{R}_+} \mathbb{1}_{B^{\sigma,\omega}}(u, \tilde{\omega})\, du\, \nu(\omega)(d\tilde{\omega})\, \mathbb{P}(d\omega) = 0$$

and therefore $\bar{\mathbb{P}} \in \mathcal{P}_x^\Theta(s, \bar{\omega})$, as required.

At last, we show the validity of the measurability condition in Lemma 4.30: Note that

$$\{(x, s, \omega, \theta) \in \mathbb{R}^d \times \mathbb{R}_+ \times D(\mathbb{R}_+) \times \mathfrak{S} \mid \omega \in D_x(\mathbb{R}_+)\}$$

is a closed set and therefore Borel-measurable, and hence

$$B := \{(x, s, \omega, \theta) \in \mathbb{R}^d \times \mathbb{R}_+ \times D(\mathbb{R}_+) \times \mathfrak{S} \mid \omega \in D_x(\mathbb{R}_+) \text{ and } \theta \notin \Theta_s(\omega)\}$$
$$= \{(x, s, \omega, \theta) \in \mathbb{R}^d \times \mathbb{R}_+ \times D(\mathbb{R}_+) \times \mathfrak{S} \mid \omega \in D_x(\mathbb{R}_+)\} \cap (\mathbb{R}^d \times A^C)$$
$$\in \mathcal{B}(\mathbb{R}^d) \otimes \mathcal{B}(\mathbb{R}_+) \otimes \mathcal{B}(D(\mathbb{R}_+)) \otimes \mathcal{B}(\mathfrak{S}) = \mathcal{B}(\mathbb{R}^d \times \mathbb{R}_+ \times D(\mathbb{R}_+) \times \mathfrak{S})$$

using $A := \{(s, \omega, \theta) \in \mathbb{R}_+ \times D(\mathbb{R}_+) \times \mathfrak{S} \mid \theta \in \Theta_s(\omega)\} \in \mathcal{B}(\mathbb{R}_+) \otimes \mathcal{B}(D(\mathbb{R}_+)) \otimes \mathcal{B}(\mathfrak{S})$ and the separability of the related spaces (cf. [30, Proposition 8.1.7, p. 243]). In particular,

$$\mathbb{R}^d \times \mathbb{R}_+ \times D(\mathbb{R}_+) \times \mathfrak{P}_{\text{sem}}^{\text{ac}}(D(\mathbb{R}_+)) \ni (x, s, \omega, \mathbb{P}) \longmapsto \mathbb{1}_B\left(x, s, \omega, \left(b_s^\mathbb{P}(\omega), c_s^\mathbb{P}(\omega), F_s^\mathbb{P}(\omega)\right)\right)$$

is Borel-measurable due to Theorem 4.20, which implies that

$$\mathbb{R}^d \times \mathfrak{P}_{\text{sem}}^{\text{ac}}(D(\mathbb{R}_+)) \ni (x, \mathbb{P}) \longmapsto \int_{D(\mathbb{R}_+)} \int_{\mathbb{R}_+} \mathbb{1}_B\left(x, s, \omega, \left(b_s^\mathbb{P}(\omega), c_s^\mathbb{P}(\omega), F_s^\mathbb{P}(\omega)\right)\right) ds\, \mathbb{P}(d\omega)$$

is Borel-measurable as well, according to [16, Proposition 7.29, p. 144]. Altogether,

$$\{(x, \mathbb{P}) \in \mathbb{R}^d \times \mathfrak{P}(D(\mathbb{R}_+)) \mid x \in \mathbb{R}^d \text{ and } \mathbb{P}|_{D_x(\mathbb{R}_+)} \in \mathcal{P}_x^\Theta\}$$
$$= \{(x, \mathbb{P}) \in \mathbb{R}^d \times \mathfrak{P}_{\text{sem}}^{\text{ac}}(D(\mathbb{R}_+)) \mid \mathbb{P}(D_x(\mathbb{R}_+)) = 1, (b_s^\mathbb{P}, c_s^\mathbb{P}, F_s^\mathbb{P})(\omega) \in \Theta_s(\omega)\ (\lambda \times \mathbb{P})\text{-a.e.}\}$$
$$= \{(x, \mathbb{P}) \in \mathbb{R}^d \times \mathfrak{P}_{\text{sem}}^{\text{ac}}(D(\mathbb{R}_+)) \mid \int \int \mathbb{1}_B(x, s, \omega, (b_s^\mathbb{P}(\omega), c_s^\mathbb{P}(\omega), F_s^\mathbb{P}(\omega))) ds\, \mathbb{P}(d\omega) = 0\}$$

is Borel-measurable and thus analytic, as required in order to apply Lemma 4.30. $\square$

Lemma 4.32 (Markov Property). *Suppose that* $\bar{\Theta}: \mathbb{R}^d \to 2^\mathfrak{S} = 2^{\mathbb{R}^d \times \mathbb{S}_+^{d \times d} \times \mathcal{L}(\mathbb{R}^d)}$ *has a Borel-measurable graph in that* $\{(x, \theta) \in \mathbb{R}^d \times \mathfrak{S} \mid \theta \in \bar{\Theta}(x)\} \in \mathcal{B}(\mathbb{R}^d) \otimes \mathcal{B}(\mathfrak{S})$, *then*

$$\Theta: D(\mathbb{R}_+; \mathbb{R}^d) \times \mathbb{R}_+ \longrightarrow 2^\mathfrak{S} = 2^{\mathbb{R}^d \times \mathbb{S}_+^{d \times d} \times \mathcal{L}(\mathbb{R}^d)}$$
$$(\omega, t) \longmapsto \Theta(\omega, t) := \bar{\Theta}(\omega_t)$$

is adapted to the natural filtration $\mathcal{F} = (\mathcal{F}_t)_{t \geq 0}$ *as in Proposition 4.31, Moreover, the corresponding sublinear conditional expectation from Theorem 4.29 satisfies*

$$E^x(E^{X_t}(\xi)) = E^x(\xi \circ \vartheta_t)$$

for every upper semi-analytic $\xi: D(\mathbb{R}_+) \to \mathbb{R}$, $t \geq 0$ *and* $x \in \mathbb{R}^d$, *where* $(X_t)_{t \geq 0}$ *is the* coordinate-mapping process and $\vartheta_t : D(\mathbb{R}_+) \to D(\mathbb{R}_+)$ with $\vartheta_t(\omega) = (\omega_{t+s})_{s\geq 0}$ is the shift operator on the path space $D(\mathbb{R}_+)$. In particular,

$$E^x(E^{X_t}(\varphi(X_s))) = E^x(\varphi(X_{t+s}))$$

for all upper semi-analytic $\varphi : \mathbb{R}^d \to \mathbb{R}$ and $x \in \mathbb{R}^d$.

Proof. Regarding the adaptedness, note that $\phi_t : [0,t] \times D(\mathbb{R}_+) \times \mathfrak{S} \to \mathbb{R}^d \times \mathfrak{S}$ with $\phi_t(s,\omega,\theta) := (\omega_s, \theta)$ is obviously measurable with respect to $\mathcal{B}([0,t]) \otimes \mathcal{F}_t \otimes \mathcal{B}(\mathfrak{S})$ for every $t \geq 0$, which implies that the preimage

$$\{(s,\omega,\theta) \in [0,t] \times D(\mathbb{R}_+) \times \mathfrak{S} \mid \theta \in \Theta(\omega, s)\} = \phi_t^{-1}\left(\{(x,\theta) \in \mathbb{R}^d \times \mathfrak{S} \mid \theta \in \bar{\Theta}(x)\}\right)$$

is in $\mathcal{B}([0,t]) \otimes \mathcal{F}_t \otimes \mathcal{B}(\mathfrak{S})$ for every $t \geq 0$, as required.

Regarding the Markov property, note that it is easy to check that the construction $\hat{\mathbb{P}}(d\hat{\omega}) := \mathbb{P}(d\hat{\omega} - \omega_t + x)$ for $\mathbb{P} \in \mathcal{P}_x^{\Theta}(t,\omega) \subset \mathfrak{P}_{\text{sem}}^{\text{ac}}(D_x(\mathbb{R}_+))$ defines $\hat{\mathbb{P}} \in \mathfrak{P}_{\text{sem}}^{\text{ac}}(D_{\omega(t)}(\mathbb{R}_+))$ with differential characteristics

$$(b_s^{\hat{\mathbb{P}}}, c_s^{\hat{\mathbb{P}}}, F_s^{\hat{\mathbb{P}}})(\hat{\omega}) = (b_s^{\mathbb{P}}, c_s^{\mathbb{P}}, F_s^{\mathbb{P}})(\hat{\omega} - \omega_t + x)$$

$\lambda(ds) \times \hat{\mathbb{P}}(d\hat{\omega})$-almost everywhere. Hence, $\hat{\mathbb{P}} \in \mathcal{P}_{\omega(t)}^{\Theta}(0, \omega(t))$, since the inclusion

$$\begin{aligned}(b_s^{\hat{\mathbb{P}}}, c_s^{\hat{\mathbb{P}}}, F_s^{\hat{\mathbb{P}}})(\hat{\omega}) = (b_s^{\mathbb{P}}, c_s^{\mathbb{P}}, F_s^{\mathbb{P}})(\hat{\omega} - \omega_t + x) &\in \Theta(\omega \otimes_t (\hat{\omega} - \omega_t + x), t+s)\\ &= \bar{\Theta}\left((\omega \otimes_t (\hat{\omega} - \omega_t + x))(t+s)\right)\\ &= \bar{\Theta}(\hat{\omega}_s - \hat{\omega}_0 + \omega_t)\\ &= \Theta(\omega_t \otimes_0 \hat{\omega}, s)\end{aligned}$$

holds $\lambda(ds) \times \hat{\mathbb{P}}(d\hat{\omega})$-almost everywhere, where $\omega_t = \omega(t)$ represents both the starting point in $\mathbb{R}^d$ and the constant path in $D_{\omega(t)}(\mathbb{R}_+)$. Analogously, one can show that

$$\mathbb{P}(d\hat{\omega}) := \hat{\mathbb{P}}(d\hat{\omega} + \omega_t - x)$$

for $\hat{\mathbb{P}} \in \mathcal{P}_{\omega(t)}^{\Theta}(0, \omega(t))$ defines $\mathbb{P} \in \mathcal{P}_x^{\Theta}(t,\omega)$. In particular,

$$\begin{aligned}E_t^x(\xi \circ \vartheta_t)(\omega) = \sup_{\mathbb{P} \in \mathcal{P}_x^{\Theta}(t,\omega)} \mathbb{E}_{\mathbb{P}}\left((\xi \circ \vartheta_t)^{t,\omega}\right) &= \sup_{\mathbb{P} \in \mathcal{P}_x^{\Theta}(t,\omega)} \int_{\Omega} (\xi \circ \vartheta_t)(\omega \otimes_t \bar{\omega})\, \mathbb{P}(d\bar{\omega})\\ &= \sup_{\mathbb{P} \in \mathcal{P}_x^{\Theta}(t,\omega)} \int_{\Omega} \xi(\bar{\omega}(\cdot) - \bar{\omega}(0) + \omega(t))\, \mathbb{P}(d\bar{\omega})\\ &= \sup_{\hat{\mathbb{P}} \in \mathcal{P}_{\omega_t}^{\Theta}(0,\omega_t)} \int_{\Omega} \xi(\hat{\omega})\, \hat{\mathbb{P}}(d\hat{\omega}) = E^{X_t(\omega)}(\xi)(\omega),\end{aligned}$$

since $\vartheta_t(\omega \otimes_t \bar{\omega})(s) = (\omega \otimes_t \bar{\omega})(t+s) = \bar{\omega}(s) - \bar{\omega}(0) + \omega(t)$. This implies

$$E^x(\xi \circ \vartheta_t) = E^x(E_t^x(\xi \circ \vartheta_t)) = E^x(E^{X_t}(\xi)),$$

using the aggregation property from Theorem 4.29. $\square$

4. Stochastic Processes

Remark 4.33 (Construction of Markov Semigroups). Suppose that the uncertainty coefficients $(b_\alpha, c_\alpha, F_\alpha) : \mathbb{R}^d \to \mathfrak{S} := \mathbb{R}^d \times \mathbb{S}_+^{d \times d} \times \mathcal{L}(\mathbb{R}^d)$ with $\alpha \in \mathcal{A}$ satisfy

$$\left\{ (x, \theta) \in \mathbb{R}^d \times \mathfrak{S} \;\middle|\; \theta \in \bigcup_{\alpha \in \mathcal{A}} (b_\alpha, c_\alpha, F_\alpha)(x) \right\} \in \mathcal{B}(\mathbb{R}^d \times \mathfrak{S}).$$

A combination of Theorem 4.29, Lemma 4.30, Proposition 4.31 and Lemma 4.32 show that for $t \geq 0$, $x \in \mathbb{R}^d$ and bounded, upper semi-analytic $\varphi : \mathbb{R}^d \to \mathbb{R}$

$$T_t \varphi(x) := \sup_{\mathbb{P} \in \mathcal{P}_x} \mathbb{E}_\mathbb{P}(\varphi(X_t))$$

defines a sublinear Markov semigroup $(T_t)_{t \geq 0}$ as in Section 4.1, where $(X_t)_{t \geq 0}$ is the coordinate-mapping process on $D(\mathbb{R}_+; \mathbb{R}^d)$ and the uncertainty subsets are given by

$$\mathcal{P}_x := \left\{ \mathbb{P} \in \mathfrak{P}_{\text{sem}}^{\text{ac}}(D_x(\mathbb{R}_+)) \;\middle|\; (b_s^\mathbb{P}, c_s^\mathbb{P}, F_s^\mathbb{P})(\omega) \in \bigcup_{\alpha \in \mathcal{A}} (b_\alpha, c_\alpha, F_\alpha)(\omega_s) \;\; \lambda(ds) \times \mathbb{P}(d\omega)\text{-a.e.} \right\}.$$

This nontrivial construction suggests to interpret the resulting Markov semigroups as a generalization of classical (linear) Markov semigroups under uncertainty in their characteristics. Note, however, that $\mathbb{P} \in \mathcal{P}_x$ for $x \in \mathbb{R}^d$ are not necessarily Markovian, in the sense that their characteristics $(b_t^\mathbb{P}, c_t^\mathbb{P}, F_t^\mathbb{P})$ are in general not a deterministic function of the process X_t at time $t \geq 0$.

Eventually, we want to apply our results from Section 4.1 to connect the sublinear Markov semigroups in Remark 4.33 with viscosity solutions of their generator equations. As a first step, we determine the explicit form of their sublinear generators (in terms of integro-differential operators) under some additional regularity assumptions on the uncertainty coefficients. The proof of this statement relies on the following intermediate result on the regularity in time of sublinear Markov semigroups from Remark 4.33.

Remark 4.34 (Quadratic Variations). The square bracket $[X, Y]$ and $[X] = [X, X]$ denotes the *quadratic (co-)variation* of semimartingales X and Y as in [73, Chapter 1, Definition 4.45, p. 51], as opposed to the angle bracket $\langle X, Y \rangle$ and $\langle X \rangle = \langle X, X \rangle$ for the *predictable quadratic (co-)variation* of locally square-integrable martingales X and Y as in [73, Chapter 1, Theorem 4.2, p. 38]. Recall that for martingales with jumps those two notions do in general not coincide, cf. e.g. [73, Chapter 1, Theorem 4.52, p. 55].

Lemma 4.35 (Burkholder-Davis-Gundy Inequality for Increments). *For every exponent $p \geq 1$ there exist constants $0 < c_p \leq C_p < \infty$ such that the inequalities*

$$c_p \cdot \mathbb{E}_\mathbb{P}\left(([X]_t - [X]_s)^{p/2} \right) \leq \mathbb{E}_\mathbb{P}\left(\sup_{u \in [s,t]} |X_u - X_s|^p \right) \leq C_p \cdot \mathbb{E}_\mathbb{P}\left(([X]_t - [X]_s)^{p/2} \right)$$

hold for all càdlàg $\mathbb{P}$-$\mathcal{F}$-local martingales $(X_t)_{t \geq 0}$ and every $t \geq s \geq 0$.

Proof. It is easy to check (using its definition directly) that for every $s \geq 0$ the process

$$(Y_t)_{t \geq 0} := (X_{s+t} - X_s)_{t \geq 0}$$

is a càdlàg $\mathbb{P}$-$\bar{\mathcal{F}}$-local martingale with respect to the filtration $(\bar{\mathcal{F}}_t)_{t \geq 0} := (\mathcal{F}_{s+t})_{t \geq 0}$. Hence, [110, Chapter 4, Theorem 48, p. 193] implies for $p \geq 1$ the existence of constants $0 < c_p \leq C_p < \infty$ (independent of $(Y_t)_{t \geq 0}$) such that

$$c_p \cdot \mathbb{E}_{\mathbb{P}}\left([Y]_{t-s}^{p/2}\right) \leq \mathbb{E}_{\mathbb{P}}\left(\sup_{u \in [0,t-s]} |Y_u|^p\right) = \mathbb{E}_{\mathbb{P}}\left(\sup_{u \in [s,t]} |X_u - X_s|^p\right) \leq C_p \cdot \mathbb{E}_{\mathbb{P}}\left([Y]_{t-s}^{p/2}\right)$$

holds for all $t \geq s$. It remains to show that $\mathbb{P}([Y]_{t-s} = [X]_t - [X]_s) = 1$ for $t \geq s$: According to [110, Chapter 2, Theorem 23, p. 66], we have uniform in probability

$$\sum_{i=0}^{k_n - 1} (X^{T_{i+1}^n} - X^{T_i^n})(Y^{T_{i+1}^n} - Y^{T_i^n}) \longrightarrow [X, Y]$$

as $n \to \infty$, where $0 = T_0^n \leq T_1^n \leq \ldots \leq T_{k_n}^n$ is an arbitrary sequence of stopping times with $\lim_{n \to \infty} T_{k_n}^n = \infty$ and $\lim_{n \to \infty} \sup_{i \in \{1, \ldots, k_n\}} |T_i^n - T_{i-1}^n| = 0$ $\mathbb{P}$-almost surely. Thus,

$$[X_{s+\cdot}]_{t-s} = \mathbb{P}\text{-}\lim_{n \to \infty} \sum_{i=0}^{n-1} \left|X_{s+\frac{i+1}{n}(t-s)} - X_{s+\frac{i}{n}(t-s)}\right|^2$$

$$[X]_t = \mathbb{P}\text{-}\lim_{n \to \infty} \left(\sum_{i=0}^{n-1} \left|X_{\frac{i+1}{n}s} - X_{\frac{i}{n}s}\right|^2 + \sum_{i=0}^{n-1} \left|X_{s+\frac{i+1}{n}(t-s)} - X_{s+\frac{i}{n}(t-s)}\right|^2\right)$$

$$[X]_s = \mathbb{P}\text{-}\lim_{n \to \infty} \sum_{i=0}^{n-1} \left|X_{\frac{i+1}{n}s} - X_{\frac{i}{n}s}\right|^2,$$

which implies $[X_{s+\cdot}]_{t-s} = [X]_t - [X]_s$ $\mathbb{P}$-almost surely. In particular,

$$[Y]_{t-s} = [X_{s+\cdot} - X_s]_{t-s} = [X_{s+\cdot}]_{t-s} = [X]_t - [X]_s$$

$\mathbb{P}$-almost surely, using the translation invariance of the quadratic variation. $\square$

Proposition 4.36 (Continuity in Time). *If* $\mathcal{P} \subset \mathfrak{P}_{sem}^{ac}(D(\mathbb{R}_+; \mathbb{R}^d))$ *with*

$$\sup_{\mathbb{P} \in \mathcal{P}} \left(\operatorname*{ess\,sup}^{\mathbb{P}}_{(s,\omega) \in \mathbb{R}_+ \times D(\mathbb{R}_+)} \left(|b_s^{\mathbb{P}}(\omega)| + |c_s^{\mathbb{P}}(\omega)| + \int_{\mathbb{R}^d} (|z| \wedge |z|^2) F_s^{\mathbb{P}}(dz)(\omega)\right)\right) < \infty,$$

then there exists a constant $C > 0$ (depending only on the preceding quantity and the cutoff function $h : \mathbb{R}^d \to \mathbb{R}^d$ for the characteristics) such that

$$\sup_{\mathbb{P} \in \mathcal{P}} \mathbb{E}_{\mathbb{P}}\left(\sup_{u \in [s,t]} |X_u - X_s|\right) \leq C\left(|t - s| + |t - s|^{1/2}\right)$$

for all $t \geq s \geq 0$.

4. Stochastic Processes

Proof. In order to avoid cluttered notation, we use $C > 0$ as a universal constant that might change from line to line, but only depends on the bound for the characteristics and the truncation function. According to [73, Chapter 2, Theorem 2.34, p. 84],

$$X_t - X_0 = \int_0^t b_s^{\mathbb{P}} \, ds + X_t^{c,\mathbb{P}} + X_t^{d,\mathbb{P}} + \int_0^t \int_{\mathbb{R}^d} (z - h(z)) \, \mu^X(ds, dz)$$

under $\mathbb{P} \in \mathcal{P}$, where $\mu^X = \sum_{s \geq 0} \mathbb{1}_{\Delta X_s \neq 0} \delta_{(s, \Delta X_s)}$ is the associated jump measure,

$$X_t^{d,\mathbb{P}} = \int_0^t \int_{\mathbb{R}^d} h(z) \left(\mu^X(ds, dz) - F_s^{\mathbb{P}}(dz) ds \right)$$

is the purely-discontinuous and $(X_t^{c,\mathbb{P}})_{t \geq 0}$ the continuous local martingale part. Together with the triangle inequality, this representation allows us consider the terms separately: The bounded variation part can be treated as

$$\mathbb{E}_{\mathbb{P}} \left(\sup_{u \in [s,t]} \left| \int_s^u b_v^{\mathbb{P}} \, dv \right| \right) \leq \mathbb{E}_{\mathbb{P}} \left(\int_s^t |b_v^{\mathbb{P}}| \, dv \right) \leq C(t - s),$$

and the sum of the large jumps can be bounded as

$$\mathbb{E}_{\mathbb{P}} \left(\sup_{u \in [s,t]} \left| \int_s^u \int_{\mathbb{R}^d} (z - h(z)) \, \mu^X(dv, dz) \right| \right) \leq \mathbb{E}_{\mathbb{P}} \left(\int_s^t \int_{\mathbb{R}^d} |z - h(z)| \, \mu^X(dv, dz) \right)$$

$$\leq C \cdot \mathbb{E}_{\mathbb{P}} \left(\int_s^t \int_{\mathbb{R}^d} (|z| \wedge |z|^2) \, F_v^{\mathbb{P}}(dz) \, dv \right)$$

$$\leq C(t - s),$$

using $|z - h(z)| \leq C_h(|z| \wedge |z|^2)$ for some constant $C_h > 0$ only depending on $h : \mathbb{R}^d \to \mathbb{R}^d$, and that $F_s(dz)ds$ is the compensator of $\mu^X(ds, dz)$ in the sense of [73, Chapter 2, Section 2, p. 66]. For the local martingale parts, an application of the Burkholder-Davis-Gundy inequality in the form of Lemma 4.35 implies

$$\mathbb{E}_{\mathbb{P}} \left(\sup_{u \in [s,t]} |X_u^{c,\mathbb{P}} - X_s^{c,\mathbb{P}}| \right) \leq C \cdot \mathbb{E}_{\mathbb{P}} \left(([X^{c,\mathbb{P}}]_t - [X^{c,\mathbb{P}}]_s)^{1/2} \right)$$

$$= C \cdot \mathbb{E}_{\mathbb{P}} \left(\left| \int_s^t c_v \, dv \right|^{1/2} \right) \leq C(t - s)^{1/2}$$

for the continuous part, and similarly for the purely-discontinuous part

$$\mathbb{E}_{\mathbb{P}} \left(\sup_{u \in [s,t]} |X_u^{d,\mathbb{P}} - X_s^{d,\mathbb{P}}| \right) \leq C \cdot \mathbb{E}_{\mathbb{P}} \left(([X^{d,\mathbb{P}}]_t - [X^{d,\mathbb{P}}]_s)^{1/2} \right)$$

$$= C \cdot \mathbb{E}_{\mathbb{P}} \left(\left(\int_s^t \int_{\mathbb{R}^d} |h(z)|^2 \, F_v^{\mathbb{P}}(dz) \, dv \right)^{1/2} \right)$$

$$\leq C \cdot \mathbb{E}_{\mathbb{P}} \left(\left(\int_s^t \int_{\mathbb{R}^d} (1 \wedge |z|^2) \, F_v^{\mathbb{P}}(dz) \, dv \right)^{1/2} \right) \leq C(t - s)^{1/2},$$

using $|h(z)|^2 \leq C_h(1 \wedge |z|^2)$ for some constant $C_h > 0$ only depending on $h : \mathbb{R}^d \to \mathbb{R}^d$. $\square$

4.2. Construction of Processes

Theorem 4.37 (Form of Generators). *Suppose that the uncertainty coefficients*

$$(b_\alpha, c_\alpha, F_\alpha) : \mathbb{R}^d \longrightarrow \mathbb{R}^d \times \mathbb{S}_+^{d\times d} \times \mathfrak{L}(\mathbb{R}^d)$$

are uniformly bounded in space and $\alpha \in \mathcal{A}$, i.e.

$$\sup_{\alpha \in \mathcal{A}} \sup_{x \in \mathbb{R}^d} \left(|b_\alpha(x)| + |c_\alpha(x)| + \int_{\mathbb{R}^d} (|z| \wedge |z|^2) \, F_\alpha(x, dz) \right) < \infty,$$

and uniformly Lipschitz-continuous in $\alpha \in \mathcal{A}$, i.e. there exists $L > 0$ such that

$$\sup_{\alpha \in \mathcal{A}} \left(|b_\alpha(x) - b_\alpha(y)| + |c_\alpha(x) - c_\alpha(y)| \right) \leq L|x - y|$$

$$\sup_{\alpha \in \mathcal{A}} \left| \int_{\mathbb{R}^d} g(z) \, F_\alpha(x, dz) - \int_{\mathbb{R}^d} g(z) \, F_\alpha(y, dz) \right| \leq \sup_{z \in \mathbb{R}^d \setminus \{0\}} \left(\frac{|g(z)|}{1 \wedge |z|^2} \right) L|x - y|$$

for all $x, y \in \mathbb{R}^d$ and $g \in C_b(\mathbb{R}^d)$. If $E^x(\cdot) = \sup_{\mathbb{P} \in \mathcal{P}_x} \mathbb{E}_\mathbb{P}(\cdot)$ for some uncertainty subsets

$$\mathcal{P}_x \subset \left\{ \mathbb{P} \in \mathfrak{P}_{sem}^{ac}(D_x(\mathbb{R}_+)) \; \middle| \; (b_s^\mathbb{P}, c_s^\mathbb{P}, F_s^\mathbb{P})(\omega) \in \bigcup_{\alpha \in \mathcal{A}} (b_\alpha, c_\alpha, F_\alpha)(\omega_s) \quad \lambda(ds) \times \mathbb{P}(d\omega)\text{-a.e.} \right\}$$

that contain $\mathbb{P}_\alpha \in \mathcal{P}_x$ for all $\alpha \in \mathcal{A}$ and $x \in \mathbb{R}^d$ with differential characteristics

$$(b_s^{\mathbb{P}_\alpha}, c_s^{\mathbb{P}_\alpha}, F_s^{\mathbb{P}_\alpha})(\omega) = (b_\alpha, c_\alpha, F_\alpha)(\omega_s)$$

$\lambda(ds) \times \mathbb{P}_\alpha(d\omega)$-*almost everywhere, then $G\phi(x) := \lim_{\delta \downarrow 0} \delta^{-1} \left(E^x(\phi(X_\delta)) - \phi(x) \right)$ converges for all $\phi \in C_b^\infty(\mathbb{R}^d)$ and $x \in \mathbb{R}^d$ with limits $G\phi(x) = \sup_{\alpha \in \mathcal{A}} G_\alpha \phi(x)$, where*

$$G_\alpha \phi(x) := b_\alpha(x) D\phi(x) + \frac{1}{2} \text{tr} \left(c_\alpha(x) D^2 \phi(x) \right)$$
$$+ \int_{\mathbb{R}^d} \left(\phi(x+z) - \phi(x) - D\phi(x)h(z) \right) F_\alpha(x, dz).$$

Proof. The proof has two parts: In the first half, we will show that

$$\lim_{\delta \downarrow 0} \frac{E^x(\phi(X_\delta)) - \phi(x)}{\delta}$$
$$= \lim_{\delta \downarrow 0} \sup_{\mathbb{P} \in \mathcal{P}_x} \frac{1}{\delta} \mathbb{E}_\mathbb{P} \left(\int_0^\delta b_s^\mathbb{P} D\phi(x) \, ds + \frac{1}{2} \int_0^\delta \text{tr} \left(c_s^\mathbb{P} D^2 \phi(x) \right) ds \right.$$
$$\left. + \int_0^\delta \int_{\mathbb{R}^d} \left(\phi(x+z) - \phi(x) - D\phi(x)h(z) \right) F_s^\mathbb{P}(dz) \, ds \right),$$

using Itô's formula for semimartingales and the regularity of the function $\phi \in C_b^\infty(\mathbb{R}^d)$. In the second half, we will use this representation together with the structure of $(\mathcal{P}_x)_{x \in \mathbb{R}^d}$ and the regularity of $x \mapsto (b_\alpha, c_\alpha, F_\alpha)(x)$, in order to obtain the desired result.

4. Stochastic Processes

First of all, [73, Chapter 2, Theorem 2.34, p. 84] shows that every semimartingale under $\mathbb{P}$ has a canonical representation of the form

$$X_t - X_0 = \int_0^t b_s^{\mathbb{P}}\, ds + X_t^{c,\mathbb{P}} + X_t^{d,\mathbb{P}} + \int_0^t \int_{\mathbb{R}^d} (z - h(z))\, \mu^X(ds, dz)$$

with the associated jump measure $\mu^X = \sum_{s \geq 0} \mathbb{1}_{\Delta X_s \neq 0} \delta_{(s, \Delta X_s)}$, the purely-discontinuous

$$X_t^{d,\mathbb{P}} = \int_0^t \int_{\mathbb{R}^d} h(z) \left(\mu^X(ds, dz) - F_s^{\mathbb{P}}(dz) ds \right)$$

and continuous local martingale part $(X_t^{c,\mathbb{P}})_{t \geq 0}$. According to Itô's formula for general semimartingales (cf. [73, Chapter 1, Theorem 4.57, p. 57]),

$$\phi(X_\delta) - \phi(X_0)$$
$$= \int_0^\delta D\phi(X_{s-})\, dX_s + \frac{1}{2} \int_0^\delta D^2\phi(X_{s-})\, d\langle X^{c,\mathbb{P}} \rangle_s$$
$$+ \sum_{s \leq \delta} \left(\phi(X_s) - \phi(X_{s-}) - D\phi(X_{s-}) \Delta X_s \right)$$
$$= \int_0^\delta b_s^{\mathbb{P}} D\phi(X_{s-})\, ds + \int_0^\delta D\phi(X_{s-})\, d(X_s^{c,\mathbb{P}} + X_s^{d,\mathbb{P}}) + \frac{1}{2} \int_0^\delta \operatorname{tr}\left(c_s^{\mathbb{P}} D^2 \phi(X_{s-}) \right) ds$$
$$+ \int_0^\delta \int_{\mathbb{R}^d} \left(\phi(X_{s-} + z) - \phi(X_{s-}) - D\phi(X_{s-})h(z) \right) \mu^X(ds, dz)$$

under each $\mathbb{P} \in \mathfrak{P}_{\text{sem}}^{\text{ac}}(D(\mathbb{R}_+))$. Note that the second term

$$\delta \longmapsto \int_0^\delta D\phi(X_{s-})\, d(X_s^{c,\mathbb{P}} + X_s^{d,\mathbb{P}})$$

is a local martingale (cf. [110, Chapter 2, Theorem 20, p. 63]) and even a martingale by [110, Chapter 2, Corollary 3, p. 73], since for all $t \geq 0$ and $\mathbb{P} \in \mathcal{P}_x$

$$\mathbb{E}_{\mathbb{P}}\left(\left[\int_0^\cdot D\phi(X_{s-})\, d(X_s^{c,\mathbb{P}} + X_s^{d,\mathbb{P}}) \right]_t \right) = \mathbb{E}_{\mathbb{P}}\left(\int_0^t |D\phi(X_{s-})|^2\, d\left[X^{c,\mathbb{P}} + X^{d,\mathbb{P}} \right]_s \right)$$
$$\leq \sup_{x \in \mathbb{R}^d} |D\phi(x)|^2\, \mathbb{E}_{\mathbb{P}}\left(\left[X^{c,\mathbb{P}} + X^{d,\mathbb{P}} \right]_t \right)$$
$$\leq C\, t^{1/2} \sup_{x \in \mathbb{R}^d} |D\phi(x)|^2 < \infty,$$

using the same line of arguments as in the proof of Proposition 4.36. Hence,

$$\mathbb{E}_{\mathbb{P}}\left(\phi(X_\delta) - \phi(X_0) \right)$$
$$= \mathbb{E}_{\mathbb{P}}\bigg(\int_0^\delta b_s^{\mathbb{P}} D\phi(X_{s-})\, ds + \frac{1}{2} \int_0^\delta \operatorname{tr}\left(c_s^{\mathbb{P}} D^2 \phi(X_{s-}) \right) ds$$
$$+ \int_0^\delta \int_{\mathbb{R}^d} \left(\phi(X_{s-} + z) - \phi(X_{s-}) - D\phi(X_{s-})h(z) \right) \mu^X(ds, dz) \bigg)$$

for all $\mathbb{P} \in \mathcal{P}_x$ and $\delta > 0$. The uniform bound for the differential characteristics together with the Lipschitz-continuity of $D\phi$ and $D^2\phi$ (and $\operatorname{tr}(A) \leq d\,\|A\|$ for $A \in \mathbb{S}^{d\times d}$) imply

$$\left|\int_0^\delta b_s^{\mathbb{P}} D\phi(X_{s-})\,ds - \int_0^\delta b_s^{\mathbb{P}} D\phi(x)\,ds\right| \leq \delta C \sup_{s\in[0,\delta]} |X_s - x|$$

$$\left|\int_0^\delta \operatorname{tr}\left(c_s^{\mathbb{P}} D^2\phi(X_{s-})\right)\,ds - \int_0^\delta \operatorname{tr}\left(c_s^{\mathbb{P}} D^2\phi(x)\right)\,ds\right| \leq \delta C \sup_{s\in[0,\delta]} |X_s - x|$$

$\mathbb{P}$-almost surely for some constant $C > 0$ (independent of $\mathbb{P} \in \mathcal{P}_x$ and $\delta > 0$). Similarly, there exists a constant $C > 0$ (independent of $\mathbb{P} \in \mathcal{P}_x$ and $s > 0$) such that

$$|(\phi(X_{s-} + z) - \phi(X_{s-}) - D\phi(X_{s-})h(z)) - (\phi(x+z) - \phi(x) - D\phi(x)h(z))|$$
$$\leq C \cdot |X_{s-} - x| \cdot |z| \cdot |h(z)| + C \cdot |X_{s-} - x| \cdot |z - h(z)| \leq C|X_{s-} - x|\left(|z| \wedge |z|^2\right)$$

using Taylor's theorem repeatedly and $|z| \cdot |h(z)| + |z - h(z)| \leq C_h(|z| \wedge |z|^2)$ for some constant $C_h > 0$ only depending on $h : \mathbb{R}^d \to \mathbb{R}^d$. Therefore,

$$\mathbb{E}_{\mathbb{P}}\left(\left|\int_0^\delta \int_{\mathbb{R}^d} \left(\phi(X_{s-} + z) - \phi(X_{s-}) - D\phi(X_{s-})h(z)\right) \mu^X(ds, dz)\right.\right.$$
$$\left.\left. - \int_0^\delta \int_{\mathbb{R}^d} \left(\phi(x + z) - \phi(x) - D\phi(x)h(z)\right) \mu^X(ds, dz)\right|\right)$$
$$\leq C \mathbb{E}_{\mathbb{P}}\left(\sup_{s\in[0,\delta]} |X_s - x| \int_0^\delta \int_{\mathbb{R}^d} \left(|z| \wedge |z|^2\right) \mu^X(ds, dz)\right)$$
$$= C \mathbb{E}_{\mathbb{P}}\left(\sup_{s\in[0,\delta]} |X_s - x| \int_0^\delta \int_{\mathbb{R}^d} \left(|z| \wedge |z|^2\right) F_s^{\mathbb{P}}(dz)\,ds\right) \leq \delta C \mathbb{E}_{\mathbb{P}}\left(\sup_{s\in[0,\delta]} |X_s - x|\right)$$

with Tonelli's theorem and the uniform bound for the characteristics. Altogether,

$$\lim_{\delta \downarrow 0} \frac{E^x(\phi(X_\delta)) - \phi(x)}{\delta} = \lim_{\delta \downarrow 0} \sup_{\mathbb{P}\in\mathcal{P}_x} \frac{\mathbb{E}_{\mathbb{P}}(\phi(X_\delta) - \phi(x))}{\delta}$$
$$= \lim_{\delta \downarrow 0} \sup_{\mathbb{P}\in\mathcal{P}_x} \frac{1}{\delta} \mathbb{E}_{\mathbb{P}}\left(\int_0^\delta b_s^{\mathbb{P}} D\phi(X_{s-})\,ds + \frac{1}{2}\int_0^\delta \operatorname{tr}\left(c_s^{\mathbb{P}} D^2\phi(X_{s-})\right)\,ds\right.$$
$$\left. + \int_0^\delta \int_{\mathbb{R}^d} \left(\phi(X_{s-} + z) - \phi(X_{s-}) - D\phi(X_{s-})h(z)\right) \mu^X(ds, dz)\right)$$
$$= \lim_{\delta \downarrow 0} \sup_{\mathbb{P}\in\mathcal{P}_x} \frac{1}{\delta} \mathbb{E}_{\mathbb{P}}\left(\int_0^\delta b_s^{\mathbb{P}} D\phi(x)\,ds + \frac{1}{2}\int_0^\delta \operatorname{tr}\left(c_s^{\mathbb{P}} D^2\phi(x)\right)\,ds\right.$$
$$\left. + \int_0^\delta \int_{\mathbb{R}^d} \left(\phi(x + z) - \phi(x) - D\phi(x)h(z)\right) F_s^{\mathbb{P}}(dz)\,ds\right),$$

since Proposition 4.36 implies the existence of a constant $C > 0$ such that

$$\sup_{\mathbb{P}\in\mathcal{P}_x} \mathbb{E}_{\mathbb{P}}\left(\sup_{s\in[0,\delta]} |X_s - x|\right) \leq C(\delta + \delta^{1/2})$$

holds for all $\delta > 0$. This concludes the first half of the proof.

4. Stochastic Processes

For the second half, note that the structural assumptions on $\mathcal{P}_x \subset \mathfrak{P}^{ac}_{sem}(D_x(\mathbb{R}_+))$ together with the uniform Lipschitz-continuity of $x \mapsto (b_\alpha, c_\alpha, F_\alpha)(x)$ imply that

$$\int_0^\delta b_s^\mathbb{P} D\phi(x)\, ds + \frac{1}{2}\int_0^\delta \mathrm{tr}\left(c_s^\mathbb{P} D^2\phi(x)\right) ds$$
$$+ \int_0^\delta \int_{\mathbb{R}^d} \left(\phi(x+z) - \phi(x) - D\phi(x)h(z)\right) F_s^\mathbb{P}(dz)\, ds$$
$$\leq \sup_{\alpha \in \mathcal{A}} \left(\int_0^\delta b_\alpha(X_{s-}) D\phi(x)\, ds + \frac{1}{2}\int_0^\delta \mathrm{tr}\left(c_\alpha(X_{s-}) D^2\phi(x)\right) ds \right.$$
$$\left. + \int_0^\delta \int_{\mathbb{R}^d} \left(\phi(x+z) - \phi(x) - D\phi(x)h(z)\right) F_\alpha(X_{s-})(dz)\, ds \right)$$
$$\leq \sup_{\alpha \in \mathcal{A}} G_\alpha \phi(x) + \delta\, C\, L \sup_{s \in [0,\delta]} |X_s - x|$$

holds $\mathbb{P}$-almost surely for all $\mathbb{P} \in \mathcal{P}_x$. Similarly, we find that

$$\int_0^\delta b_s^{\mathbb{P}_\alpha} D\phi(x)\, ds + \frac{1}{2}\int_0^\delta \mathrm{tr}\left(c_s^{\mathbb{P}_\alpha} D^2\phi(x)\right) ds$$
$$+ \int_0^\delta \int_{\mathbb{R}^d} \left(\phi(x+z) - \phi(x) - D\phi(x)h(z)\right) F_s^{\mathbb{P}_\alpha}(dz)\, ds$$
$$= \int_0^\delta b_\alpha(X_{s-}) D\phi(x)\, ds + \frac{1}{2}\int_0^\delta \mathrm{tr}\left(c_\alpha(X_{s-}) D^2\phi(x)\right) ds$$
$$+ \int_0^\delta \int_{\mathbb{R}^d} \left(\phi(x+z) - \phi(x) - D\phi(x)h(z)\right) F_\alpha(X_{s-})(dz)\, ds$$
$$\geq G_\alpha \phi(x) - \delta\, C\, L \sup_{s \in [0,\delta]} |X_s - x|$$

holds $\mathbb{P}_\alpha$-almost surely for all $\alpha \in \mathcal{A}$. Since $\mathbb{P}_\alpha \in \mathcal{P}_x$ for all $\alpha \in \mathcal{A}$ and $x \in \mathbb{R}^d$, this shows

$$\lim_{\delta \downarrow 0} \frac{E^x(\phi(X_\delta)) - \phi(x)}{\delta} = \sup_{\alpha \in \mathcal{A}} G_\alpha \phi(x)$$

for all $\phi \in C_b(\mathbb{R}^d)$ and $x \in \mathbb{R}^d$, using the representation from the first half and $\lim_{\delta \downarrow 0} E^x(\sup_{s \in [0,\delta]} |X_s - x|) = 0$ from Proposition 4.36. $\square$

In the special case of spatially homogeneous coefficients $(b_\alpha, c_\alpha, F_\alpha)_{\alpha \in \mathcal{A}}$, our results allow us to reconstruct the conclusions from [93] on Lévy processes for sublinear expectations (which in turn are generalizations of the findings in [67]):

Remark 4.38 (Lévy Processes). Suppose that the coefficients

$$(b_\alpha, c_\alpha, F_\alpha)_{\alpha \in \mathcal{A}} \subset \mathbb{R}^d \times \mathbb{S}_+^{d \times d} \times \mathfrak{L}(\mathbb{R}^d)$$

are spatially homogeneous (i.e. not dependent on space) and uniformly bounded

$$\sup_{\alpha \in \mathcal{A}} \left(|b_\alpha| + |c_\alpha| + \int_{\mathbb{R}^d} \left(|z| \wedge |z|^2\right) F_\alpha(dz) \right) < \infty.$$

4.2. Construction of Processes

Let $(X_t)_{t\geq 0}$ denote the coordinate-mapping process on $D(\mathbb{R}_+;\mathbb{R}^d)$ and set for $x \in \mathbb{R}^d$

$$\mathcal{P}_x := \left\{ \mathbb{P} \in \mathfrak{P}^{\mathrm{ac}}_{\mathrm{sem}}(D_x(\mathbb{R}_+)) \;\middle|\; (b^{\mathbb{P}}_s, c^{\mathbb{P}}_s, F^{\mathbb{P}}_s)(\overline{\omega}) \in \bigcup_{\alpha \in \mathcal{A}} (b_\alpha, c_\alpha, F_\alpha) \; ds \times \mathbb{P}(d\overline{\omega})\text{-a.e.} \right\}.$$

A straightforward combination of Theorem 4.29, Lemma 4.30, Proposition 4.31 and Lemma 4.32 shows that $T_t\varphi(x) := E^x(\varphi(X_t))$ defines a sublinear Markov semigroup $(T_t)_{t\geq 0}$ on the convex cone $\mathcal{H}$ of bounded, upper semi-analytic functions, where

$$E^x_\tau(\xi)(\omega) := \sup_{\mathbb{P} \in \mathcal{P}_x} \mathbb{E}_\mathbb{P}(\xi^{\tau(\omega),\omega})$$

for upper semi-analytic functions $\xi : D(\mathbb{R}_+) \to \overline{\mathbb{R}}$, stopping times $\tau : D(\mathbb{R}_+) \to [0,\infty]$ and $x \in \mathbb{R}^d$. Moreover, the construction of the uncertainty subsets $(\mathcal{P}_x)_{x\in\mathbb{R}^d}$ and the spatial homogeneity of the uncertainty coefficients $(b_\alpha, c_\alpha, F_\alpha)_{\alpha \in \mathcal{A}}$ imply

$$T_t\varphi(x) = E^x(\varphi(X_t)) = E^0(\varphi(x + X_t))$$

for all $x \in \mathbb{R}^d$ and upper semi-analytic $\varphi : \mathbb{R}^d \to \mathbb{R}$. Hence, $(T_t)_{t\geq 0}$ also defines a sublinear Markov semigroup on the linear space of bounded, Lipschitz-continuous functions

$$C^{\mathrm{Lip}}_b(\mathbb{R}^d) := \{\varphi : \mathbb{R}^d \to \mathbb{R} \mid \varphi \text{ is bounded and Lipschitz-continuous}\}.$$

Since $C^{\mathrm{Lip}}_b(\mathbb{R}^d)$ is a space of integrands (cf. Definition 3.1), this shows that $(X_t)_{t\geq 0}$ is a Markov process for sublinear expectations (cf. Definition 4.6) on $(\Omega, \hat{\mathcal{H}}, E^x)_{x\in\mathbb{R}^d}$ with $\Omega := D(\mathbb{R}_+; \mathbb{R}^d)$, the linear space $\hat{\mathcal{H}}$ of bounded, Borel-measurable functions (on $\Omega = D(\mathbb{R}_+; \mathbb{R}^d)$) and $\mathcal{T} := C^{\mathrm{Lip}}_b(\mathbb{R}^d)$. In this thesis, we refer to such processes $(X_t)_{t\geq 0}$ as *Lévy processes for sublinear expectations* (whereas [67] calls them *G-Lévy processes*, in analogy to the continuous-paths *G-Brownian motions* from [102]).

Lévy processes for sublinear expectations have independent and stationary increments: The Markov property and spatial homogeneity show that

$$E^x\big(\varphi(X_{t+h} - X_t)\big) = E^x\big(E^{X_t}(\varphi(X_h - X_0))\big) = E^x\big(E^x(\varphi(X_h - X_0))\big)$$
$$= E^x\big(\varphi(X_h - X_0)\big)$$

for $\varphi \in C^{\mathrm{Lip}}_b(\mathbb{R}^d)$, $x \in \mathbb{R}^d$ and $t, h \geq 0$, i.e. $X_{t+h} - X_t \sim X_h - X_0$ with respect to $C^{\mathrm{Lip}}_b(\mathbb{R}^d)$ (in terms of Definition 3.11). Similarly,

$$E^x\big(\varphi(X_{t+h} - X_t, X_t - X_0)\big) = E^x\big(E^y(\varphi(X_h - X_0, y - x))\big|_{y=X_t}\big)$$
$$= E^x\big(E^x(\varphi(X_h - X_0, z))\big|_{z=X_t-X_0}\big)$$
$$= E^x\big(E^x(\varphi(X_{t+h} - X_t, z))\big|_{z=X_t-X_0}\big)$$

for $\varphi \in C^{\mathrm{Lip}}_b(\mathbb{R}^d \times \mathbb{R}^d)$, $x \in \mathbb{R}^d$ and $t, h \geq 0$, i.e. $X_{t+h} - X_t \perp\!\!\!\perp X_t - X_0$ with respect to $C^{\mathrm{Lip}}_b(\mathbb{R}^d \times \mathbb{R}^d)$ (in terms of Definition 3.13). Using the same arguments iteratively, we can show for all $t_n \geq \ldots \geq t_0 \geq 0$ and $n \in \mathbb{N}$ that

$$X_{t_n} - X_{t_{n-1}} \perp\!\!\!\perp X_{t_{n-2}} - X_{t_{n-3}}, \ldots, X_{t_1} - X_{t_0}$$

with respect to $C^{\mathrm{Lip}}_b(\mathbb{R}^{nd})$.

179

4. Stochastic Processes

Lévy processes for sublinear expectations have a strong connection to their sublinear generator equations: For fixed $\varphi \in C_b^{\mathrm{Lip}}(\mathbb{R}^d)$, define $u^\varphi : [0, \infty) \times \mathbb{R}^d \to \mathbb{R}$ by

$$u^\varphi(t, x) := T_t\varphi(x) = E^x(\varphi(X_t)) = E^0(\varphi(x + X_t))$$

for $(t, x) \in [0, \infty) \times \mathbb{R}^d$. According to Proposition 4.36 and the spatial homogeneity,

$$(t, x) \longmapsto u^\varphi(t, x)$$

is uniformly bounded by $\|\varphi\|_\infty = \sup_{x \in \mathbb{R}^d} |\varphi(x)|$ and (global) Hölder-continuous, because

$$|u^\varphi(t, x) - u^\varphi(s, y)| \leq |u^\varphi(t, x) - u^\varphi(t, y)| + |u^\varphi(t, y) - u^\varphi(s, y)|$$
$$\leq L|x - y| + LC\left(|t - s| + |t - s|^{1/2}\right)$$

for all $t, s \geq 0$ and $x, y \in \mathbb{R}^d$ (where $L > 0$ is the Lipschitz-constant of $\varphi \in C_b^{\mathrm{Lip}}(\mathbb{R}^d)$ and $C > 0$ the Hölder constant from Proposition 4.36). Since $(X_t)_{t \geq 0}$ is a Markov process for sublinear expectations with respect to $\mathcal{T} = C_b^{\mathrm{Lip}}(\mathbb{R}^d)$, Proposition 4.10 implies that $u^\varphi \in C_b([0, \infty) \times \mathbb{R}^d)$ is a viscosity solution in $(0, \infty) \times \mathbb{R}^d$ of the generator equation

$$\partial_t u^\varphi(t, x) - A(u^\varphi(t, \cdot))(x) = 0$$

for $u^\varphi(0, \cdot) = \varphi(\cdot) \in C_b^{\mathrm{Lip}}(\mathbb{R}^d)$. Furthermore, Theorem 4.37 shows that the sublinear generator (as introduced in Definition 4.9) is of the form

$$A(\psi)(x) = \lim_{\delta \downarrow 0} \frac{T_\delta \psi(x) - T_0 \psi(x)}{\delta} = \lim_{\delta \downarrow 0} \frac{E^x(\psi(X_\delta)) - E^x(\psi(X_0))}{\delta}$$
$$= \sup_{\alpha \in \mathcal{A}} \left(b_\alpha D\psi(x) + \frac{1}{2} \operatorname{tr}\left(c_\alpha D^2 \psi(x)\right) \right.$$
$$\left. + \int_{\mathbb{R}^d} (\psi(x + z) - \psi(x) - D\psi(x)h(z))\, F_\alpha(dz) \right)$$

for $\psi \in C_b^\infty(\mathbb{R}^d)$, where $h : \mathbb{R}^d \to \mathbb{R}^d$ is the truncation function for semimartingale characteristics (cf. Definition 4.13), which is used to define the uncertainty subsets $(\mathcal{P}_x)_{x \in \mathbb{R}^d}$. In order to apply Theorem 4.37, we have used (cf. [117, Chapter 2, Corollary 11.6, p. 63]) that there exists $\mathbb{P}_\alpha \in \mathcal{P}_x$ for every $\alpha \in \mathcal{A}$ such that $(X_t)_{t \geq 0}$ is a classical Lévy process under $\mathbb{P}_\alpha$ with Lévy triplet $(b_\alpha, c_\alpha, F_\alpha)$ (with respect to $h : \mathbb{R}^d \to \mathbb{R}^d$), and hence

$$(b_t^{\mathbb{P}_\alpha}, c_t^{\mathbb{P}_\alpha}, F_t^{\mathbb{P}_\alpha})(\omega) = (b_\alpha, c_\alpha, F_\alpha)$$

$\lambda(dt) \times \mathbb{P}_\alpha(d\omega)$-almost everywhere according to [73, Chapter 4, Corollary 4.19, p. 107]. In particular, the generator equations of Lévy processes for sublinear expectations are Hamilton-Jacobi-Bellman equations from Section 2.3: It is easy to check that

$$-A(\psi)(x) = G(x, D\psi(x), D^2\psi(x), \psi) := \inf_{\alpha \in \mathcal{A}} G_\alpha(x, D\psi(x), D^2\psi(x), \psi)$$

for $\psi \in C_b^\infty(\mathbb{R}^d)$ and $x \in \mathbb{R}^d$ with the linear operators

$$G_\alpha(x, p, X, \phi) := -\mathcal{L}_\alpha(x, p, X) - \mathcal{I}_\alpha(x, \phi)$$

$$\mathcal{L}_\alpha(x, p, X) := b'_\alpha p + \frac{1}{2} \operatorname{tr}\left(\sigma_\alpha \sigma_\alpha^T X\right)$$

$$\mathcal{I}_\alpha(x, \phi) := \int_{|z| \leq 1} (\phi(x+z) - \phi(x) - D\phi(x)z) \, F_\alpha(dz)$$

$$+ \int_{|z| > 1} (\phi(x+z) - \phi(x)) \, F_\alpha(dz)$$

for $(x, p, X, \phi) \in \mathbb{R}^d \times \mathbb{R}^d \times \mathbb{S}^{d \times d} \times C_b^2(\mathbb{R}^d)$, where

$$b'_\alpha := b_\alpha + \int_{|z| \leq 1} (z - h(z)) \, F_\alpha(dz) - \int_{|z| > 1} h(z) \, F_\alpha(dz)$$

and $\sigma_\alpha := c_\alpha^{1/2}$ is the unique square root of a positive semi-definite, symmetric matrix (cf. [89, Section 6.1.4, p. 88]). Moreover, if the tightness condition

$$\lim_{\kappa \to 0} \sup_{\alpha \in \mathcal{A}} \int_{|z| \leq \kappa} |z|^2 \, F_\alpha(dz) = 0$$

holds, the generator equation satisfies all assumptions of Remark 2.28. Thus, Corollary 2.34 shows that $u^\varphi \in C_b(\mathbb{R}^d)$ is the unique viscosity solution in $(0, \infty) \times \mathbb{R}^d$ of

$$\partial_t u^\varphi(t, x) - \sup_{\alpha \in \mathcal{A}} \left(b_\alpha D u^\varphi(t, x) + \frac{1}{2} \operatorname{tr}\left(c_\alpha D^2 u^\varphi(t, x)\right) \right.$$

$$\left. + \int_{\mathbb{R}^d} (u^\varphi(t, x+z) - u^\varphi(t, x) - D u^\varphi(t, x) h(z)) \, F_\alpha(dz) \right) = 0$$

with $u^\varphi(0, \cdot) = \varphi(\cdot) \in C_b^{\text{Lip}}(\mathbb{R}^d)$. In order to apply the comparison principle from Corollary 2.34, we have to use Lemma 2.4, Remark 2.5 and Lemma 2.6 first, to show that the two notions of viscosity solutions in Definition 2.1 and Proposition 4.10 coincide.

One of the main objectives for the rest of this thesis is to construct and characterize Markov processes similar as in Remark 4.38 for spatially inhomogeneous coefficients

$$(b_\alpha, c_\alpha, F_\alpha) : \mathbb{R}^d \longrightarrow \mathbb{R}^d \times \mathbb{S}_+^{d \times d} \times \mathfrak{L}(\mathbb{R}^d)$$

with $\alpha \in \mathcal{A}$. In contrast to sublinear Markov semigroups for spatially homogeneous coefficients, however, it is much more involved for general Markov semigroups $(T_t)_{t \geq 0}$ (constructed as in Remark 4.33) to show that the functions

$$T_t \varphi : \mathbb{R}^d \longrightarrow \mathbb{R}$$

are continuous for $\varphi \in C_b(\mathbb{R}^d)$ and $t \geq 0$ - or equivalently $T_t C_b(\mathbb{R}^d) \subset C_b(\mathbb{R}^d)$ for $t \geq 0$, which is referred to as the C_b-Feller property for linear Markov semigroups (cf. e.g. [21, Definition 1.6, p. 19]). As a first step towards this objective, we obtain conditions for the compactness of the corresponding uncertainty subsets $(\mathcal{P}_x)_{x \in \mathbb{R}^d} \subset \mathfrak{P}(\Omega)$ and relate these findings to the C_b-Feller property of our sublinear Markov semigroups from Remark 4.33:

4. Stochastic Processes

Remark 4.39 (Assumptions on $(b_\alpha, c_\alpha, F_\alpha)$). In order to show that the spatially inhomogeneous uncertainty coefficients $(b_\alpha, c_\alpha, F_\alpha) : \mathbb{R}^d \to \mathbb{R}^d \times \mathbb{S}_+^{d \times d} \times \mathfrak{L}(\mathbb{R}^d)$ with $\alpha \in \mathcal{A}$ lead to compact uncertainty subsets, we have to impose the following assumptions:

(D1) (Boundedness & Tightness) The uncertainty coefficients $(b_\alpha, c_\alpha, F_\alpha)_{\alpha \in \mathcal{A}}$ are uniformly bounded and uniformly tight in $\alpha \in \mathcal{A}$ and $x \in \mathbb{R}^d$, i.e.

$$\sup_{\alpha \in \mathcal{A}} \sup_{x \in \mathbb{R}^d} \left(|b_\alpha(x)| + |c_\alpha(x)| + \int_{\mathbb{R}^d} (1 \wedge |z|^2) \, F_\alpha(x, dz) \right) < \infty$$

$$\lim_{R \to \infty} \sup_{\alpha \in \mathcal{A}} \sup_{x \in \mathbb{R}^d} \left(\int_{|z| < R^{-1}} (1 \wedge |z|^2) \, F_\alpha(x, dz) + \int_{|z| > R} (1 \wedge |z|^2) \, F_\alpha(x, dz) \right) = 0.$$

(D2) (Continuity) There exists a (Lipschitz) constant $L > 0$ such that

$$\sup_{\alpha \in \mathcal{A}} (|b_\alpha(x) - b_\alpha(y)| + |c_\alpha(x) - c_\alpha(y)|) \leq L|x - y|$$

$$\sup_{\alpha \in \mathcal{A}} \left| \int_{\mathbb{R}^d} g(z) \, F_\alpha(x, dz) - \int_{\mathbb{R}^d} g(z) \, F_\alpha(y, dz) \right| \leq \sup_{z \in \mathbb{R}^d \setminus \{0\}} \left(\frac{|g(z)|}{1 \wedge |z|^2} \right) L|x - y|$$

holds for all $x, y \in \mathbb{R}^d$ and $g \in C_b(\mathbb{R}^d)$.

(D3) (Convexity & Closedness) The range of the uncertainty coefficients in $\alpha \in \mathcal{A}$

$$\bigcup_{\alpha \in \mathcal{A}} (b_\alpha, c_\alpha, F_\alpha)(x) = \{ (b_\alpha, c_\alpha, F_\alpha)(x) \mid \alpha \in \mathcal{A} \} \subset \mathbb{R}^d \times \mathbb{S}_+^{d \times d} \times \mathfrak{L}(\mathbb{R}^d)$$

is convex and closed for every $x \in \mathbb{R}^d$.

Lemma 4.40 (Properties of Uncertainty Coefficients). *Suppose that the spatially inhomogeneous uncertainty coefficients $(b_\alpha, c_\alpha, F_\alpha) : \mathbb{R}^d \to \mathbb{R}^d \times \mathbb{S}_+^{d \times d} \times \mathfrak{L}(\mathbb{R}^d)$ with $\alpha \in \mathcal{A}$ satisfy all assumptions from Remark 4.39, then*

$$D(\mathbb{R}_+; \mathbb{R}^d) \ni \omega \longmapsto \left(\bigcup_{\alpha \in \mathcal{A}} (b_\alpha, c_\alpha, g \circ F_\alpha)(\omega_s) \right)_{s \geq 0} \subset L^1_{loc}(\mathbb{R}_+; \mathbb{R}^d \times \mathbb{S}_+^{d \times d} \times \mathbb{R})$$

is continuous, with respect to the Hausdorff distance $d_H^{L^1_{loc}}$ (as introduced in Remark A.15) on the power set of $L^1_{loc}(\mathbb{R}_+; \mathbb{R}^d \times \mathbb{S}_+^{d \times d} \times \mathbb{R})$, for every

$$g \in \mathcal{C} := \left\{ \varphi \in C_b(\mathbb{R}^d) \mid \exists \delta > 0 \, \forall z \in B_\delta(0) : g(z) = 0 \right\},$$

where $g \circ \nu := \int_{\mathbb{R}^d} g(z) \, \nu(dz)$ is defined for all Lévy measures $\nu \in \mathfrak{L}(\mathbb{R}^d)$. Moreover, for every $\omega \in D(\mathbb{R}_+; \mathbb{R}^d)$ the set (of locally bounded, absolutely continuous functions)

$$\left\{ \int_0^\cdot \theta_s \, ds \in C(\mathbb{R}_+; \mathbb{R}^d \times \mathbb{S}_+^{d \times d} \times \mathbb{R}) \,\middle|\, \forall s \geq 0 : \theta_s \in \bigcup_{\alpha \in \mathcal{A}} (b_\alpha, c_\alpha, g \circ F_\alpha)(\omega_s) \right\}$$

is closed in $C(\mathbb{R}_+; \mathbb{R}^d \times \mathbb{S}_+^{d \times d} \times \mathbb{R})$ with respect to locally uniform convergence.

Proof. For the first part, suppose that $(\omega^n)_{n\in\mathbb{N}} \subset D(\mathbb{R}_+;\mathbb{R}^d)$ such that

$$\lim_{n\to\infty} \omega^n = \omega \in D(\mathbb{R}_+;\mathbb{R}^d)$$

as in Lemma 4.2, i.e. there exists a sequence $(\lambda_n)_{n\in\mathbb{N}}$ of strictly increasing functions $\lambda_n : \mathbb{R}_+ \to \mathbb{R}_+$ with $\lim_{n\to\infty}\sup_{t\geq 0}|\lambda_n(t) - t| = \lim_{n\to\infty}\sup_{s\geq 0}|s - \lambda_n^{-1}(s)| = 0$ and

$$\lim_{n\to\infty}\sup_{t\in[0,T]}|\omega^n(\lambda_n(t)) - \omega(t)| = 0$$

for all $T > 0$. In particular, $T_s := \sup_{n\in\mathbb{N}} \lambda_n^{-1}(s) < \infty$ for all $s \geq 0$. Hence, for continuity points $s \in \mathbb{R}_+ \setminus \Delta(\omega)$ with $\Delta(\omega) := \{t \in \mathbb{R}_+ \mid \Delta\omega_t \neq 0\}$ we find

$$|\omega^n(s) - \omega(s)| \leq |\omega^n(s) - \omega(\lambda_n^{-1}(s))| + |\omega(\lambda_n^{-1}(s)) - \omega(s)|$$
$$\leq \sup_{t\in[0,T_s]}|\omega^n(\lambda_n(t)) - \omega(t)| + |\omega(\lambda_n^{-1}(s)) - \omega(s)| \longrightarrow 0$$

as $n \to \infty$. According to continuity assumption (D2), $(\hat{L}|\omega^n(s) - \omega(s)|)_{s\geq 0}$ dominates

$$\left(\sup_{\alpha\in\mathcal{A}}(|b_\alpha(\omega_s^n) - b_\alpha(\omega_s)| + |c_\alpha(\omega_s^n) - c_\alpha(\omega_s)| + |g\circ F_\alpha(\omega_s^n,dz) - g\circ F_\alpha(\omega_s,dz)|)\right)_{s\geq 0}$$

with $\hat{L} := L\left(1 + \sup_{z\in\mathbb{R}^d\setminus\{0\}}(1 \wedge |z|^2)^{-1}|g(z)|\right) < \infty$, which implies the pointwise convergence to zero on $\mathbb{R}_+ \setminus \Delta(\omega)$. Since the processes are locally bounded in $s \in \mathbb{R}_+$ (uniformly in $n \in \mathbb{N}$) according to boundedness assumption (D1), and since $\Delta(\omega) \subset \mathbb{R}_+$ is countable, the dominated convergence theorem implies that for every $T > 0$

$$\int_0^T \sup_{\alpha\in\mathcal{A}}(|b_\alpha(\omega_s^n) - b_\alpha(\omega_s)| + |c_\alpha(\omega_s^n) - c_\alpha(\omega_s)| + |g\circ F_\alpha(\omega_s^n,dz) - g\circ F_\alpha(\omega_s,dz)|)\,ds$$

vanishes as $n \to \infty$. Hence, the definition of the Hausdorff distance implies that

$$d_H^{L^1}\left(\bigcup_{\alpha\in\mathcal{A}}(b_\alpha,c_\alpha,g\circ F_\alpha)(\omega^n|_{[0,T]}), \bigcup_{\alpha\in\mathcal{A}}(b_\alpha,c_\alpha,g\circ F_\alpha)(\omega|_{[0,T]})\right)$$
$$= \max\Big\{\sup_{\alpha\in\mathcal{A}}\inf_{\beta\in\mathcal{A}} d_{L^1}((b_\alpha,c_\alpha,g\circ F_\alpha)(\omega^n|_{[0,T]}),(b_\beta,c_\beta,g\circ F_\beta)(\omega|_{[0,T]})),$$
$$\sup_{\beta\in\mathcal{A}}\inf_{\alpha\in\mathcal{A}} d_{L^1}((b_\alpha,c_\alpha,g\circ F_\alpha)(\omega^n|_{[0,T]}),(b_\beta,c_\beta,g\circ F_\beta)(\omega|_{[0,T]}))\Big\}$$
$$\leq \sup_{\alpha\in\mathcal{A}} d_{L^1}((b_\alpha,c_\alpha,g\circ F_\alpha)(\omega^n|_{[0,T]}),(b_\alpha,c_\alpha,g\circ F_\alpha)(\omega|_{[0,T]}))$$
$$\leq \int_0^T \sup_{\alpha\in\mathcal{A}}(|b_\alpha(\omega_s^n) - b_\alpha(\omega_s)| + |c_\alpha(\omega_s^n) - c_\alpha(\omega_s)| + |g\circ F_\alpha(\omega_s^n,dz) - g\circ F_\alpha(\omega_s,dz)|)\,ds$$

also vanishes as $n \to \infty$ for every $T > 0$, as required.

4. Stochastic Processes

For the second part, denote by $\Upsilon : L^1_{\text{loc}}(\mathbb{R}_+; \mathbb{R}^d \times \mathbb{S}^{d\times d}_+ \times \mathbb{R}) \to C(\mathbb{R}_+; \mathbb{R}^d \times \mathbb{S}^{d\times d}_+ \times \mathbb{R})$ the (Bochner) integration operator from Remark A.13 with

$$\Upsilon(\theta) := \int_0^{\cdot} \theta_s \, ds$$

for $\theta \in L^1_{\text{loc}}(\mathbb{R}_+; \mathbb{R}^d \times \mathbb{S}^{d\times d}_+ \times \mathbb{R})$. Suppose that for $\omega \in D(\mathbb{R}_+; \mathbb{R}^d)$ the sequence

$$(B^n, C^n, \Gamma^n)_{n \in \mathbb{N}} \subset \Upsilon\left(\left(\bigcup_{\alpha \in \mathcal{A}} (b_\alpha, c_\alpha, g \circ F_\alpha)(\omega_s)\right)_{s \geq 0}\right) \subset C(\mathbb{R}_+; \mathbb{R}^d \times \mathbb{S}^{d\times d}_+ \times \mathbb{R})$$

is so that $\lim_{n \to \infty}(B^n, C^n, \Gamma^n) = (B, C, \Gamma)$ locally uniformly in $C(\mathbb{R}_+; \mathbb{R}^d \times \mathbb{S}^{d\times d}_+ \times \mathbb{R})$. In order to prove the second part of the statement, it suffices to show that

$$(B, C, \Gamma) \in \Upsilon\left(\left(\bigcup_{\alpha \in \mathcal{A}} (b_\alpha, c_\alpha, g \circ F_\alpha)(\omega_s)\right)_{s \geq 0}\right)$$

holds. First, we prove that (B, C, Γ) is absolutely continuous: By definition, there exists

$$(b^n, c^n, \gamma^n)_{n \in \mathbb{N}} \subset \left(\bigcup_{\alpha \in \mathcal{A}} (b_\alpha, c_\alpha, g \circ F_\alpha)(\omega_s)\right)_{s \geq 0} \subset L^1_{\text{loc}}(\mathbb{R}_+; \mathbb{R}^d \times \mathbb{S}^{d\times d}_+ \times \mathbb{R})$$

such that $\Upsilon((b^n, c^n, \gamma^n)) = (B^n, C^n, \gamma^n)$ for all $n \in \mathbb{N}$. The uniform boundedness assumption (D1) implies the existence of $\hat{L} > 0$ (independent of $n \in \mathbb{N}$) such that

$$|B^n_t - B^n_s| + |C^n_t - C^n_s| + |\Gamma^n_t - \Gamma^n_s| \leq \int_s^t |b^n_u| \, du + \int_s^t |c^n_u| \, du + \int_s^t |\gamma^n_u| \, du \leq \hat{L}|t - s|$$

holds for all $t > s \geq 0$ and every $n \in \mathbb{N}$. Since $(B^n, C^n, \Gamma^n) \to (B, C, \Gamma)$ locally uniformly, the limit (B, C, Γ) is also Lipschitz-continuous and hence absolutely continuous. The bijectivity of Υ from $L^1_{\text{loc}}(\mathbb{R}_+)$ onto

$$\{\varphi \in C(\mathbb{R}_+) : \varphi \text{ is absolutely continuous}\}$$

(as in Remark A.13) therefore implies that $(b, c, \gamma) := \Upsilon^{-1}((B, C, \Gamma))$ is well-defined. It remains to show that $(b, c, \gamma) \in (\bigcup_{\alpha \in \mathcal{A}}(b_\alpha, c_\alpha, g \circ F_\alpha)(\omega_s))_{s \geq 0}$ holds, i.e.

$$\lambda\left(\left\{s \in \mathbb{R}_+ \;\middle|\; (b_s, c_s, \gamma_s) \notin \bigcup_{\alpha \in \mathcal{A}} (b_\alpha, c_\alpha, g \circ F_\alpha)(\omega_s)\right\}\right) = 0,$$

where λ is the Lebesgue measure on $\mathbb{R}$: According to the boundedness assumption (D1), the sequence $(b^n, c^n, \gamma^n)_{n \in \mathbb{N}} \subset L^1_{\text{loc}}(\mathbb{R}_+; \mathbb{R}^d \times \mathbb{S}^{d\times d}_+ \times \mathbb{R})$ is uniformly integrable. The Dunford-Pettis theorem (cf. [38, Chapter 2, Theorem 25, p. 27]) thus implies the existence of a weakly convergent subsequence $(b^{n_k}, c^{n_k}, \gamma^{n_k})_{k \in \mathbb{N}} \subset L^1_{\text{loc}}(\mathbb{R}_+; \mathbb{R}^d \times \mathbb{S}^{d \times d}_+ \times \mathbb{R})$. Since the integration operator Υ is a closed operator with respect to weak convergence in $L^1_{\text{loc}}(\mathbb{R}_+)$ and the local uniform convergence in $C(\mathbb{R}_+)$, and since

$$\lim_{k \to \infty}(B^{n_k}, C^{n_k}, \Gamma^{n_k}) = \lim_{k \to \infty} \Upsilon((b^{n_k}, c^{n_k}, \gamma^{n_k})) = (B, C, \Gamma) = \Upsilon((b, c, \gamma))$$

locally uniformly in $C(\mathbb{R}_+)$, we have $\lim_{k \to \infty}(b^{n_k}, c^{n_k}, \gamma^{n_k}) = (b, c, \gamma)$ weakly in $L^1_{\text{loc}}(\mathbb{R}_+)$. In particular,

$$\lim_{k \to \infty} m \int_s^{s+1/m} (b^{n_k}_u, c^{n_k}_u, \gamma^{n_k}_u)\, du = m \int_s^{s+1/m} (b_u, c_u, \gamma_u)\, du$$

for all $s \in \mathbb{R}_+$ and $m \in \mathbb{N}$. According to Lemma A.12,

$$m \int_s^{s+1/m} (b^{n_k}_u, c^{n_k}_u, \gamma^{n_k}_u)\, du \in \overline{\text{conv}}\left(\bigcup_{u \in [s, s+1/m]} \bigcup_{\alpha \in \mathcal{A}} (b_\alpha, c_\alpha, g \circ F_\alpha)(\omega_u) \right)$$

using $(b^n, c^n, \gamma^n)_{n \in \mathbb{N}} \subset (\bigcup_{\alpha \in \mathcal{A}}(b_\alpha, c_\alpha, g \circ F_\alpha)(\omega_s))_{s \geq 0}$ and therefore

$$m \int_s^{s+1/m} (b_u, c_u, \gamma_u)\, du \in \overline{\text{conv}}\left(\bigcup_{u \in [s, s+1/m]} \bigcup_{\alpha \in \mathcal{A}} (b_\alpha, c_\alpha, g \circ F_\alpha)(\omega_u) \right)$$

for all $s \in \mathbb{R}_+$ and $m \in \mathbb{N}$, since the right-hand side is closed. Lebsgue's differentiation theorem (cf. [119, Theorem 19.20, p. 218] or [114, Theorem 7.7, p. 138]) then leads to

$$\lambda\left(\left\{ s \in \mathbb{R}_+ \,\middle|\, (b_s, c_s, \gamma_s) \notin \bigcap_{m \in \mathbb{N}} \overline{\text{conv}}\left(\bigcup_{u \in [s, s+1/m]} \bigcup_{\alpha \in \mathcal{A}} (b_\alpha, c_\alpha, g \circ F_\alpha)(\omega_u) \right) \right\} \right) = 0,$$

which shows that it suffices to prove that the inclusion

$$\bigcap_{m \in \mathbb{N}} \overline{\text{conv}}\left(\bigcup_{u \in [s, s+1/m]} \bigcup_{\alpha \in \mathcal{A}} (b_\alpha, c_\alpha, g \circ F_\alpha)(\omega_u) \right) \subset \bigcup_{\alpha \in \mathcal{A}} (b_\alpha, c_\alpha, g \circ F_\alpha)(\omega_s)$$

holds for all $s \in \mathbb{R}_+$: The right-continuity of $\omega \in D(\mathbb{R}_+; \mathbb{R}^d)$ and the uniform Lipschitz-continuity assumption (D2) imply

$$\bigcap_{m \in \mathbb{N}} \overline{\text{conv}}\left(\bigcup_{u \in [s, s+1/m]} \bigcup_{\alpha \in \mathcal{A}} (b_\alpha, c_\alpha, g \circ F_\alpha)(\omega_u) \right)$$

$$\subset \bigcap_{\varepsilon > 0} \overline{\text{conv}}\left(\bigcup_{\alpha \in \mathcal{A}} (b_\alpha, c_\alpha, g \circ F_\alpha)\left(\omega_s + \overline{B_\varepsilon(0)}\right) \right)$$

$$\subset \bigcap_{\varepsilon > 0} \overline{\text{conv}}\left(\bigcup_{\alpha \in \mathcal{A}} (b_\alpha, c_\alpha, g \circ F_\alpha)(\omega_s) + \overline{B_{\hat{L}\varepsilon}(0)} \right)$$

with some finite (Lipschitz) constant $\hat{L} := L\left(1 + \sup_{z \in \mathbb{R}^d \setminus \{0\}}(1 \wedge |z|^2)^{-1}|g(z)|\right) < \infty$.

4. Stochastic Processes

Since L^1_{loc}-convergence on σ-finite measure spaces implies the existence of an almost everywhere convergent subsequence (cf. e.g. [56, Theorem 2.30, p. 61]), the convexity-and-closedness assumption (D3) implies that

$$\bigcup_{\alpha \in \mathcal{A}} (b_\alpha, c_\alpha, g \circ F_\alpha)(\omega_s) + \overline{B_{\hat{L}\varepsilon}(0)}$$

are convex and closed subsets of $L^1_{\text{loc}}(\mathbb{R}_+; \mathbb{R}^d \times \mathbb{S}^{d \times d}_+ \times \mathbb{R})$ for $\varepsilon > 0$. Therefore,

$$\bigcap_{m \in \mathbb{N}} \overline{\text{conv}} \left(\bigcup_{u \in [s, s+1/m]} \bigcup_{\alpha \in \mathcal{A}} (b_\alpha, c_\alpha, g \circ F_\alpha)(\omega_u) \right)$$
$$\subset \bigcap_{\varepsilon > 0} \overline{\text{conv}} \left(\bigcup_{\alpha \in \mathcal{A}} (b_\alpha, c_\alpha, g \circ F_\alpha)(\omega_s) + \overline{B_{\hat{L}\varepsilon}(0)} \right)$$
$$= \bigcap_{\varepsilon > 0} \left(\bigcup_{\alpha \in \mathcal{A}} (b_\alpha, c_\alpha, g \circ F_\alpha)(\omega_s) + \overline{B_{\hat{L}\varepsilon}(0)} \right) = \bigcup_{\alpha \in \mathcal{A}} (b_\alpha, c_\alpha, g \circ F_\alpha)(\omega_s)$$

holds for all $s \in \mathbb{R}_+$, which implies the desired result. □

Theorem 4.41 (Compactness of Uncertainty Subsets). *Suppose that*

$$(b_\alpha, c_\alpha, F_\alpha) : \mathbb{R}^d \longrightarrow \mathbb{R}^d \times \mathbb{S}^{d \times d}_+ \times \mathfrak{L}(\mathbb{R}^d)$$

with $\alpha \in \mathcal{A}$ are given (spatially inhomogeneous) uncertainty coefficients and let

$$\mathcal{P}_x := \left\{ \mathbb{P} \in \mathfrak{P}^{ac}_{sem}(D_x(\mathbb{R}_+)) \ \middle| \ (b^\mathbb{P}_s, c^\mathbb{P}_s, F^\mathbb{P}_s)(\omega) \in \bigcup_{\alpha \in \mathcal{A}} (b_\alpha, c_\alpha, F_\alpha)(\omega_s) \ \lambda(ds) \times \mathbb{P}(d\omega)\text{-a.e.} \right\}$$

for $x \in \mathbb{R}^d$ be the corresponding uncertainty subsets. If the set $K \subset \mathbb{R}^d$ is bounded and the boundedness-and-tightness assumption (D1) from Remark 4.39 holds, then

$$\bigcup_{x \in K} \mathcal{P}_x \subset \mathfrak{P}(D(\mathbb{R}_+; \mathbb{R}^d))$$

is relatively compact (with respect to the weak convergence of probability measures). If the set $K \subset \mathbb{R}^d$ is compact and all assumptions from Remark 4.39 hold, then

$$\bigcup_{x \in K} \mathcal{P}_x \subset \mathfrak{P}(D(\mathbb{R}_+; \mathbb{R}^d))$$

is compact and sequentially compact in itself (with respect to the weak convergence of probability measures, cf. Remark 4.16).

Proof. Before we start, note that, since $\mathfrak{P}(D(\mathbb{R}_+;\mathbb{R}^d))$ is a metric space according to Remark 4.16, every relatively compact (compact) subset is sequentially compact (sequentially compact in itself, respectively) and vice versa, cf. [50, Theorem 4.1.17, p. 256]. For the first half of the statement, it therefore suffices to show that every

$$(\mathbb{P}_n)_{n\in\mathbb{N}} \subset \bigcup_{x\in K} \mathcal{P}_x \subset \mathfrak{P}(D(\mathbb{R}_+;\mathbb{R}^d))$$

has a convergent subsequence $(\mathbb{P}_{n_k})_{k\in\mathbb{N}}$ (with respect to weak convergence of probability measures) if $K \subset \mathbb{R}^d$ is bounded and the boundedness-and-tightness assumption (D1) holds. For the second half, we have to show that the weak limit satisfies

$$\lim_{k\to\infty} \mathbb{P}_{n_k} \in \bigcup_{x\in \overline{K}} \mathcal{P}_x$$

under the additional assumptions from Remark 4.39.

Suppose now that $(\mathbb{P}_n)_{n\in\mathbb{N}} \subset \bigcup_{x\in K}\mathcal{P}_x \subset \mathfrak{P}(D(\mathbb{R}_+;\mathbb{R}^d))$ for $K \subset \mathbb{R}^d$ bounded and let

$$(B_t^n, C_t^n, \nu^n(dt,dx)) := (B_t^{\mathbb{P}_n}, C_t^{\mathbb{P}_n}, \nu^{\mathbb{P}_n}(dt,dx)) = (b_t^n, c_t^n, F_t^{\mathbb{P}_n}(dx))\, dt$$
$$(b_t^n, c_t^n, F_t^n(dx)) := (b_t^{\mathbb{P}_n}, c_t^{\mathbb{P}_n}, F_t^{\mathbb{P}_n}(dx))$$

be the measurable (differential) characteristics (from Theorem 4.18 and Theorem 4.20) with respect to one fixed truncation function $h : \mathbb{R}^d \to \mathbb{R}$. Moreover, define

$$\widetilde{C}^{n,ij} := C^{n,ij} + (h^i h^j) * \nu^n + \sum_{s \leq \cdot} \Delta B_s^{n,i} \Delta B_s^{n,j}$$

for $i, j \in \{1,\ldots,d\}$ and $n \in \mathbb{N}$, and note that $\widetilde{C}^n$ is the predictable quadratic covariation process (cf. Remark 4.34) of the truncated process $\widehat{M}$ from Remark 4.14, according to [73, Chapter 3, Theorem 2.7, p. 153]. The proof is split into three parts: In the first part, we will show that under the boundedness-and-tightness assumption (D1)

$$\left(\mathbb{P}_n \circ (X, B^n, \widetilde{C}^n, g * \nu^n)^{-1}\right)_{n\in\mathbb{N}}$$

is a tight sequence of probability measures (cf. [73, Chapter 6, Section 3a, p. 347]) on the Skorokhod space $D(\mathbb{R}_+;\mathbb{R}^d \times \mathbb{R}^d \times \mathbb{R}^{d\times d} \times \mathbb{R})$ for fixed

$$g \in \mathcal{C} := \{\varphi \in C_b(\mathbb{R}^d) \mid \varphi \equiv 0 \text{ around } z = 0\},$$

where $X = (X_t)_{t\geq 0}$ is the coordinate-mapping process. Since tightness is equivalent to relative compactness according to Prokhorov's theorem (cf. e.g. [73, Chater 6, Section 3a, p. 347]), this proves the first half of the statement. In the second part, we will prove that all limit points of $(\mathbb{P}_n)_{n\in\mathbb{N}} = (\mathbb{P}_n \circ X^{-1})_{n\in\mathbb{N}}$ are in $\mathfrak{P}_{\text{sem}}^{\text{ac}}(D_x(\mathbb{R}^d))$ for some $x \in \overline{K} \subset \mathbb{R}^d$. In the third part, we will show that all limit points of $(\mathbb{P}_n)_{n\in\mathbb{N}}$ are in $\bigcup_{x\in\overline{K}}\mathcal{P}_x$ under the additional (continuity and convexity) assumptions from Remark 4.39.

4. Stochastic Processes

In order to show that $(Y^n)_{n\in\mathbb{N}} := (X, B^n, \tilde{C}^n, g*\nu^n)_{n\in\mathbb{N}}$ is tight with respect to $(\mathbb{P}_n)_{n\in\mathbb{N}}$ for the first part of the proof, we will use that Y^n is a $\mathbb{P}_n$-semimartingale for all $n \in \mathbb{N}$: Note that (by assumption) B^n and C^n are not only bounded variation processes, but even have absolutely continuous paths. Therefore, for all $i,j \in \{1,\dots,d\}$

$$\tilde{C}^{n,ij} = C^{n,ij} + (h^i h^j)*\nu^n + \sum_{s\leq \cdot} \Delta B_s^{n,i} \Delta B_s^{n,j} = C^{n,ij} + \int_0^\cdot \int_{\mathbb{R}^d} h^i(z) h^j(z)\, \nu^n(ds,dz)$$

$$= C^{n,ij} + \int_0^\cdot \int_{\mathbb{R}^d} h^i(z) h^j(z)\, F_s^n(dz)\, ds$$

is also absolutely continuous, since the second terms is Lipschitz-continuous, due to

$$\left| \int_s^t \int_{\mathbb{R}^d} h^i(z) h^j(z)\, F_u^n(dz)\, du \right| \leq |t-s| \sup_{\alpha \in \mathcal{A}} \sup_{x \in \mathbb{R}^d} \int_{\mathbb{R}^d} \left(|h^i(z)| \cdot |h^j(z)| \right) F_\alpha(x, dz)$$

$$\leq |t-s| \sup_{\alpha \in \mathcal{A}} \sup_{x \in \mathbb{R}^d} \int_{\mathbb{R}^d} C(1 \wedge |z|^2)\, F_\alpha(x, dz)$$

for all $t \geq s \geq 0$ and some $C > 0$ only depending on the truncation function $h : \mathbb{R}^d \to \mathbb{R}^d$, and due to the boundedness assumption (D1). Similarly, $g*\nu^n$ is also Lipschitz, due to

$$\left| \int_s^t \int_{\mathbb{R}^d} g(z)\, F_u^n(dz)\, du \right| \leq |t-s| \sup_{\alpha \in \mathcal{A}} \sup_{x \subset \mathbb{R}^d} \int_{\mathbb{R}^d} |g(z)|\, F_\alpha(x, dz)$$

$$\leq |t-s| \sup_{\alpha \in \mathcal{A}} \sup_{x \in \mathbb{R}^d} \int_{\mathbb{R}^d} C(1 \wedge |z|^2)\, F_\alpha(x, dz)$$

for all $t \geq s \geq 0$ and some $C > 0$ only depending on the function $g \in \mathcal{C}$, and due to the boundedness assumption (D1). This shows that the process $Y_n = (X, B^n, \tilde{C}^n, g\circ\nu^n)$ is a $\mathbb{P}_n$-semimartingale with differential characteristics

$$\left((b^n, b^n, c^n + (hh^T)\circ F^n, g\circ F^n), (c^n, 0, 0, 0), (F^n, 0, 0, 0)\right),$$

using the construction of characteristics from Remark 4.14 and the fact that all bounded variation processes are $\mathbb{P}_n$-semimartingales. If the sequence $(D^n)_{n\in\mathbb{N}}$ with

$$D^n := \sum_{i\leq d} \left(\operatorname{var}(B^{n,i}) + C^{n,ii} \right) + \left(1 \wedge |z|^2\right) * \nu^n + \sum_{i\leq d} \operatorname{var}(B^{n,i})$$

$$+ \sum_{i,j\leq d} \operatorname{var}\left(C^{n,ij} + (h^i h^j)*\nu^n \right) + \sum_{i\leq d} \operatorname{var}(g*\nu^n)$$

is C-tight (i.e. tight and all limit points are laws of continuous processes), if the origin $(Y_0^n)_{n\in\mathbb{N}} = (X_0^n, 0, 0, 0)_{n\in\mathbb{N}}$ is tight (in $\mathbb{R}^d \times \mathbb{R}^d \times \mathbb{R}^{d\times d} \times \mathbb{R}$) for $(\mathbb{P}_n)_{n\in\mathbb{N}}$ and if

$$\lim_{r\to\infty} \limsup_{n\to\infty} \mathbb{P}_n\left(\nu^n\big([0,N] \times \{z \in \mathbb{R}^d : |z| > r\}\big) > \varepsilon\right) = 0$$

for all $N, \varepsilon > 0$, then $(Y^n)_{n\in\mathbb{N}}$ is tight with respect to $(\mathbb{P}_n)_{n\in\mathbb{N}}$ according to [73, Chapter 6,

Theorem 4.18, p. 359] and [73, Chapter 6, Remark 4.20, p. 359]. Since

$$Y_0^n = (X_0^n, 0, 0, 0) \in K \times \{0\} \times \{0\} \times \{0\}$$

$\mathbb{P}_n$-almost surely for all $n \in \mathbb{N}$ with $K \subset \mathbb{R}^d$ bounded, and since for all $N, \varepsilon > 0$

$$\mathbb{P}_n\left(\nu^n\big([0,N] \times \{z \in \mathbb{R}^d : |z| > r\}\big) > \varepsilon\right) \leq \varepsilon^{-1} \mathbb{E}_{\mathbb{P}_n}\left(\nu^n\big([0,N] \times \{z \in \mathbb{R}^d : |z| > r\}\big)\right)$$

$$= \varepsilon^{-1} \mathbb{E}_{\mathbb{P}_n}\left(\int_0^N \int_{\mathbb{R}^d} \mathbb{1}_{\{|z|>r\}} F_s(dz)\, ds\right)$$

$$\leq \varepsilon^{-1} N \sup_{\alpha \in \mathcal{A}} \sup_{x \in \mathbb{R}^d} \int_{|z|>r} \left(1 \wedge |z|^2\right) F_\alpha(x)(dz)$$

together with the tightness assumption (D1), it remains to show that $(D^n)_{n \in \mathbb{N}}$ is C-tight: Because D^n is a non-decreasing process for all $n \in \mathbb{N}$ (since predictable quadratic variations and total variations are always non-decreasing), [73, Chapter 6, Proposition 3.35, p. 354] shows that $(D^n)_{n \in \mathbb{N}}$ is C-tight with respect to $(\mathbb{P}_n)_{n \in \mathbb{N}}$ if we can find a constant $C > 0$ (independent of $n \in \mathbb{N}$) such that

$$D_t^n(\omega) - D_s^n(\omega) \leq C(t-s)$$

for all $t \geq s \geq 0$ and $\mathbb{P}_n$-almost every $\omega \in D(\mathbb{R}_+; \mathbb{R}^d)$, or in other words $(D_t^n - Ct)_{t \geq 0}$ are ($\mathbb{P}_n$-almost surely) non-decreasing processes for all $n \in \mathbb{N}$. But this holds due to

$$D_t^n(\omega) - D_s^n(\omega) = \sum_{i \leq d} \left(\int_s^t |b_u^{n,i}|\, du + \int_s^t c_u^{n,ii}\, du\right) + \int_s^t \int_{\mathbb{R}^d} \left(1 \wedge |z|^2\right) F_u(dz)\, du$$

$$+ \sum_{i \leq d} \int_s^t |b_u^{n,i}|\, du \qquad + \int_s^t \int_{\mathbb{R}^d} |g(z)|\, F_u(dz)\, du$$

$$+ \sum_{i,j \leq d} \int_s^t \left| c_u^{n,ij} + \int_{\mathbb{R}^d} \left(h^i(z) h^j(z)\right) F_u(dz) \right| du$$

$$\leq (t-s)\, C \sup_{\alpha \in \mathcal{A}} \sup_{x \in \mathbb{R}^d} \left(|b_\alpha(x)| + |c_\alpha(x)| + \int_{\mathbb{R}^d} (1 \wedge |z|^2)\, F_\alpha(x, dz)\right)$$

for all $t \geq s \geq 0$ and some constant $C > 0$ depending only on the truncation functions $h : \mathbb{R}^d \to \mathbb{R}^d$ and $g \in \mathcal{C}$, and due to the boundedness assumption (D1). In particular, we finished the first part of the proof.

For the second part, suppose that $(\mathbb{P}_{n_k})_{k \in \mathbb{N}}$ is a weakly convergent subsequence of $(\mathbb{P}_n)_{n \in \mathbb{N}}$ and $\mathbb{P} := \lim_{k \to \infty} \mathbb{P}_{n_k}$. Analogously to [73, Chapter 7, Section 2a, p. 394], Corollary 1.29 shows how to construct a countable subfamily $\mathcal{C}_0 \subset \mathcal{C}$ such that for every function $g \in C_b(\mathbb{R}^d)$ with $\sup_{z \in \mathbb{R}^d \setminus \{0\}} (1 \wedge |z|^2)^{-1} |g(z)| < \infty$, there exists a sequence $(g_n)_{n \in \mathbb{N}} \subset \mathcal{C}_0$ with $\sup_{n \in \mathbb{N}} \sup_{z \in \mathbb{R}^d \setminus \{0\}} (1 \wedge |z|^2)^{-1} |g_n(z)| < \infty$ and for all $r > 0$

$$\lim_{n \to \infty} \sup_{z \in B_r(0) \setminus \{0\}} (1 \wedge |z|^2)^{-1} |g_n(z) - g(z)| = 0.$$

4. Stochastic Processes

The intermediate result from the first part (of this proof) implies that we can pick a convergent subsequence (using a diagonal argument) such that

$$\left(\mathbb{P}_{n_k} \circ (X, B^{n_k}, \tilde{C}^{n_k}, g * \nu^{n_k})^{-1}\right)_{k \in \mathbb{N}}$$

converges for all $g \in \mathcal{C}_0$. Denote by $(B, \tilde{C})$ the weak limit of $(B^{n_k}, \tilde{C}^{n_k})_{k \in \mathbb{N}}$ with respect to $(\mathbb{P}_{n_k})_{k \in \mathbb{N}}$, and let ν be the compensator of the random measure μ^X with respect to $\mathbb{P}$ as in [73, Chapter 2, Theorem 1.8, p. 66], then the uniqueness of weak limits leads to

$$\mathbb{P}_{n_k} \circ (X, B^{n_k}, \tilde{C}^{n_k}, g * \nu^{n_k})^{-1} \longrightarrow \mathbb{P} \circ (X, B, \tilde{C}, g * \nu)^{-1}$$

weakly for all $g \in \mathcal{C}_0$. Hence, [73, Chapter 9, Theorem 2.4, p. 528] implies that X is a $\mathbb{P}$-semimartingale with characteristics (B, C, ν), where

$$C^{ij} := \tilde{C}^{ij} - (h^i h^j) * \nu^n - \sum_{s \leq \cdot} \Delta B_s^i \Delta B_s^j.$$

Similar to the first part of this proof, the boundedness assumption (D1) implies that there exists $L > 0$ (only depending on the truncation function $h : \mathbb{R}^d \to \mathbb{R}^d$) such that

$$\mathbb{P}_n\left(\Lambda(B^n) \leq L\right) = 1 = \mathbb{P}_n\left(\Lambda(C^n) \leq L\right) = 1 = \mathbb{P}_n\left(\Lambda(g * \nu^n) \leq L \cdot \sup_{z \neq 0} \frac{|g(z)|}{1 \wedge |z|^2}\right)$$

holds for all $n \in \mathbb{N}$ and $g \in \mathcal{C}_0$, where the operator $\Lambda : D(\mathbb{R}_+) \to [0, \infty]$ assigns the optimal Lipschitz constant

$$\Lambda(\omega) := \sup\left\{\frac{|\omega(t) - \omega(s)|}{|t - s|} : t > s \geq 0\right\}$$

to a càdlàg path $\omega \in D(\mathbb{R}_+)$. According to Lemma A.14, the map $\Lambda(\cdot)$ is lower semicontinuous, which implies that sets of the form $\{\omega \in D(\mathbb{R}_+) : \Lambda(\omega) \leq L\}$ are closed with respect to the Skorokhod topology. Hence, we obtain

$$\mathbb{P}\left(\Lambda(B) \leq L\right) = \left(\mathbb{P} \circ B^{-1}\right)(\Lambda \leq L) \geq \limsup_{k \to \infty}\left(\mathbb{P}_{n_k} \circ (B^{n_k})^{-1}\right)(\Lambda \leq L)$$

$$= \limsup_{k \to \infty} \mathbb{P}_{n_k}\left(\Lambda(B^{n_k}) \leq L\right) = 1$$

using the portmanteau theorem, and similarly

$$\mathbb{P}\left(\Lambda(C) \leq L\right) = 1 = \mathbb{P}\left(\Lambda(g * \nu) \leq L \cdot \sup_{z \neq 0} \frac{|g(z)|}{1 \wedge |z|^2}\right).$$

The countability paired with the approximation properties of the family $\mathcal{C}_0$ then lead to

$$1 = \mathbb{P}\left(\forall g \in \mathcal{C}_0 : \Lambda(g * \nu) \leq L \cdot \sup_{z \neq 0} \frac{|g(z)|}{1 \wedge |z|^2}\right) \leq \mathbb{P}\left(\Lambda((1 \wedge |z|^2) * \nu) \leq L\right)$$

and $\mathbb{P}((B, C, \nu) \ll dt) = \mathbb{P}(\text{var}(B) + \text{var}(C) + (1 \wedge |z|^2) * \nu \ll dt) = 1$ (cf. Remark 4.19).

This implies $\mathbb{P} \in \mathfrak{P}^{ac}_{sem}(D_x(\mathbb{R}_+))$ for some $x \in \overline{K} \subset \mathbb{R}^d$, and in particular
$$(B_t, C_t, \nu(dt, dx))(\omega) = (b_t\, dt, c_t\, dt, F_t(dx)\, dt)(\omega)$$
holds $\lambda(dt) \times \mathbb{P}(\omega)$-almost everywhere for the differential characteristics $(b_t, c_t, F_t)_{t \geq 0}$ of $(X_t)_{t \geq 0}$ under $\mathbb{P} \in \mathfrak{P}^{ac}_{sem}(D_x(\mathbb{R}_+))$.

For the third part of the proof, it remains to show that $\mathbb{P} \in \cup_{x \in K} \mathcal{P}_x$ under the additional assumptions from Remark 4.39, i.e.
$$(b_t, c_t, F_t)(\omega) \in \bigcup_{\alpha \in \mathcal{A}} (b_\alpha, c_\alpha, F_\alpha)(\omega_t)$$
holds $\lambda(dt) \times \mathbb{P}(d\omega)$-almost everywhere. First of all, we are going to prove that the map $\Phi : D(\mathbb{R}_+; \mathbb{R}^d) \times C(\mathbb{R}_+; \mathbb{R}^d \times \mathbb{S}^{d \times d}_+ \times \mathbb{R}) \to \{0, 1\} \subset \mathbb{R}$ defined by
$$\Phi(\omega, \beta, \gamma, \zeta) := \mathbb{1}_{\Upsilon(\cup_{\alpha \in \mathcal{A}}(b_\alpha, c_\alpha, g \circ F_\alpha)(\omega))}(\beta, \gamma, \zeta)$$
for $(\omega, \beta, \gamma, \zeta) \in D(\mathbb{R}_+; \mathbb{R}^d) \times C(\mathbb{R}_+; \mathbb{R}^d \times \mathbb{S}^{d \times d}_+ \times \mathbb{R})$ is upper semicontinuous (with respect to the Skorokhod topology on $D(\mathbb{R}_+; \mathbb{R}^d)$ and the locally uniform convergence on $C(\mathbb{R}_+; \mathbb{R}^d \times \mathbb{S}^{d \times d}_+ \times \mathbb{R}))$, where
$$\Upsilon : L^1_{loc}(\mathbb{R}_+; \mathbb{R}^d \times \mathbb{S}^{d \times d}_+ \times \mathbb{R}) \longrightarrow C(\mathbb{R}_+; \mathbb{R}^d \times \mathbb{S}^{d \times d}_+ \times \mathbb{R})$$
$$\theta \longmapsto \Upsilon(\theta) := \int_0^\cdot \theta_s\, ds$$
is the Bochner integration operator (as defined in Remark A.13) and $g \in \mathcal{C}_0$: Suppose that $(\omega^n, \beta^n, \gamma^n, \zeta^n), (\omega, \beta, \gamma, \zeta) \in D(\mathbb{R}_+; \mathbb{R}^d) \times C(\mathbb{R}_+; \mathbb{R}^d \times \mathbb{S}^{d \times d}_+ \times \mathbb{R})$ for $n \in \mathbb{N}$ with
$$\lim_{n \to \infty} (\omega^n, \beta^n, \gamma^n, \zeta^n) = (\omega, \beta, \gamma, \zeta).$$
On the one hand, if $(\beta, \gamma, \zeta) \in \Upsilon(\cup_{\alpha \in \mathcal{A}}(b_\alpha, c_\alpha, g \circ F_\alpha)(\omega))$, then
$$\limsup_{n \to \infty} \Phi(\omega^n, \beta^n, \gamma^n, \zeta^n) \leq 1 = \mathbb{1}_{\Upsilon(\cup_{\alpha \in \mathcal{A}}(b_\alpha, c_\alpha, g \circ F_\alpha)(\omega))}(\beta, \gamma, \zeta) = \Phi(\omega, \beta, \gamma, \zeta)$$
holds, since indicator functions only attain values less than one. On the other hand, if $(\beta, \gamma, \zeta) \notin \Upsilon(\cup_{\alpha \in \mathcal{A}}(b_\alpha, c_\alpha, g \circ F_\alpha)(\omega))$, the triangle inequality for Hausdorff distances (cf. Remark A.15) implies
$$d_C\Big((\beta^n, \gamma^n, \zeta^n), \Upsilon\big(\cup_{\alpha \in \mathcal{A}}(b_\alpha, c_\alpha, g \circ F_\alpha)(\omega^n)\big)\Big)$$
$$\geq d_C\Big((\beta, \gamma, \zeta), \Upsilon\big(\cup_{\alpha \in \mathcal{A}}(b_\alpha, c_\alpha, g \circ F_\alpha)(\omega)\big)\Big) - d_C\Big((\beta^n, \gamma^n, \zeta^n), (\beta, \gamma, \zeta)\Big)$$
$$- d^C_H\Big(\Upsilon\big(\cup_{\alpha \in \mathcal{A}}(b_\alpha, c_\alpha, g \circ F_\alpha)(\omega^n)\big), \Upsilon\big(\cup_{\alpha \in \mathcal{A}}(b_\alpha, c_\alpha, g \circ F_\alpha)(\omega)\big)\Big)$$
for all $n \in \mathbb{N}$. Since the Bochner integration operator $\Upsilon(\cdot)$ is Lipschitz in that
$$d_C\Big(\Upsilon(\theta_1), \Upsilon(\theta_2)\Big) = \sum_{n \in \mathbb{N}} 2^{-n} \left| \int_0^n \theta_1(s)\, ds - \int_0^n \theta_2(s)\, ds \right| \leq d_{L^1_{loc}}(\theta_1, \theta_2)$$
for all $\theta_1, \theta_2 \in L^1_{loc}(\mathbb{R}_+)$, the continuity of the map $\omega \mapsto \cup_{\alpha \in \mathcal{A}}(b_\alpha, c_\alpha, g \circ F_\alpha)(\omega)$ with

4. Stochastic Processes

respect to the Hausdorff distance $d_H^{L^1_{\text{loc}}}$ from Lemma 4.40 leads to

$$d_H^C\Big(\Upsilon\big(\cup_{\alpha\in\mathcal{A}}(b_\alpha,c_\alpha,g\circ F_\alpha)(\omega^n)\big), \Upsilon\big(\cup_{\alpha\in\mathcal{A}}(b_\alpha,c_\alpha,g\circ F_\alpha)(\omega)\big)\Big)$$
$$\leq d_H^{L^1_{\text{loc}}}\Big(\cup_{\alpha\in\mathcal{A}}(b_\alpha,c_\alpha,g\circ F_\alpha)(\omega^n), \cup_{\alpha\in\mathcal{A}}(b_\alpha,c_\alpha,g\circ F_\alpha)(\omega)\Big) \longrightarrow 0$$

as $n\to\infty$. Moreover, $\Upsilon(\cup_{\alpha\in\mathcal{A}}(b_\alpha,c_\alpha,g\circ F_\alpha)(\omega))$ is closed according to Lemma 4.40 and hence $(\beta,\gamma,\zeta)\notin\Upsilon(\cup_{\alpha\in\mathcal{A}}(b_\alpha,c_\alpha,g\circ F_\alpha)(\omega))$ implies

$$d_C\Big((\beta,\gamma,\zeta),\Upsilon\big(\cup_{\alpha\in\mathcal{A}}(b_\alpha,c_\alpha,g\circ F_\alpha)(\omega)\big)\Big) > 0.$$

Altogether, this shows that $d_C((\beta^n,\gamma^n,\zeta^n),\Upsilon(\cup_{\alpha\in\mathcal{A}}(b_\alpha,c_\alpha,g\circ F_\alpha)(\omega^n))) > 0$ and thus $(\beta^n,\gamma^n,\zeta^n)\notin\Upsilon(\cup_{\alpha\in\mathcal{A}}(b_\alpha,c_\alpha,g\circ F_\alpha)(\omega^n))$ for large $n\in\mathbb{N}$, which implies

$$\limsup_{n\to\infty}\Phi(\omega^n,\beta^n,\gamma^n,\zeta^n) = \limsup_{n\to\infty}\mathbf{1}_{\Upsilon(\cup_{\alpha\in\mathcal{A}}(b_\alpha,c_\alpha,g\circ F_\alpha)(\omega^n))}(\beta^n,\gamma^n,\zeta^n)$$
$$= 0 = \mathbf{1}_{\Upsilon(\cup_{\alpha\in\mathcal{A}}(b_\alpha,c_\alpha,g\circ F_\alpha)(\omega))}(\beta,\gamma,\zeta) = \Phi(\omega,\beta,\gamma,\zeta).$$

In particular, we have proved the upper semicontinuity of $(\omega,\beta,\gamma,\zeta)\mapsto\Phi(\omega,\beta,\gamma,\zeta)$. The portmanteau theorem now implies that for all $g\in\mathcal{C}_0$

$$\mathbb{P}\Big((B,C,g*\nu)\in\Upsilon\big(\cup_{\alpha\in\mathcal{A}}(b_\alpha,c_\alpha,g\circ F_\alpha)(X)\big)\Big) = \mathbb{E}_\mathbb{P}\Big(\Phi(X,B,C,g*\nu)\Big)$$
$$\geq \limsup_{k\to\infty} E_{\mathbb{P}_{n_k}}\Big(\Phi(X,B^{n_k},C^{n_k},g*\nu^{n_k})\Big) = 1,$$

using the semicontinuity of $\Phi(\cdot)$ and $\mathbb{P}_{n_k}\in\cup_{x\in K}\mathcal{P}_x$ for all $k\in\mathbb{N}$. This shows that

$$(B,C,g*\nu)\in\Upsilon\left(\bigcup_{\alpha\in\mathcal{A}}(b_\alpha,c_\alpha,g\circ F_\alpha)(X)\right)$$

for all $g\in\mathcal{C}_0$ $\mathbb{P}$-almost surely, using the countability of $\mathcal{C}_0$. Since Υ is bijective from $L^1_{\text{loc}}(\mathbb{R}_+)$ onto $\{\varphi\in C(\mathbb{R}_+):\varphi$ is absolutely continuous$\}$ (cf. Remark A.13) and differential characteristics are unique up to $\mathbb{P}$-indistinguishability according to Remark 4.19, this implies that $\lambda(dt)\times\mathbb{P}(d\omega)$-almost everywhere

$$(b_t^\mathbb{P},c_t^\mathbb{P},g\circ F_t^\mathbb{P})(\omega)\in\bigcup_{\alpha\in\mathcal{A}}(b_\alpha,c_\alpha,g\circ F_\alpha)(\omega_t)$$

holds for all $g\in\mathcal{C}_0$. Finally, the approximation properties of $\mathcal{C}_0$ and an application of the hyperplane separation theorem from Theorem A.10 (together with the convexity and closedness of $\cup_{\alpha\in\mathcal{A}}(b_\alpha,c_\alpha,F_\alpha)(x)$ for every $x\in\mathbb{R}^d$) show that the inclusion

$$(b_t^\mathbb{P},c_t^\mathbb{P},F_t^\mathbb{P})(\omega)\in\bigcup_{\alpha\in\mathcal{A}}(b_\alpha,c_\alpha,F_\alpha)(\omega_t)$$

holds $\lambda(dt)\times\mathbb{P}(d\omega)$-almost everywhere, as required. (Note that it suffices to check the last implication for fixed $(t,\omega)\in\mathbb{R}_+\times\Omega$ and then apply Theorem A.10.) $\square$

4.2. Construction of Processes

As indicated earlier, we will now discuss, how the assumptions on the uncertainty coefficients in Remark 4.39 and the compactness result in Theorem 4.41 relate to the C_b-Feller property of the sublinear Markov semigroups $(T_t)_{t\geq 0}$ from Remark 4.33: It is easy to construct counterexamples, which show that the continuity of

$$x \longmapsto T_t\varphi(x) = E^x(\varphi(X_t)) = \sup_{\mathbb{P}\in\mathcal{P}_x} \mathbb{E}_\mathbb{P}(\varphi(X_t))$$

for all $\varphi \in C_b(\mathbb{R}^d)$ requires some form of continuity of the uncertainty coefficients

$$x \longmapsto \bigcup_{\alpha \in \mathcal{A}} (b_\alpha, c_\alpha, F_\alpha)(x),$$

similar to (D2) from Remark 4.39. In order to prove the converse, i.e. the continuity of

$$x \longmapsto T_t\varphi(x) = E^x(\varphi(X_t)) = \sup_{\mathbb{P}\in\mathcal{P}_x} \mathbb{E}_\mathbb{P}(\varphi(X_t))$$

for all $\varphi \in C_b(\mathbb{R}^d)$ under the assumptions in Remark 4.39, it seems to be necessary to relate the elements of the uncertainty subsets $\mathcal{P}_x$ to the elements of $\mathcal{P}_y$ for $x \in \mathbb{R}^d$ and $y \in \mathbb{R}^d$ close to each other. In the special case of spatially homogeneous uncertainty coefficients (i.e. Lévy processes for sublinear expectations from Remark 4.17),

$$\mathcal{P}_y = \{\mathbb{P}(\,\cdot\, + (y-x)) \mid \mathbb{P} \in \mathcal{P}_x\}$$

holds for all $x, y \in \mathbb{R}^d$, which allowed us to obtain related continuity results easily. In the rest of this section, we will study the upper and lower semicontinuity (in space) of sublinear Markov semigroups (from Remark 4.33) with spatially inhomogeneous uncertainty coefficients $(b_\alpha, c_\alpha, F_\alpha)_{\alpha \in \mathcal{A}}$. In particular, we will relate the semicontinuity to the compactness result from Theorem 4.41.

Lemma 4.42 (Upper Semicontinuity in Space). *Suppose that $X = (X_t)_{t\geq 0}$ is the coordinate-mapping process on $\Omega = D(\mathbb{R}_+; \mathbb{R}^d)$, and that*

$$\mathcal{P}_x \subset \mathfrak{P}(D_x(\mathbb{R}_+; \mathbb{R}^d))$$

for all $x \in \mathbb{R}^d$. If $\cup_{x \in K} \mathcal{P}_x$ is sequentially compact in itself (with respect to weak convergence of measures) for compact $K \subset \mathbb{R}^d$, then

$$\mathbb{R}^d \ni x \longmapsto E^x(\varphi(X_t)) := \sup_{\mathbb{P}\in\mathcal{P}_x} \mathbb{E}_\mathbb{P}(\varphi(X_t)) \in \mathbb{R}$$

is upper semicontinuous for every $t \geq 0$ and $\varphi \in USC_b(\mathbb{R}^d)$.

Proof. Suppose that $t \geq 0$, $\varphi \in USC_b(\mathbb{R}^d)$ and $(x_n)_{n\in\mathbb{N}} \subset \mathbb{R}^d$ with $\lim_{n\to\infty} x_n = x \in \mathbb{R}^d$. The definition of the supremum implies the existence of $\mathbb{P}_n \in \mathcal{P}_{x_n}$ for $n \in \mathbb{N}$ with

$$\mathbb{E}_{\mathbb{P}_n}(\varphi(X_t)) \leq E^{x_n}(\varphi(X_t)) \leq \mathbb{E}_{\mathbb{P}_n}(\varphi(X_t)) + n^{-1}.$$

4. Stochastic Processes

Moreover, there exists a subsequence $(\mathbb{P}_{n_k})_{k\in\mathbb{N}}$ such that $\mathbb{P} := \lim_{k\to\infty} \mathbb{P}_{n_k}$ weakly and

$$\limsup_{n\to\infty} E^{x_n}(\varphi(X_t)) = \limsup_{n\to\infty} \mathbb{E}_{\mathbb{P}_n}(\varphi(X_t)) = \lim_{k\to\infty} \mathbb{E}_{\mathbb{P}_{n_k}}(\varphi(X_t)),$$

because of the sequential compactness of $\cup_{x\in K}\mathcal{P}_x \subset \mathfrak{P}(D(\mathbb{R}_+;\mathbb{R}^d))$ with

$$K := \{x\} \cup \{x_n : n \in \mathbb{N}\},$$

and the characterization of the limit superior as the largest accumulation point. In fact,

$$\mathbb{P} \in \mathcal{P}_x = \mathfrak{P}(D_x(\mathbb{R}_+;\mathbb{R}^d)) \cap \left(\bigcup_{y\in K} \mathcal{P}_y\right),$$

since $\lim_{n\to\infty} x_n = x$ holds and $\omega \mapsto X_0(\omega)$ is a continuous mapping. Altogether,

$$\limsup_{n\to\infty} E^{x_n}(\varphi(X_t)) = \lim_{k\to\infty} \mathbb{E}_{\mathbb{P}_{n_k}}(\varphi(X_t)) \leq \mathbb{E}_{\mathbb{P}}(\varphi(X_t)) \leq E^x(\varphi(X_t))$$

by the generalized portmanteau theorem for semicontinuous functions (cf. [41, Theorem 2.7.5, p. 222]), and the upper semicontinuity of $\omega \mapsto X_t(\omega)$. $\square$

One is lead to assume, that the relative compactness of the uncertainty subsets $\cup_{x\in K}\mathcal{P}_x$ for compact $K \subset \mathbb{R}^d$ would suffice to prove the upper semicontinuity of

$$\mathbb{R}^d \ni x \longmapsto E^x(\varphi(X_t)) := \sup_{\mathbb{P}\in\mathcal{P}_x} \mathbb{E}_{\mathbb{P}}(\varphi(X_t)) \in \mathbb{R}$$

for continuous (instead of semicontinuous in Lemma 4.42) $\varphi \in C_b(\mathbb{R}^d)$, since

$$E^x(\varphi(X_t)) = \sup_{\mathbb{P}\in\mathcal{P}_x} \mathbb{E}_{\mathbb{P}}(\varphi(X_t)) = \sup_{\mathbb{P}\in\overline{\mathcal{P}_x}} \mathbb{E}_{\mathbb{P}}(\varphi(X_t))$$

for $t \geq 0$ and $x \in \mathbb{R}^d$. Here, $\overline{\mathcal{P}_x} \subset \mathfrak{P}(D_x(\mathbb{R}_+))$ denotes the closure of $\mathcal{P}_x$ with respect to the weak convergence of measures. However, the proof of Lemma 4.42 is not easily adapted to this generalized situation: The reason for this is, that one had to show that

$$\lim_{n\to\infty} \mathbb{P}_n \circ X_t^{-1} \in \overline{\mathcal{P}_x} \circ X_t^{-1} \tag{4.3}$$

for any weakly convergent sequence $(\mathbb{P}_n)_{n\in\mathbb{N}}$ with $\mathbb{P}_n \in \mathcal{P}_{x_n}$ and $\lim_{n\to\infty} x_n = x$. If

$$\mathcal{P}_x = \left\{\mathbb{P} \in \mathfrak{P}^{ac}_{sem}(D_x(\mathbb{R}_+)) \,\middle|\, (b^{\mathbb{P}}_s, c^{\mathbb{P}}_s, F^{\mathbb{P}}_s)(\omega) \in \bigcup_{\alpha\in\mathcal{A}} (b_\alpha, c_\alpha, F_\alpha)(\omega_s)\ \lambda(ds) \times \mathbb{P}(d\omega)\text{-a.e.}\right\}$$

for (spatially inhomogeneous) uncertainty coefficients satisfying the boundedness-and-tightness assumption (D1) from Remark 4.39, then the proof of Theorem 4.41 shows

$$\overline{\mathcal{P}_x} \subset \mathfrak{P}^{ac}_{sem}(D_x(\mathbb{R}_+))$$

and $\lim_{n\to\infty} \mathbb{P}_n \in \mathfrak{P}^{ac}_{sem}(D_x(\mathbb{R}_+))$, for convergent $(\mathbb{P}_n)_{n\in\mathbb{N}}$ with $\mathbb{P}_n \in \mathcal{P}_{x_n}$ and $x_n \to x$.

4.2. Construction of Processes

The condition in Equation (4.3) hence reduces to analyzing the structure of $\overline{\mathcal{P}_x}$ in terms of the differential characteristics $(b^{\mathbb{P}}, c^{\mathbb{P}}, F^{\mathbb{P}})$ for $\mathbb{P} \in \overline{\mathcal{P}_x}$: Under continuity assumption (D2), a minor modification of (part three of) the proof of Theorem 4.41 shows that

$$(b_s^{\mathbb{P}}, c_s^{\mathbb{P}}, F_s^{\mathbb{P}})(\omega) \in \overline{\mathrm{conv}}\left(\bigcup_{\alpha \in \mathcal{A}} (b_\alpha, c_\alpha, F_\alpha)(\omega_s)\right) \tag{4.4}$$

holds $\lambda(ds) \times \mathbb{P}(d\omega)$-almost everywhere for $\mathbb{P} = \lim_{n \to \infty} \mathbb{P}_n$ with $\mathbb{P}_n \in \mathcal{P}_{x_n}$ and $x_n \to x$. However, the converse inclusion, i.e. $\overline{\mathcal{P}_x}$ contains all

$$\mathbb{P} \in \mathfrak{P}_{\mathrm{sem}}^{\mathrm{ac}}(D_x(\mathbb{R}_+; \mathbb{R}^d))$$

such that Equation (4.4) holds $\lambda(ds) \times \mathbb{P}(d\omega)$-almost everywhere, is essentially a special case of the martingale problem (cf. [73, Chapter 3, Section 2a, p. 152]) and does not seem easy to prove in general.

Remark 4.43 (Lower Semicontinuity in Space). Suppose that $X = (X_t)_{t \geq 0}$ is the coordinate-mapping process on $\Omega = D(\mathbb{R}_+; \mathbb{R}^d)$, and that

$$\mathcal{P}_x \subset \mathfrak{P}(D_x(\mathbb{R}_+; \mathbb{R}^d))$$

for all $x \in \mathbb{R}^d$. Dual to Lemma 4.42, we would like to prove that

$$\mathbb{R}^d \ni x \longmapsto E^x(\varphi(X_t)) := \sup_{\mathbb{P} \in \mathcal{P}_x} E_{\mathbb{P}}(\varphi(X_t)) \in \mathbb{R}$$

is lower semicontinuous for fixed $t \geq 0$ and $\varphi \in \mathrm{LSC}_b(\mathbb{R}^d)$, under some additional conditions on the uncertainty subsets $\mathcal{P}_x$ with $x \in \mathbb{R}^d$. In other words, we want

$$\liminf_{n \to \infty} E^{x_n}(\varphi(X_t)) = \liminf_{n \to \infty} \sup_{\mathbb{P} \in \mathcal{P}_{x_n}} E_{\mathbb{P}}(\varphi(X_t)) \geq \sup_{\mathbb{P} \in \mathcal{P}_x} E_{\mathbb{P}} = E^x(\varphi(X_t))$$

for $(x_n)_{n \in \mathbb{N}} \subset \mathbb{R}^d$ with $\lim_{n \to \infty} x_n = x$. Unfortunately, it seems that the arguments for the upper semicontinuity from Lemma 4.42 cannot be adapted to this situation: In fact, we had to show that for every $\mathbb{P} \in \mathcal{P}_x$ there exist some $\mathbb{P}_n \in \mathcal{P}_{x_n}$ for $n \in \mathbb{N}$ such that

$$\lim_{n \to \infty} \mathbb{P}_n \circ X_t^{-1} = \mathbb{P} \circ X_t^{-1}$$

weakly (at least along a subsequence). In case the uncertainty subsets are of the form

$$\mathcal{P}_x = \left\{ \mathbb{P} \in \mathfrak{P}_{\mathrm{sem}}^{\mathrm{ac}}(D_x(\mathbb{R}_+)) \;\middle|\; (b_s^{\mathbb{P}}, c_s^{\mathbb{P}}, F_s^{\mathbb{P}})(\omega) \in \bigcup_{\alpha \in \mathcal{A}} (b_\alpha, c_\alpha, F_\alpha)(\omega_s) \;\; \lambda(ds) \times \mathbb{P}(d\omega)\text{-a.e.} \right\}$$

for (spatially inhomogeneous) uncertainty coefficients, this leads naturally to the question, if we can find $\mathbb{P}_n \in \mathfrak{P}_{\mathrm{sem}}^{\mathrm{ac}}(D_{x_n}(\mathbb{R}^d))$ for a triplet of potential differential characteristics. However, this question seems not to be easy to answer, since it is a special case of the martingale problem (cf. [73, Chapter 3, Section 2a, p. 152]).

4. Stochastic Processes

In order to avoid the difficulties mentioned in the last two paragraphs (i.e. related to the continuity of $x \mapsto E^x(\varphi(X_t))$ for fixed $\varphi \in C_b(\mathbb{R}^d)$ and $t \geq 0$), we will present another approach to introduce spatially inhomogeneous jump-type processes for sublinear expectations in the following two sections: We will start off with a Lévy process $(X_t)_{t\geq 0}$ on a sublinear expectation space $(\Omega, \mathcal{H}, E)$ as in Remark 4.38, construct stochastic integrals for sublinear expectations, and solve stochastic differential equations of the form

$$Z_t^x = x + \int_0^t f(Z_{s-}^x)\, dX_s$$

for globally Lipschitz-continuous $f : \mathbb{R}^k \to \mathbb{R}^{k \times d}$. The resulting processes $(Z^x)_{t\geq 0}$ will be referred to as Lévy-type processes for sublinear expectations, and generate sublinear Markov semigroups $(T_t)_{t\geq 0}$ by

$$T_t\varphi(x) = E(\varphi(Z_t^x))$$

on the convex cone of bounded, upper semi-analytic functions. Furthermore, we will show that these sublinear Markov semigroups always satisfy the C_b-Feller property. Therefore, we can establish a strong connection (similar to Remark 4.38) between Lévy-type processes for sublinear expectations and their associated sublinear generator equations (based on the general theory in Section 4.1).

4.3. Stochastic Integration

Stochastic integration theory, as a general framework to define integrals of stochastic processes with respect to stochastic processes, plays an important role in the majority of applications of stochastic processes. The most commonly used notion (in classical probability theory) of Itô integrals and the related Itô calculus for semimartingales was originally developed by Kiyosi Itô in [71]. For a historical overview of stochastic integration theory for linear expectations and its applications in mathematical finance, we refer the interested reader to [76]. Due to its relevance for applications in mathematical finance, Shige Peng already included a canonical construction of stochastic integrals (for continuous-path processes) for sublinear expectations in his original article [102] on G-Brownian motion. This intrinsic approach was later extended to G-Lévy processes (with uniformly bounded Lévy measures) by Krzysztof Paczka in [101].

In this section, we will introduce a very general notion of stochastic integrals for sublinear expectations. Instead of generalizing Shige Peng's intrinsic approach from [102], which is based on the classical construction via elementary functions, we will employ a pathwise construction of stochastic integrals from classical probability theory due to Marcel Nutz in [97]. This will allow us to introduce stochastic integrals for a large class of integrands driven by general stochastic processes in sublinear expectation spaces, which are semimartingales (with jumps) for each element of the uncertainty subset. Furthermore, we are going to study the measurability of the resulting random variables, which will play an essential role in Section 4.4, in order to construct Lévy-type processes for sublinear expectations.

4.3. Stochastic Integration

Remark 4.44 (Usual Conditions). Many of the following results on classical stochastic integration assume that we work with a probability space $(\Omega, \mathcal{A}, \mathbb{P})$ together with a filtration $\mathcal{F} = (\mathcal{F}_t)_{t \geq 0}$ satisfying the *usual conditions* (or *usual hypotheses*): The σ-algebras $\mathcal{A}$ and $\mathcal{F}_t$ for $t \geq 0$ are complete (i.e. they contain all subsets of $\mathbb{P}$-null sets) and the filtration $\mathcal{F}$ is right-continuous (i.e. $\mathcal{F}_t = \cap_{s \geq t} \mathcal{F}_s$ for all $t \geq 0$). It is easy to check (cf. e.g. [73, Chapter 1, Section 1a, p. 2]) that every completed probability space $(\Omega, \mathcal{A}^{\mathbb{P}}, \overline{\mathbb{P}})$ together with the augmented filtration

$$\mathcal{F}_+^{\mathbb{P}} := \left(\mathcal{F}_{t+}^{\mathbb{P}}\right)_{t \geq 0} = \left(\left(\cap_{s \geq t} \mathcal{F}_s\right)^{\mathbb{P}}\right)_{t \geq 0}$$

satisfies the usual conditions, where $\mathcal{A}^{\mathbb{P}}$ is the completion of $\mathcal{A}$ under $\mathbb{P}$, and $\overline{\mathbb{P}}$ the corresponding unique extension of $\mathbb{P}$. Moreover, if $X = (X_t)_{t \geq 0}$ is a $\mathbb{P}$-$\mathcal{F}$-semimartingale, then it is also a $\overline{\mathbb{P}}$-$\mathcal{F}_+^{\mathbb{P}}$-semimartingale with the same characteristics, cf. Lemma 4.15. This procedure allows us to start with a filtered probability space without the usual conditions, by extending the space and augmenting the filtration before applying the results from classical stochastic integration.

For some of the results in Section 4.2, it is important that we work with the original canonical filtration generated by the stochastic process $X = (X_t)_{t \geq 0}$, since they rely on characterizations of the canonical filtration such as Galmarino's test (cf. [38, Chapter 4, Theorem 100, p. 149]). Therefore, we will strictly differentiate between the original filtration $\mathcal{F}$ and its augmentation $\mathcal{F}_+^{\mathbb{P}}$ in the following two sections. However, in order to avoid cluttered notation, we will occasionally use $\mathbb{P}$ for both the original probability measure as well as its extension to the completed σ-algebra. Since the extension is unique, this abuse of notation should not cause any confusion. Moreover, although we have used the common notation based on the integral sign for Itô integrals from classical probability theory in some of the proofs in Section 4.2, we will use a different notation in this section to stress the dependence on the probability measures.

Since our stochastic integrals will almost surely coincide with classical Itô integrals (for linear expectations) under each probability measure in the uncertainty subset, let us quickly recall the notation and some important properties:

Remark 4.45 (Classical Stochastic Integration). Suppose that $X = (X_t)_{t \geq 0}$ is a d-dimensional $\mathbb{P}$-$\mathcal{F}$-semimartingale on a classical probability space $(\Omega, \mathcal{A}, \mathbb{P})$ adapted to a filtration $\mathcal{F} = (\mathcal{F}_t)_{t \geq 0}$, then there exists (cf. for example [73, Chapter 3, Section 6, p. 203] or [110, Chapter 4, p. 153]) a linear space $L_{\mathbb{P}}(X) = L_{\mathbb{P}}(X; \mathbb{R}^d)$ of d-dimensional predictable processes (referred to as the *family of integrable processes*) and a mapping

$$L_{\mathbb{P}}(X) \ni (H_t)_{t \geq 0} \longmapsto \mathcal{J}_{\mathbb{P}}(H, X) = (\mathcal{J}_{\mathbb{P}}(H, X)_t)_{t \geq 0}$$

(referred to as *stochastic integral* or *Itô integral* with respect to X) that assigns to every integrand $H = (H_t)_{t \geq 0} \in L_{\mathbb{P}}(X)$ a one-dimensional $\mathbb{P}$-$\mathcal{F}_+^{\mathbb{P}}$-semimartingale $\mathcal{J}_{\mathbb{P}}(H, X)$, such that the following properties hold (up to $\mathbb{P}$-indistinguishability):

4. Stochastic Processes

(i) (Construction) For *elementary functions* $H \in L_{\mathbb{P}}(X)$, i.e. integrands of the form

$$H = \xi_0 \mathbf{1}_{\{0\}} + \sum_{j=1}^{m} \xi_j \mathbf{1}_{(t_j, t_{j+1}]}$$

for some (deterministic) times $0 = t_0 \leq t_1 < \ldots < t_m < \infty$ with $m \in \mathbb{N}$ and some d-dimensional, $\mathcal{F}_{t_j}$-measurable random variables ξ_j for $j \in \{0, \ldots, m\}$,

$$\mathcal{J}_{\mathbb{P}}(H, X)_t = \sum_{j=1}^{m} \xi_j \left(X_{t_{j+1} \wedge t} - X_{t_j \wedge t} \right)$$

holds for all $t \geq 0$.

(ii) (Locally Bounded Integrands) If the d-dimensional predictable process $(H_t)_{t \geq 0}$ is locally bounded, i.e. there exists a sequence of stopping times $(\tau_n)_{n \in \mathbb{N}}$ with $\lim_{n \to \infty} \tau_n = \infty$ $\mathbb{P}$-almost surely and $(K_n)_{n \in \mathbb{N}} \subset \mathbb{R}$ such that

$$\mathbb{P}\left(\forall t \geq 0 : |H_{\tau_n \wedge t}| \leq K_n \right) = 1$$

for all $n \in \mathbb{N}$, then $H = (H_t)_{t \geq 0} \in L_{\mathbb{P}}(X)$ and $\mathcal{J}_{\mathbb{P}}(H, X) = \sum_{i \leq d} \mathcal{J}_{\mathbb{P}}(H^i, X^i)$.

(iii) (Linearity) The stochastic integration map $L_{\mathbb{P}}(X) \ni H \mapsto \mathcal{J}_{\mathbb{P}}(H, X)$ is linear. Moreover, if $H \in L_{\mathbb{P}}(X) \cap L_{\mathbb{P}}(Y)$ for two $\mathbb{P}$-$\mathcal{F}$-semimartingales X and Y, then $H \in L_{\mathbb{P}}(aX + bY)$ and

$$\mathcal{J}_{\mathbb{P}}(H, aX + bY) = a \mathcal{J}_{\mathbb{P}}(H, X) + b \mathcal{J}_{\mathbb{P}}(H, Y)$$

holds for all constants $a, b \in \mathbb{R}$.

(iv) (Localization) For all stopping times $\tau : \Omega \to [0, \infty]$ and integrands $H \in L_{\mathbb{P}}(X)$, the stopped integrands $H \mathbf{1}_{[0, \tau]} \in L_{\mathbb{P}}(X)$ and

$$\mathcal{J}_{\mathbb{P}}(H, X^\tau) = \mathcal{J}_{\mathbb{P}}(H \mathbf{1}_{[0, \tau]}, X) = \mathcal{J}_{\mathbb{P}}(H, X)^\tau$$

holds, where $Y^\tau := (Y_{\tau \wedge t})_{t \geq 0}$ for random processes $Y = (Y_t)_{t \geq 0}$.

(v) (Continuous Martingale Part & Jumps) The continuous martingale part and the jumps of a stochastic integral are given by

$$(\mathcal{J}_{\mathbb{P}}(H, X))^C = \mathcal{J}_{\mathbb{P}}(H, X^C)$$
$$\Delta \mathcal{J}_{\mathbb{P}}(H, X) = \sum_{i \leq d} H^i \Delta X^i$$

for every $H \in L_{\mathbb{P}}(X)$, where Y^C is the continuous martingale part of a semimartingale $Y = (Y_t)_{t \geq 0}$ as defined in [73, Chapter 1, Proposition 4.27, p. 45].

(vi) (Associativity) If $H \in L_P(X)$ and $K \in L_{\mathbb{P}}(\mathcal{J}_{\mathbb{P}}(H,X))$, then $KH \in L_{\mathbb{P}}(X)$ and

$$\mathcal{J}_{\mathbb{P}}(K, \mathcal{J}_{\mathbb{P}}(H,X)) = \mathcal{J}_{\mathbb{P}}(KH, X)$$

holds, where $(KH)(\omega) := K(\omega)H(\omega)$ for all $\omega \in \Omega$.

(vii) (Translation Invariance) Stochastic integrals start at the origin and are invariant under translations of the integrator, i.e. $\mathcal{J}_{\mathbb{P}}(H,X)_0 = 0$ and

$$\mathcal{J}_{\mathbb{P}}(H, X - X_0) = \mathcal{J}_{\mathbb{P}}(H, X)$$

for every $H \in L_{\mathbb{P}}(X)$.

(viii) (Stability) Suppose that $(H^n)_{n \in \mathbb{N}} \subset L_{\mathbb{P}}(X)$, $H \in L_{\mathbb{P}}(X)$ and that $K = (K_t)_{t \geq 0}$ is a one-dimensional predictable, locally bounded process as in property (ii). If

$$\sup_{n \in \mathbb{N}} |H_t^n(\omega) - H_t(\omega)| \leq K_t(\omega)$$

and $\lim_{n \to \infty} H_t^n(\omega) = H_t(\omega)$ for all $(t, \omega) \in \mathbb{R}_+ \times \Omega$, then

$$\sup_{s \leq t} |\mathcal{J}_{\mathbb{P}}(H^n, X)_s - \mathcal{J}_{\mathbb{P}}(H, X)_s| \longrightarrow 0$$

in $\mathbb{P}$-probability for all $t \geq 0$.

Lemma 4.46 (Characterization of $L_{\mathbb{P}}(X)$). *Suppose that $(X_t)_{t \geq 0}$ is a d-dimensional $\mathbb{P}$-$\mathcal{F}$-semimartingale on a classical probability space $(\Omega, \mathcal{A}, \mathbb{P})$ and*

$$(B_t, C_t, \nu(dt, dx)) = (b_t \, dA_t, c_t \, dA_t, F_t(dx) \, dA_t)$$

is the differential representation of its characteristics with respect to a truncation function $h : \mathbb{R}^d \to \mathbb{R}^d$ as in Remark 4.19. If $H = (H_t)_{t \geq 0}$ is a predictable d-dimensional process, then the Lebesgue-Stieltjes integral

$$\int_0^t \left(H_s^T c_s H_s + \int_{|z| \leq 1} \left(1 \wedge |H_s^T z|^2 \right) F_s(dz) \right.$$
$$- \left| \int_{|z| \leq 1, |H_s^T z| \leq 1} H_s^T z \, F_s(dz) \right|^2 \Delta A_s$$
$$\left. + \left| H_s^T b_s + \int \left((H_s^T z) \mathbf{1}_{|z| \leq 1, |H_s^T z| \leq 1} - H_s^T h(z) \right) F_s(dz) \right| \right) dA_s$$

is finite $\mathbb{P}$-almost surely for all $t \geq 0$ if and only if $H \in L_{\mathbb{P}}(X)$.

Proof. $\hookrightarrow$ [73, Chapter 3, Theorem 6.30, p. 212] $\square$

4. Stochastic Processes

Remark 4.47 (Scalar & Matrix-Valued Integrands). If $X = (X_t)_{t \geq 0}$ is a d-dimensional $\mathbb{P}$-$\mathcal{F}$-semimartingale on a classical probability space $(\Omega, \mathcal{A}, \mathbb{P})$, then the (classical) stochastic integral from Remark 4.45 can be generalized to

$$L_{\mathbb{P}}(X; \mathbb{R}^{k \times d}) := \left\{ \hat{H} = (\hat{H}^i)_{i \leq k} : \mathbb{R}_+ \times \Omega \to \mathbb{R}^{k \times d} \mid \forall i \in \{1, \ldots, k\} : \hat{H}^i \in L_{\mathbb{P}}(X; \mathbb{R}^d) \right\}$$
$$L_{\mathbb{P}}(X; \mathbb{R}) := \left\{ \check{H} : \mathbb{R}_+ \times \Omega \to \mathbb{R} \mid \forall i \in \{1, \ldots, d\} : \check{H} \in L_{\mathbb{P}}(X^i; \mathbb{R}) \right\}$$
$$= \left\{ \check{H} : \mathbb{R}_+ \times \Omega \to \mathbb{R} \mid \mathrm{diag}(\check{H}, \ldots, \check{H}) \in L_{\mathbb{P}}(X; \mathbb{R}^{d \times d}) \right\},$$

the family of *matrix-valued* or *scalar integrable processes*, by defining

$$\mathcal{J}_{\mathbb{P}}\left(\hat{H}, X\right) := \left(\mathcal{J}_{\mathbb{P}}\left(\hat{H}^1, X\right), \ldots, \mathcal{J}_{\mathbb{P}}\left(\hat{H}^k, X\right) \right)^T$$
$$\mathcal{J}_{\mathbb{P}}\left(\check{H}, X\right) := \left(\mathcal{J}_{\mathbb{P}}\left(\check{H}, X^1\right), \ldots, \mathcal{J}_{\mathbb{P}}\left(\check{H}, X^d\right) \right)^T = \mathcal{J}_{\mathbb{P}}(\mathrm{diag}(\check{H}, \ldots, \check{H}), X)$$

for $\hat{H} \in L_{\mathbb{P}}(X; \mathbb{R}^{k \times d})$ and $\check{H} \in L_{\mathbb{P}}(X; \mathbb{R})$. Note that the predictability assumption for the integrands is inherited from the definitions of $L_{\mathbb{P}}(X; \mathbb{R}^d)$ and $L_{\mathbb{P}}(X^i; \mathbb{R})$, respectively.

Lemma 4.48 (Characteristics of Stochastic Integrals). *If the process $X = (X_t)_{t \geq 0}$ is a d-dimensional $\mathbb{P}$-$\mathcal{F}$-semimartingale on a classical probability space $(\Omega, \mathcal{A}, \mathbb{P})$ with characteristics (B, C, ν) for a truncation function $h : \mathbb{R}^d \to \mathbb{R}^d$ as in Definition 4.13, $H \in L_{\mathbb{P}}(X; \mathbb{R}^{k \times d})$ is an integrable (matrix-valued) process with $k \in \mathbb{N}$ and*

$$\hat{B} := HB + (\hat{h}(Hz) - Hh(z)) * \nu$$
$$= \left(\sum_{j \leq d} H^{(i,j)} B^j + \int_{[0, \cdot]} \int_{\mathbb{R}^d} \left(\hat{h}^i(H_s z) - \sum_{j \leq d} H_s^{i,j} h^j(z) \right) \nu(ds, dz) \right)_{i \leq k}$$
$$\hat{C} := HCH^T = \left(\sum_{m, n \leq d} H^{i, m} C^{m, n} H^{j, n} \right)_{i, j \leq k}$$
$$\hat{\nu}(B) := \int_{\mathbb{R}_+} \int_{\mathbb{R}^d} \mathbb{1}_B(s, H_s z) \, \nu(ds, dz)$$

for $B \in \mathcal{B}(\mathbb{R}_+ \times \mathbb{R}^k)$, then $(\hat{B}, \hat{C}, \hat{\nu})$ are the characteristics of the $\mathbb{P}$-$\mathcal{F}_+^{\mathbb{P}}$-semimartingale $\mathcal{J}_{\mathbb{P}}(H, X)$ with respect to the truncation function $\hat{h} : \mathbb{R}^k \to \mathbb{R}^k$.

Proof. $\hookrightarrow$ [73, Chapter 9, Proposition 5.3, p. 565] □

In the light of the connection between the continuity of sublinear Markov semigroups and the classical martingale problem (as discussed in Remark 4.43), it seems interesting to note that it is possible to prove a converse of the preceding statement (cf. [73, Chapter 9, Proposition 5.6, p. 566] for details).

We now present a result due to Marcel Nutz in [97] on the pathwise construction of classical stochastic integrals (as defined in Remark 4.45), and show how to apply it, in order to introduce stochastic integration for sublinear expectations. The main idea of the proof of Marcel Nutz's result is to combine an approximation argument for classical stochastic integrals with medial limits:

4.3. Stochastic Integration

Theorem 4.49 (Medial Limits). *Suppose that $(\Omega, \mathcal{A})$ is a measurable space and*

$$\mathcal{A}^* = \cap_{\mathbb{P} \in \mathfrak{P}(X)} \mathcal{A}^{\mathbb{P}}$$

its universal completion as in Remark 4.25 (i.e. $\mathcal{A}^{\mathbb{P}}$ is the smallest σ-algebra that contains $\mathcal{A}$ and all subsets of $\mathbb{P}$-null sets). If $(X_n)_{n \in \mathbb{N}}$ is a sequence of $\mathcal{A}$-measurable random variables and Cantor's continuum hypothesis holds (i.e. there exists a bijection $\pi : A \to \mathbb{R}$ for every uncountable $A \subset \mathbb{R}$), then there exists a $\mathcal{A}^*$-measurable

$$\lim_{n \to \infty} \operatorname{med} X_n : \Omega \longrightarrow \overline{\mathbb{R}} = \mathbb{R} \cup \{\pm \infty\},$$

referred to as (Mokobodzki's) medial limit of the sequence $(X_n)_{n \in \mathbb{N}}$, such that

$$\lim_{n \to \infty} \operatorname{med} X_n = X^{\mathbb{P}}$$

holds $\mathbb{P}$-almost surely for every $\mathbb{P} \in \mathfrak{P}(\Omega)$ with $\lim_{n \to \infty} X_n = X^{\mathbb{P}}$ in probability.

Proof. $\hookrightarrow$ [39, Chapter 10, Section 2, Theorem 57, p. 114] □

Theorem 4.50 (Pathwise Stochastic Integration). *Suppose that $X = (X_t)_{t \geq 0}$ is a stochastic process on a measurable space $(\Omega, \mathcal{A})$ adapted to a filtration $\mathcal{F} = (\mathcal{F}_t)_{t \geq 0}$. Furthermore, assume that Cantor's continuum hypothesis holds,*

$$\mathcal{P} \subset \mathfrak{P}_{sem}(\Omega) = \{\mathbb{P} \in \mathfrak{P}(\Omega) \mid (X_t)_{t \geq 0} \text{ is a } \mathbb{P}\text{-}\mathcal{F}\text{-semimartingale}\}$$

and that $A = (A_t)_{t \geq 0}$ is a predictable (with respect to the filtration $\mathcal{F}$), increasing stochastic process with càdlàg paths such that

$$\operatorname{var}(B^{\mathbb{P}}) + \operatorname{var}(C^{\mathbb{P}}) + \int_{[0, \cdot]} \int_{\mathbb{R}^d} (1 \wedge |z|^2) \, \nu^{\mathbb{P}}(ds, dz) \ll A$$

holds $\mathbb{P}$-almost surely for all $\mathbb{P} \in \mathcal{P}$, where $(B^{\mathbb{P}}, C^{\mathbb{P}}, \nu^{\mathbb{P}})$ are the characteristics of X under $\mathbb{P} \in \mathfrak{P}_{sem}(\Omega)$ as in Definition 4.13. For every

$$H = (H_t)_{t \geq 0} \in \bigcap_{\mathbb{P} \in \mathcal{P}} L_{\mathbb{P}}(X),$$

there exists a stochastic process $Y = (Y_t)_{t \geq 0}$ adapted to the $\mathcal{P}$-universally augmented filtration $\mathcal{F}_+^{\mathcal{P}} = (\cap_{\mathbb{P} \in \mathcal{P}} \mathcal{F}_{t+}^{\mathbb{P}})_{t \geq 0}$ with càdlàg paths, such that

$$Y = \mathcal{J}_{\mathbb{P}}(H, X)$$

holds $\mathbb{P}$-almost surely for all $\mathbb{P} \in \mathcal{P}$. Moreover, the construction of the path $(Y_t(\omega))_{t \geq 0}$ for any $\omega \in \Omega$ only depends on the paths $(X_t(\omega))_{t \geq 0}$, $(A_t(\omega))_{t \geq 0}$ and $(H_t(\omega))_{t \geq 0}$.

Proof. $\hookrightarrow$ [97, Theorem 2.2] □

4. Stochastic Processes

Definition 4.51 (Stochastic Integrals). Suppose that $X = (X_t)_{t\geq 0}$ is a stochastic process on a measurable space $(\Omega, \mathcal{A})$ adapted to a filtration $\mathcal{F} = (\mathcal{F}_t)_{t\geq 0}$, and that Cantor's continuum hypothesis holds. For every uncertainty subset $\mathcal{P} \subset \mathfrak{P}^{ac}_{sem}(\Omega)$ (i.e. a family consisting of probability measures $\mathbb{P} \in \mathfrak{P}(\Omega)$ such that $(X_t)_{t\geq 0}$ is a $\mathbb{P}$-$\mathcal{F}$-semimartingale with absolutely continuous characteristics, cf. Theorem 4.20) and

$$H = (H_t)_{t\geq 0} \in L_{\mathcal{P}}(X) := L_{\mathcal{P}}(X; \mathbb{R}^d) := \bigcap_{\mathbb{P}\in\mathcal{P}} L_{\mathbb{P}}(X; \mathbb{R}^d),$$

the *stochastic integral* of the integrand $(H_t)_{t\geq 0}$ with respect to integrator $(X_t)_{t\geq 0}$

$$\mathcal{J}_{\mathcal{P}}(H, X) = (\mathcal{J}_{\mathcal{P}}(H, X)_t)_{t\geq 0} : \Omega \longrightarrow D(\mathbb{R}_+)$$

is defined as the stochastic process adapted to the $\mathcal{P}$-universally augmented filtration

$$\mathcal{F}^{\mathcal{P}}_+ := \left(\bigcap_{\mathbb{P}\in\mathcal{P}} \mathcal{F}^{\mathbb{P}}_{t+}\right)_{t\geq 0}$$

with càdlàg paths (constructed by Theorem 4.50) such that

$$\mathcal{J}_{\mathcal{P}}(H, X) = \mathcal{J}_{\mathbb{P}}(H, X)$$

holds $\mathbb{P}$-almost surely for all $\mathbb{P} \in \mathcal{P}$. For scalar and matrix-valued integrands

$$\hat{H} \in L_{\mathcal{P}}(X; \mathbb{R}^{k\times d}) := \bigcap_{\mathbb{P}\in\mathcal{P}} L_{\mathbb{P}}(X; \mathbb{R}^{k\times d})$$

$$\check{H} \in L_{\mathcal{P}}(X; \mathbb{R}^1) := \bigcap_{\mathbb{P}\in\mathcal{P}} L_{\mathbb{P}}(X; \mathbb{R}^1),$$

define the *matrix-valued* or *scalar stochastic integral* component-wise as in Remark 4.47.

It is well-known that the space of integrands $L^{\mathbb{P}}(X)$ for a fixed $\mathbb{P} \in \mathfrak{P}_{sem}(\Omega)$ is the largest class of integrands satisfying some natural conditions, cf. [29, Theorem 12.3.22, p. 276]. Therefore, the space of integrands $L^{\mathcal{P}}(X)$ for $\mathcal{P} \subset \mathfrak{P}^{ac}_{sem}(\Omega)$ from Definition 4.51 is maximal in a similar sense, since the stochastic integral for $\mathcal{P}$ coincides with the classical stochastic integral for $\mathbb{P}$ (from Remark 4.45) almost surely under each $\mathbb{P} \in \mathcal{P}$. However, the price for this general definition is its non-constructive nature, in that it is based on medial limits from Theorem 4.49 and thus relies on the validity of Cantor's continuity hypothesis. Fortunately, Rajeeva Karandikar introduced in [80] another construction of pathwise stochastic integrals for integrands with *càglàd paths* (continue à gauche, limite à droite - i.e. left continuous with right limits) without relying on Cantor's continuity hypothesis. In the rest of this section, we extend Karandikar's result by showing that the resulting stochastic integrals are adapted to the completed filtration $(\mathcal{F}^{\mathbb{P}}_t)_{t\geq 0}$ instead of its augmentation $(\mathcal{F}^{\mathbb{P}}_{t+})_{t\geq 0}$. As remarked earlier, this measurability is essential for our construction of Lévy-type processes in Section 4.4. Before we prove this supplement, let us recall that almost sure limits in Polish spaces can chosen to be measurable with respect to the original σ-algebra instead of its completion:

4.3. Stochastic Integration

Lemma 4.52 (Measurability of Limits). *Suppose that $(\Omega, \mathcal{A})$ is a measurable space, $(\mathcal{S}, d)$ is a Polish space (i.e. a complete, separable metric space) with $x_0 \in \mathcal{S}$, and that the maps $X_n : \Omega \to \mathcal{S}$ are $\mathcal{A}/\mathcal{B}(\mathcal{S})$-measurable for $n \in \mathbb{N}$, then $X : \Omega \to \mathcal{S}$ defined by*

$$X(\omega) := \begin{cases} \lim_{n \to \infty} X_n(\omega) & \text{if } \lim_{n \to \infty} X_n(\omega) \text{ exists} \\ x_0 & \text{otherwise} \end{cases}$$

for $\omega \in \Omega$ is $\mathcal{A}/\mathcal{B}(\mathcal{S})$-measurable and $\{\omega \in \Omega : \lim_{n \to \infty} X_n(\omega) \text{ exists}\} \in \mathcal{A}$. Moreover, if

$$\lim_{n \to \infty} X_n = X^{\mathbb{P}}$$

almost surely for some probability measure $\mathbb{P} \in \mathfrak{P}(\Omega)$, then $X^{\mathbb{P}} = X$ almost surely.

Proof. It is easy to check that the metric $d : X \times X \to \mathbb{R}$ is continuous and thus $\mathcal{B}(X \times X)/\mathcal{B}(\mathbb{R})$-measurable. Since $(\mathcal{S}, d)$ is separable, we know that

$$\mathcal{B}(X \times X) = \mathcal{B}(X) \otimes \mathcal{B}(X),$$

cf. e.g. [46, Proposition 4.1.7, p. 119] and [46, Proposition 2.1.4, p. 31]. In particular,

$$d(X_n, X_m) : \Omega \longrightarrow \mathbb{R}$$

are $\mathcal{A}/\mathcal{B}(\mathbb{R})$-measurable for all $m, n \in \mathbb{N}$. Therefore, the completeness of $(\mathcal{S}, d)$ implies

$$\Omega_0 := \left\{ \omega \in \Omega : \lim_{n \to \infty} X_n(\omega) \text{ exists} \right\} = \left\{ \omega \in \Omega : (X_n(\omega))_{n \in \mathbb{N}} \subset \mathcal{S} \text{ is Cauchy} \right\}$$
$$= \left\{ \omega \in \Omega : \limsup_{n \to \infty} \sup_{k \in \mathbb{N}} d(X_n, X_{n+k}) = 0 \right\}$$

and hence the measurability of $\Omega_0 \subset \Omega$ with respect to $\mathcal{A}$. The sequence $(\hat{X}_n)_{n \in \mathbb{N}}$ with

$$\hat{X}_n(\omega) := \begin{cases} X_n(\omega) & \text{if } \omega \in \Omega_0 \\ x_0 & \text{if } \omega \in \Omega_0^C = \Omega \setminus \Omega_0 \end{cases}$$

for $\omega \in \Omega$ and $n \in \mathbb{N}$ converges to X pointwise, i.e. $\lim_{n \to \infty} \hat{X}_n(\omega) = X(\omega)$ for all $\omega \in \Omega$. According to [46, Theorem 4.2.2, p. 125], this implies X is $\mathcal{A}/\mathcal{B}(X)$-measurable, since

$$\hat{X}_n^{-1}(B) = \left(\hat{X}_n^{-1}(B) \cap \Omega_0\right) \cup \left(\hat{X}_n^{-1}(B) \cap \Omega_0^C\right)$$
$$\in \left\{ (X_n^{-1}(B) \cap \Omega_0) \cup \varnothing, (X_n^{-1}(B) \cap \Omega_0) \cup \Omega_0^C \right\} \subset \mathcal{A}$$

for $B \in \mathcal{B}(\mathcal{S})$ and $n \in \mathbb{N}$, using the $\mathcal{A}/\mathcal{B}(\mathbb{R})$-measurability of $X_n : \Omega \to \mathcal{S}$. Finally, if $\lim_{n \to \infty} X_n = X^{\mathbb{P}}$ almost surely for some $\mathbb{P} \in \mathfrak{P}(\Omega)$, then $\mathbb{P}(\Omega_0) = 1$ and therefore $X^{\mathbb{P}} = X$ almost surely. $\square$

4. Stochastic Processes

Proposition 4.53 (Measurability of Stochastic Integrals). *The mappings*

$$\tau_n^m : D(\mathbb{R}_+; \mathbb{R}^d) \times D(\mathbb{R}_+; \mathbb{R}^d) \longrightarrow [0, \infty]$$
$$\mathcal{J}^m : D(\mathbb{R}_+; \mathbb{R}^d) \times D(\mathbb{R}_+; \mathbb{R}^d) \longrightarrow D(\mathbb{R}_+; \mathbb{R}^d)$$

are Borel-measurable with respect to the Skorokhod topology for $m, n \in \mathbb{N}$, where

$$\tau_0^m(\alpha, \beta) := 0$$
$$\tau_n^m(\alpha, \beta) := \inf\left\{ t > \tau_{n-1}^m(\alpha, \beta) : \left| \alpha(t) - \alpha\left(\tau_{n-1}^m(\alpha, \beta)\right) \right| > 2^{-m} \right\}$$
$$\mathcal{J}^m(\alpha, \beta)(t) := \sum_{k=0}^{\infty} \alpha(\tau_k^m(\alpha, \beta)) \left(\beta(\tau_{k+1}^m(\alpha, \beta)) - \beta(\tau_k^m(\alpha, \beta)) \right) \mathbf{1}_{[\tau_{k+1}^m(\alpha, \beta), \infty)}(t)$$
$$+ \sum_{k=0}^{\infty} \alpha(\tau_k^m(\alpha, \beta)) \left(\beta(\quad t \quad) - \beta(\tau_k^m(\alpha, \beta)) \right) \mathbf{1}_{[\tau_k^m(\alpha, \beta), \tau_{k+1}^m(\alpha, \beta))}(t)$$

for $\alpha, \beta \in D(\mathbb{R}_+)$ and $t \geq 0$. Hence, the map $\mathcal{J} : D(\mathbb{R}_+) \times D(\mathbb{R}_+) \to D(\mathbb{R}_+)$ defined by

$$\mathcal{J}(\alpha, \beta) := \begin{cases} \lim_{m \to \infty} \mathcal{J}^m(\alpha, \beta) & \text{if } \lim_{m \to \infty} \mathcal{J}^m(\alpha, \beta) \text{ exists} \\ 0 & \text{otherwise} \end{cases}$$

is Borel-measurable as well, where the limits are taken with respect to the Skorokhod topology. Moreover, if $X = (X_t)_{t \geq 0}$ is a $\mathbb{P}$-$\mathcal{F}$-semimartingale and $H = (H_t)_{t \geq 0}$ is a stochastic process with càdlàg paths adapted to $\mathcal{F}$, then

$$\lim_{n \to \infty} \sup_{t \in [0,T]} \left| \mathcal{J}^m(H, X)(t) - \mathcal{J}_{\mathbb{P}}(H_-, X)(t) \right| = 0$$

$\mathbb{P}$-*almost surely for all $T > 0$, where $H_- := (H_{t-})_{t \geq 0}$ with $H_{0-} := H_0$. In particular,*

$$\mathcal{J}(H, X) = \mathcal{J}_{\mathbb{P}}(H_-, X)$$

$\mathbb{P}$-*almost surely, and $\mathcal{J}(H, X)$ is $(\cup_{t \geq 0} \mathcal{F}_t)$-measurable and adapted to $\mathcal{F}^{\mathbb{P}} = (\mathcal{F}_t^{\mathbb{P}})_{t \geq 0}$.*

Proof. Obviously, $\tau_0^m \equiv 0$ are Borel-measurable for all $m \in \mathbb{N}$. Inductively, for all $T \geq 0$

$$\{\tau_n^m < T\} = \bigcup_{t \in \mathbb{Q} \cap [0, T)} \left(\{\tau_{n-1}^m < t\} \cap \{(\alpha, \beta) : |\alpha(t) - \alpha(\tau_{n-1}^m(\alpha, \beta))| > 2^{-m}\} \right)$$

is Borel-measurable (and even measurable with respect to the σ-algebra $\mathcal{G}_T$ generated

$$D(\mathbb{R}_+)^2 \ni (\alpha, \beta) \mapsto (\alpha^T, \beta^T) = (\alpha_{t \wedge T}, \beta_{t \wedge T})_{t \geq 0} \in D(\mathbb{R}_+)^2$$

for $T \geq 0$) using the Borel-measurability (or $\mathcal{G}_t$-measurability, respectively) of

$$D(\mathbb{R}_+)^2 \times \mathbb{R}_+ \ni (\alpha, \beta, s) \mapsto (\alpha(s), \beta(s)) \in \mathbb{R}^d \times \mathbb{R}^d$$
$$D(\mathbb{R}_+)^2 \ni (\alpha, \beta) \mapsto (\alpha^t, \beta^t, \tau_{n-1}^m(\alpha, \beta)) \mathbf{1}_{\{\tau_{n-1}^m(\alpha, \beta) < t\}} \in D(\mathbb{R}_+)^2 \times \mathbb{R}_+$$

as well as the right-continuity of the paths iteratively.

4.3. Stochastic Integration

In order to show that
$$\mathcal{J}^m : D(\mathbb{R}_+) \times D(\mathbb{R}_+) \longrightarrow D(\mathbb{R}_+)$$
are Borel-measurable with $m \in \mathbb{N}$, it suffices to check that
$$\mathcal{J}^m(t) : D(\mathbb{R}_+) \times D(\mathbb{R}_+) \longrightarrow \mathbb{R}$$
are Borel-measurable for all $t \geq 0$, due to [18, Theorem 12.5, p. 134]. The maps
$$D(\mathbb{R}_+)^2 \times \mathbb{R}_+ \ni (\alpha, \beta, s) \longmapsto (\alpha(s), \beta(s)) \in \mathbb{R}^d \times \mathbb{R}^d$$
$$D(\mathbb{R}_+)^2 \ni (\alpha, \beta) \longmapsto (\alpha^t, \beta^t, \tau_n^m(\alpha, \beta)) \mathbf{1}_{\{\tau_n^m(\alpha,\beta) < t \leq \tau_{n+1}^m(\alpha,\beta)\}} \in D(\mathbb{R}_+)^2 \times \mathbb{R}_+$$
are Borel-measurable (and the second map even $\mathcal{G}_t$-measurable), which implies that
$$D(\mathbb{R}_+)^2 \ni (\alpha, \beta) \longmapsto \big(\alpha(\tau_n^m(\alpha, \beta)), \beta(\tau_n^m(\alpha, \beta))\big) \mathbf{1}_{(\tau_n^m(\alpha,\beta), \tau_{n+1}^m(\alpha,\beta)]}(t) \in \mathbb{R}^d \times \mathbb{R}^d$$
are Borel-measurable (and even $\mathcal{G}_t$-measurable) for all $t \geq 0$, $m \in \mathbb{N}$ and $n \in \mathbb{N}_0$. Similarly, we can show that the other terms of
$$\mathcal{J}^m(\alpha, \beta)(t) = \sum_{k=0}^{\infty} \alpha(\tau_k^m(\alpha, \beta)) \big(\beta(\tau_{k+1}^m(\alpha, \beta)) - \beta(\tau_k^m(\alpha, \beta))\big) \mathbf{1}_{(\tau_{k+1}^m(\alpha,\beta), \infty)}(t)$$
$$+ \sum_{k=0}^{\infty} \alpha(\tau_k^m(\alpha, \beta)) \big(\beta(\quad t \quad) - \beta(\tau_k^m(\alpha, \beta))\big) \mathbf{1}_{(\tau_k^m(\alpha,\beta), \tau_{k+1}^m(\alpha,\beta)]}(t)$$
are Borel-measurable (and even $\mathcal{G}_t$-measurable) for $t \geq 0$. Note that the inclusion of the endpoints of the stochastic intervals in the indicator function changed compared to the original definition of $\mathcal{J}^m(\alpha, \beta)(t)$. The Borel-measurability of the limit
$$\mathcal{J} : D(\mathbb{R}_+) \times D(\mathbb{R}_+) \longrightarrow \mathbb{R}$$
now follows from Lemma 4.52.

The $\mathbb{P}$-almost sure, local uniform convergence in the second part of the statement is a minor modification of [80, Theorem 2]: Note that we defined the limit
$$\lim_{m \to \infty} \mathcal{J}^m = \mathcal{J}$$
with respect to the Skorokhod topology instead of local uniform convergence. Since local uniform convergence is stronger than Skorokhod convergence (cf. Lemma 4.2),
$$\mathcal{J}(H, X) = \mathcal{J}_{\mathbb{P}}(H_-, X)$$
still holds $\mathbb{P}$-almost surely. Moreover, we defined the random times τ_n^m with "$> 2^{-m}$" instead of "$\geq 2^{-m}$", which does not cause any trouble, since the proof only relies on
$$\sup\left\{|\alpha(t-) - \alpha(\tau(\alpha, \beta)_k^m)| : t \in (\tau(\alpha, \beta)_k^m, \tau(\alpha, \beta)_{k+1}^m]\right\} \leq 2^{-m}$$
for all $\alpha, \beta \in D(\mathbb{R}_+)$ and $m, n \in \mathbb{N}$.

4. Stochastic Processes

It remains to show the measurability properties: The $(\cup_{t\geq 0} \mathcal{F}_t)$-measurability of

$$\mathcal{J}(H, X) : \Omega \longrightarrow D(\mathbb{R}_+)$$

follows readily from the Borel-measurability of $\mathcal{J} : D(\mathbb{R}_+) \times D(\mathbb{R}_+) \to D(\mathbb{R}_+)$ and the $(\cup_{t\geq 0} \mathcal{F}_t)$-measurability of $(H, X) : \Omega \to D(\mathbb{R}_+) \times D(\mathbb{R}_+)$. Similarly, for all $m \in \mathbb{N}$

$$\mathcal{J}^m(H, X)(t) : \Omega \longrightarrow \mathbb{R}$$

are $\mathcal{F}_t$-measurable with $t \geq 0$, since $\mathcal{J}^m(t) : D(\mathbb{R}_+) \times D(\mathbb{R}_+) \to \mathbb{R}$ are $\mathcal{G}_t$-measurable and

$$(H^t, X^t) = (H_{t \wedge s}, X_{t \wedge s})_{s \geq 0} : \Omega \longrightarrow D(\mathbb{R}_+) \times D(\mathbb{R}_+)$$

are $\mathcal{F}_t$-measurable. Finally, this implies the $\mathcal{F}_t^{\mathbb{P}}$-measurability of $\mathcal{J}(H, X)(t) : \Omega \to \mathbb{R}$ for $t \geq 0$, since $\mathcal{J}(H, X)(t) = \lim_{m \to \infty} \mathcal{J}^m(H, X)(t)$ holds $\mathbb{P}$-almost surely. □

One important consequence of Proposition 4.53 is that we can neglect the continuum hypothesis assumption from Definition 4.51 for càglàd integrands and, on top of that, work with a version of stochastic integrals that are measurable with respect to the original filtration $\cup_{t\geq 0} \mathcal{F}_t$ and adapted to the completed filtration $\mathcal{F}^{\mathbb{P}} = (\mathcal{F}_t^{\mathbb{P}})_{t \geq 0}$. Since all applications of stochastic integration in Section 4.4 only concern integrands with càglàd paths, we will tacitly use the construction of Proposition 4.53 for all stochastic integrals and, in particular, its superior measurability properties.

4.4. Stochastic Differential Equations

Stochastic differential equations (or SDEs for short) are a generalization of differential equations based on the theory of stochastic integration. Although deterministic differential equations related to stochastic processes were already studied earlier (e.g. by Albert Einstein in [48]), Kiyosi Itô introduced his stochastic integration theory in [71] to put stochastic differential equations on a rigorous mathematical foundation. For a historical account on the early history of stochastic differential equations and their applications (with a focus on mathematical finance), we refer the interested reader to [76].

In the first half of this section, we recall well-known results for stochastic differential equations in classical probability spaces. Moreover, we show how to generalize them for sublinear expectations, based on our stochastic integration theory from Section 4.3. In the second half, we demonstrate how to apply those generalized results in order to present a novel construction for spatially inhomogeneous, sublinear Markov semigroups and associated Markov processes for sublinear expectations, which can be interpreted as a generalization of classical Lévy-type processes under uncertainty in their characteristics. In particular, we relate the resulting sublinear Markov semigroups with their sublinear generator equations (in the spirit of Proposition 4.10), which is essential for potential applications of this new class of stochastic processes for sublinear expectations.

4.4. Stochastic Differential Equations

Definition 4.54 (Functional Lipschitz). Suppose that

$$\mathfrak{D}(\mathbb{R}^d) := \mathfrak{D}(\mathbb{R}^d; \mathcal{F}) := \{X : \Omega \to D(\mathbb{R}_+; \mathbb{R}^d) \mid X \text{ is adapted to } \mathcal{F}\}$$

is the set of all stochastic process on an measurable space $(\Omega, \mathcal{A})$ with càdlàg paths adapted to a filtration $\mathcal{F} = (\mathcal{F}_t)_{t \geq 0}$. An operator with $d, d' \in \mathbb{N}$

$$F : \mathfrak{D}(\mathbb{R}^d) \longrightarrow \mathfrak{D}(\mathbb{R}^{d'})$$

is called *functional Lipschitz* if $X_-^\tau = Y_-^\tau$ implies $F(X)_-^\tau = F(Y)_-^\tau$ for all stopping times $\tau : \Omega \to [0, \infty]$ and $X, Y \in \mathfrak{D}(\mathbb{R}^d)$, where $Z_-^\tau = (Z_{(\tau \wedge t)-})_{t \geq 0}$ with $Z_{0-} := Z_0$ for $Z \in \mathfrak{D}$, and if there exists a non-decreasing process $K \in \mathfrak{D}(\mathbb{R})$ such that

$$\sup_{s \leq t} |F(X)(\omega)_s - F(Y)(\omega)_s| \leq K_t(\omega) \sup_{s \leq t} |X(\omega)_s - Y(\omega)_s|$$

for every $t \geq 0$, $\omega \in \Omega$ and $X, Y \in \mathfrak{D}(\mathbb{R}^d)$.

Theorem 4.55 (Classical Stochastic Differential Equations). *Suppose $(X_t)_{t \geq 0}$ is a d-dimensional $\mathbb{P}$-$\mathcal{F}$-semimartingale on a classical probability space $(\Omega, \mathcal{A}, \mathbb{P})$ adapted to a filtration $\mathcal{F} = (\mathcal{F}_t)_{t \geq 0}$. If $Y = (Y_t)_{t \geq 0} \in \mathfrak{D}(\mathbb{R}^k)$ and $F : \mathfrak{D}(\mathbb{R}^d) \to \mathfrak{D}(\mathbb{R}^{k \times d})$ is functional Lipschitz, then the (classical) stochastic differential equation*

$$Z = Y + \mathcal{J}_\mathbb{P}(F(Z)_-, X)$$

has a solution $Z = (Z_t)_{t \geq 0} \in \mathfrak{D}(\mathbb{R}^k; \mathcal{F}_+^\mathbb{P})$, which is unique up to $\mathbb{P}$-indistinguishability. Moreover, if $(Y_t)_{t \geq 0}$ is a $\mathbb{P}$-$\mathcal{F}_+^\mathbb{P}$-semimartingale, so is the solution $(Z_t)_{t \geq 0}$.

Proof. ↪ [110, Chapter 5, Theorem 7, p. 253] □

Proposition 4.56 (Picard Iteration Method). *Suppose $(X_t)_{t \geq 0}$ is a d-dimensional $\mathbb{P}$-$\mathcal{F}$-semimartingale on a classical probability space $(\Omega, \mathcal{A}, \mathbb{P})$, $Y = (Y_t)_{t \geq 0} \in \mathfrak{D}(\mathbb{R}^k)$ and the operator $F : \mathfrak{D}(\mathbb{R}^d) \to \mathfrak{D}(\mathbb{R}^{k \times d})$ is functional Lipschitz. If $(Z^n)_{n \in \mathbb{N}} \subset \mathfrak{D}(\mathbb{R}^k; \mathcal{F}_+^\mathbb{P})$ are the Picard iterations for a starting point $Z^0 \in \mathfrak{D}(\mathbb{R}^k; \mathcal{F}_+^\mathbb{P})$, i.e. defined by*

$$Z^n := Y + \mathcal{J}_\mathbb{P}\left(F(Z^{n-1})_-, X\right) \in \mathfrak{D}(\mathbb{R}^k; \mathcal{F}_+^\mathbb{P})$$

for $n \in \mathbb{N}$, and if $Z \in \mathfrak{D}(\mathbb{R}^k; \mathcal{F}_+^\mathbb{P})$ is the solution of the stochastic differential equation

$$Z = Y + \mathcal{J}_\mathbb{P}(F(Z)_-, X),$$

then $\lim_{n \to \infty} Z^n = Z$ uniformly on compacts in probability with respect to $\mathbb{P}$, i.e.

$$\lim_{n \to \infty} \mathbb{P}\left(\sup_{s \leq t} |Z^n(s) - Z(s)| > \varepsilon\right) = 0$$

for all $t \geq 0$ and $\varepsilon > 0$. Moreover, if $F : \mathfrak{D}(\mathbb{R}^d) \to \mathfrak{D}(\mathbb{R}^{k \times d})$ is uniform Lipschitz, i.e.

$$\sup_{s \leq t} |F(X)(\omega)_s - F(Y)(\omega)_s| \leq K \sup_{s \leq t} |X(\omega)_s - Y(\omega)_s|$$

for every $t \geq 0$, $\omega \in \Omega$ and $X, Y \in \mathfrak{D}(\mathbb{R}^d)$ with some finite constant $K > 0$, then

$$\lim_{n \to \infty} \sup_{s \leq t} |Z^n(s) - Z(s)| = 0$$

$\mathbb{P}$-*almost surely for all $t \geq 0$.*

4. Stochastic Processes

Proof. Uniform convergence on compacts in probability for the functional Lipschitz case is established in [110, Chapter 5, Theorem 8, p. 255]. Almost sure convergence for the uniform Lipschitz case can be found in [17, Section 8, Theorem 8.2]. □

Theorem 4.57 (Equations with Uniform Lipschitz Coefficients). *Suppose that $X = (X_t)_{t \geq 0}$ is a stochastic process with càdlàg paths on a measurable space $(\Omega, \mathcal{A})$ adapted to a filtration $\mathcal{F} = (\mathcal{F}_t)_{t \geq 0}$, that $Y = (Y_t)_{t \geq 0} \in \mathfrak{D}(\mathbb{R}^k)$ and that the operator*

$$F: \mathfrak{D}(\mathbb{R}^d) \longrightarrow \mathfrak{D}(\mathbb{R}^{k \times d})$$

is uniform Lipschitz (cf. Proposition 4.56). If $\mathcal{P} \subset \mathfrak{P}_{sem}(\Omega)$ (i.e. a family of probability measures $\mathbb{P} \in \mathfrak{P}(\Omega)$ such that $(X_t)_{t \geq 0}$ is a $\mathbb{P}$-$\mathcal{F}$-semimartingale), then the stochastic differential equation

$$Z = Y + \mathcal{J}(F(Z), X)$$

has a solution in form of an $\mathcal{A}$-measurable stochastic process $Z = (Z_t)_{t \geq 0}$ with càdlàg paths, which is adapted to the $\mathcal{P}$-universally completed filtration

$$\mathcal{F}^{\mathcal{P}} = \left(\cap_{\mathbb{P} \in \mathcal{P}} \mathcal{F}_t^{\mathbb{P}} \right)_{t \geq 0}.$$

In $\mathfrak{D}(\mathbb{R}^k; \mathcal{F}^{\mathcal{P}})$ the solution is unique up to $\mathcal{P}$-indistinguishability, i.e. for any other solution $Z' \in \mathfrak{D}(\mathbb{R}^k; \mathcal{F}^{\mathcal{P}})$, we have $Z = Z'$ $\mathbb{P}$-almost surely for all $\mathbb{P} \in \mathcal{P}$. If $(Y_t)_{t \geq 0}$ is a $\mathbb{P}$-$\mathcal{F}$-semimartingale for $\mathbb{P} \in \mathcal{P}$, then $(Z_t)_{t \geq 0}$ is a $\mathbb{P}$-$\mathcal{F}^{\mathbb{P}}$-semimartingale. Furthermore, if $Z^0 \in \mathfrak{D}(\mathbb{R}^k)$ is $\mathcal{A}$-measurable, the Picard iterations $(Z^n)_{n \in \mathbb{N}}$ with

$$Z^n := Y + \mathcal{J}\big(F(Z^{n-1}), X\big) \in \mathfrak{D}(\mathbb{R}^k; \mathcal{F}^{\mathcal{P}})$$

for $n \in \mathbb{N}$ converge to Z locally uniform $\mathbb{P}$-almost surely for all $\mathbb{P} \in \mathcal{P}$, i.e.

$$\lim_{n \to \infty} \sup_{s \leq t} |Z^n(s) - Z(s)| = 0$$

$\mathbb{P}$-*almost surely for all $t \geq 0$.*

Proof. According to Proposition 4.53, the Picard iterations $Z^n : \Omega \to D(\mathbb{R}_+)$ are $\mathcal{A}$-measurable for all $n \in \mathbb{N}$. Therefore, Lemma 4.52 implies that the pointwise limit

$$Z(\omega) := \begin{cases} \lim_{n \to \infty} Z^n(\omega) & \text{if } \lim_{n \to \infty} Z^n(\omega) \text{ exists} \\ 0 & \text{otherwise} \end{cases}$$

for $\omega \in \Omega$ (with respect to the Skorokhod topology on $D(\mathbb{R}_+; \mathbb{R}^k)$, cf. Lemma 4.2) is $\mathcal{A}$-measurable as well. Furthermore, Proposition 4.56 shows that

$$\lim_{n \to \infty} \sup_{s \leq t} |Z^n(s) - Z(s)| = 0$$

$\mathbb{P}$-almost surely for all $t \geq 0$ and $\mathbb{P} \in \mathcal{P}$. This implies that Z is adapted to $\mathcal{F}^{\mathbb{P}}$, since the iterations Z^n are adapted to $\mathcal{F}^{\mathbb{P}}$ for $n \in \mathbb{N}$, and that Z is the (unique) solution of

$$Z = Y + \mathcal{J}_{\mathbb{P}}\left(F(Z)_-, X\right) \left(= Y + \mathcal{J}(F(Z), X) \right)$$

for all $\mathbb{P} \in \mathcal{P}$, as required. □

4.4. Stochastic Differential Equations

Remark 4.58 (Equations with Functional Lipschitz Coefficients). Suppose that $F : \mathfrak{D}(\mathbb{R}^d) \to \mathfrak{D}(\mathbb{R}^{k \times d})$ is only functional Lipschitz (as in Theorem 4.55) and Cantor's continuum hypothesis holds. It is still possible to solve the equation

$$Z = Y + \mathcal{J}(F(Z), X)$$

with a solution $Z \in \mathfrak{D}(\mathbb{R}^k; \mathcal{F}_+^{\mathcal{P}})$ adapted to the $\mathcal{P}$-universally augmented filtration

$$\mathcal{F}_+^{\mathcal{P}} = \left(\cap_{\mathbb{P} \in \mathcal{P}} \mathcal{F}_{t+}^{\mathbb{P}}\right)_{t \geq 0}$$

instead of the $\mathcal{P}$-universally completed filtration $\mathcal{F}^{\mathcal{P}}$, using a slightly different argument compared to Theorem 4.57: According to the construction of the Picard iterations and of the generalized stochastic integral in Proposition 4.53,

$$Z^n = Y + \mathcal{J}_{\mathbb{P}}\left(F(Z^{n-1})_-, X\right)$$

holds for all $n \in \mathbb{N}$ (by iteration) $\mathbb{P}$-almost surely for each $\mathbb{P} \in \mathcal{P}$. Hence, Proposition 4.56 implies $\lim_{n \to \infty} Z^n =: Z^{\mathbb{P}}$ uniformly on compacts in $\mathbb{P}$-probability with

$$Z^{\mathbb{P}} = Y + \mathcal{J}_{\mathbb{P}}\left(F(Z^{\mathbb{P}})_-, X\right)$$

for each $\mathbb{P} \in \mathcal{P}$. According to [97, Lemma 2.5], which is basically a lift of Mokobodzki's medial limit (as in Theorem 4.49) from $\overline{\mathbb{R}} = \mathbb{R} \cup \{\pm\infty\}$ to $\mathfrak{D}(\mathbb{R}^k)$ and therefore assumes the validity of Cantor's continuum hypothesis, there exists

$$Z = (Z_t)_{t \geq 0} \in \mathfrak{D}(\mathbb{R}^k; \mathcal{F}_+^{\mathcal{P}})$$

with $Z = Z^{\mathbb{P}}$ $\mathbb{P}$-almost surely for all $\mathbb{P} \in \mathcal{P}$. In particular, $\lim_{n \to \infty} Z^n = Z$ uniformly on compact in $\mathbb{P}$-probability for each $\mathbb{P} \in \mathcal{P}$ and

$$Z = Z^{\mathbb{P}} = Y + \mathcal{J}_{\mathbb{P}}\left(F(Z^{\mathbb{P}})_-\right)_{t \geq 0}, X) = Y + \mathcal{J}_{\mathbb{P}}\left(F(Z)_-, X\right)$$
$$= Y + \mathcal{J}(F(Z), X)$$

holds $\mathbb{P}$-almost surely for all $\mathbb{P} \in \mathcal{P}$, as required. It is easy to check that the solution is unique in $\mathfrak{D}(\mathbb{R}^k; \mathcal{F}_+^{\mathcal{P}})$ up to $\mathcal{P}$-indistinguishability. However, note that Z is in general only measurable with respect to the $\mathcal{P}$-universally completed filtration $\mathcal{A}^{\mathcal{P}} = \cap_{\mathbb{P} \in \mathcal{P}} \mathcal{A}^{\mathbb{P}}$.

In the rest of this section, we want to use the preceding generalization of stochastic differential equations for sublinear expectations to construct sublinear Markov semigroups. Since we want to use the special structure of the canonical filtration for the coordinate-mapping process on the Skorokhod space, we need the following generalization of Lemma 4.52 on the measurability of almost sure limits. It shows how we can apply certain measurability results (such as [38, Theorem 97, p. 147]) to the solutions of our generalization of stochastic differential equations for sublinear expectations, which are in general only adapted to the completed filtration.

4. Stochastic Processes

Lemma 4.59 (Measurable Modifications). *Suppose that $(\mathcal{S}, d)$ is a separable metric space and $(\Omega, \mathcal{A}, \mathbb{P})$ a probability space. If $X : \Omega \to \mathcal{S}$ is $\mathcal{A}^\mathbb{P}/\mathcal{B}(\mathcal{S})$-measurable, then there exists an $\mathcal{A}/\mathcal{B}(\mathcal{S})$-measurable $Y : \Omega \to \mathcal{S}$ such that $X = Y$ holds $\mathbb{P}$-almost surely.*

Proof. Since $(\mathcal{S}, d)$ is separable, there exists a countable base $(B_n)_{n \in \mathbb{N}}$ for the topology such that $\mathcal{B}(\mathcal{S})$ is generated by $(B_n)_{n \in \mathbb{N}}$ (cf. e.g. [30, Remark D.32, p. 394]). The $\mathcal{A}^\mathbb{P}/\mathcal{B}(\mathcal{S})$-measurability of $X : \Omega \to \mathcal{S}$ implies that $X^{-1}(B_n) \in \mathcal{A}^\mathbb{P}$, i.e. there exist subsets of $\mathbb{P}$-null sets $E_n \in \mathcal{A}^\mathbb{P}$ with

$$X^{-1}(B_n) \setminus E_n \in \mathcal{A}$$

for all $n \in \mathbb{N}$. Since $\cup_{n \in \mathbb{N}} E_n \in \mathcal{A}^\mathbb{P}$ is still a subset of a $\mathbb{P}$-null set, there exists a $\mathbb{P}$-null set $E \in \mathcal{A}$ with $\cup_{n \in \mathbb{N}} E_n \subset E$. Fix $x_0 \in \mathcal{S}$ and define $Y : \Omega \to \mathcal{S}$ by

$$Y(\omega) := X(\omega) \mathbb{1}_{\Omega \setminus E}(\omega) + x_0 \mathbb{1}_E(\omega)$$

for $\omega \in \Omega$. The construction entails that for all $n \in \mathbb{N}$

$$Y^{-1}(B_n) = \begin{cases} X^{-1}(B_n) \cup E = (X^{-1}(B_n) \setminus E_n) \cup E \in \mathcal{A} & \text{if } x_0 \in B_n \\ X^{-1}(B_n) \setminus E = (X^{-1}(B_n) \setminus E_n) \setminus E \in \mathcal{A} & \text{if } x_0 \notin B_n \end{cases}$$

holds. Since $(B_n)_{n \in \mathbb{N}}$ is a countable base for the topology of $\mathcal{S}$, i.e. every open set is a countable union of elements in $(B_n)_{n \in \mathbb{N}}$, this implies the $\mathcal{A}/\mathcal{B}(\mathcal{S})$-measurability of $Y : \Omega \to \mathcal{S}$. Moreover, $\mathbb{P}(E) = 0$ shows that $X = Y$ holds $\mathbb{P}$-almost surely. $\square$

Proposition 4.60 (Markov Property). *Assume that $X = (X_t)_{t \geq 0}$ is a d-dimensional Lévy process for sublinear expectations (starting in $X_0 = 0$) on a measurable space $(\Omega, \mathcal{A})$ adapted to the filtration $\mathcal{F} = (\mathcal{F}_t)_{t \geq 0}$ (as in Remark 4.38) together with the associated uncertainty subset $\mathcal{P} \subset \mathfrak{P}_{sem}(\Omega)$ and a globally Lipschitz-continuous $f : \mathbb{R}^k \to \mathbb{R}^{k \times d}$. Moreover, suppose for $(\mathcal{A} \cap \mathcal{F}_T^\mathcal{P})$-measurable maps $\xi : \Omega \to \mathbb{R}^k$ and $T > 0$ that*

$$Z^{T,\xi} : \Omega \longrightarrow D(\mathbb{R}_+; \mathbb{R}^k),$$

are the unique $\mathcal{A}$-measurable solutions (from Theorem 4.57) of the equations

$$Z^{T,\xi} = \xi + \mathcal{J}\big(f(Z^{T,\xi}), X_{T+\cdot}\big)$$

adapted to the shifted $\mathcal{P}$-universally completed filtration $\mathcal{F}_{T+\cdot}^\mathcal{P} = (\cap_{\mathbb{P} \in \mathcal{P}} \mathcal{F}_{T+t}^\mathbb{P})_{t \geq 0}$, and that

$$E = \sup_{\mathbb{P} \in \mathcal{P}} \mathbb{E}_\mathbb{P} : \bigcap_{\mathbb{P} \in \mathcal{P}} L^1(\mathbb{P}) \longrightarrow \mathbb{R}$$

is the sublinear expectation associated to $\mathcal{P} \subset \mathfrak{P}_{sem}(\Omega)$. For upper semi-analytic functions $\varphi : \mathbb{R}^k \to \mathbb{R}$, the functions $x \mapsto E\big(\varphi\big(Z_t^{0,x}\big)\big)$ are upper semi-analytic and

$$E\big(\varphi\big(Z_{s+t}^{0,x}\big)\big) = E\Big(E\big(\varphi\big(Z_t^{0,y}\big)\big)\Big|_{y=Z_s^{0,x}}\Big)$$

holds for all $t, s \geq 0$.

4.4. Stochastic Differential Equations

Proof. First of all, note that we heavily rely on the constructions of Lévy processes and its properties from Remark 4.38: In fact, the process $X = (X_t)_{t \geq 0}$ is the coordinate-mapping process on the Skorokhod space

$$\Omega = D_0(\mathbb{R}_+; \mathbb{R}^d)$$

equipped with the Borel-σ-algebra $\mathcal{A} = \mathcal{B}(D_0(\mathbb{R}_+))$ and its canonical filtration

$$\mathcal{F}_t = \sigma(X_s : s \leq t) \subset \mathcal{B}(D_0(\mathbb{R}_+))$$

for $t \geq 0$. Since $Z^{0,x} : \Omega \to D(\mathbb{R}_+)$ is $\mathcal{A}$-measurable (i.e. in this case Borel-measurable) and $\varphi : \mathbb{R}^k \to \mathbb{R}$ is upper semi-analytic, the composition

$$\varphi\left(Z^{0,x}_{s+t}\right) : \Omega \longrightarrow \mathbb{R}$$

is upper semi-analytic by [16, Lemma 7.30, p. 177]. Similar to the proof of Lemma 4.30, [16, Proposition 7.47, p. 179] and [16, Proposition 7.48, p. 180] imply that

$$\mathbb{R}^k \ni y \longmapsto E\left(\varphi\left(Z^{0,y}_t\right)\right) \in \mathbb{R}$$

is upper semi-analytic and hence is $\Omega \ni \omega \mapsto E\left(\varphi\left(Z^{0,y}_t\right)\right)\big|_{y=Z^{0,x}_s(\omega)} \in \mathbb{R}$, as a composition with a Borel-measurable function: Here, we used that

$$\mathbb{R}^k \times \Omega \ni (y, \omega) \longmapsto Z^{0,y}_t(\omega) \in \mathbb{R}$$

is upper semi-analytic for all $t \geq 0$, which follows from the pathwise construction of the solutions of stochastic differential equations from Theorem 4.57 and the Borel-measurability with respect to $(y, \omega) \in \mathbb{R}^k \in \Omega$ of the associated Picard iterations. The semi-analyticity of all functions now allows us to employ the properties of the corresponding sublinear conditional expectation $(E_t)_{t \geq 0}$ as introduced in Theorem 4.29: In fact, if we show that $Z^{0,x}_{s+t} = Z^{s,y}_t\big|_{y=Z^{0,x}_s}$ holds $\mathbb{P}$-almost surely for all $\mathbb{P} \in \mathcal{P}$ and

$$E\left(\varphi(Z^{s,y}_t)\right) = E\left(\varphi(Z^{0,y}_t)\right)$$
$$E\left(E_s\left(\varphi(Z^{s,y}_t)\right)\big|_{y=Z^{0,x}_s}\right) = E\left(E\left(\varphi(Z^{s,y}_t)\right)\right)\big|_{y=Z^{0,x}_s}$$

for $t, s \geq 0$ and $x, y \in \mathbb{R}^k$, then the aggregation property of the sublinear conditional expectation (as demonstrated in Theorem 4.29) implies that

$$E\left(\varphi(Z^{0,x}_{s+t})\right) = E\left(\varphi(Z^{s,y}_t)\big|_{y=Z^{0,x}_s}\right) = E\left(E_s\left(\varphi(Z^{s,y}_t)\big|_{y=Z^{0,x}_s}\right)\right)$$
$$= E\left(E\left(\varphi(Z^{s,y}_t)\right)\big|_{y=Z^{0,x}_s}\right)$$
$$= E\left(E\left(\varphi(Z^{0,y}_t)\right)\big|_{y=Z^{0,x}_s}\right)$$

holds. To complete the proof, we will now show that these three premises hold:

4. Stochastic Processes

As indicated before, the construction of the solution (in the proof of Theorem 4.57) as the pathwise limit of the Picard iterations implies the Borel-measurability of

$$\mathbb{R}^k \times \Omega \ni (y, \omega) \longmapsto Z^{s,y}(\omega) \in D(\mathbb{R}_+)$$

for $s \geq 0$. The uniqueness of solutions hence shows that $Z^{s,y}|_{y=Z_s^{0,x}} = Z^{s,Z_s^{0,x}}$ holds $\mathbb{P}$-almost surely for all $\mathbb{P} \in \mathcal{P}$. Moreover, the construction of the stochastic integral from Proposition 4.53 implies that for every $s \geq 0$ and $x \in \mathbb{R}^k$

$$\begin{aligned}Z^{0,x}(s + \cdot) &= x + \mathcal{J}\big(f(Z^{0,x}), X\big)(s + \cdot) \\ &= x + \mathcal{J}\big(f(Z^{0,x}), X\big)(s) \quad + \mathcal{J}\big(f(Z^{0,x}(s+\cdot)), X(s+\cdot)\big)(\cdot) \\ &= Z^{0,x}(s) \quad\quad\quad\quad\quad\quad + \mathcal{J}\big(f(Z^{0,x}(s+\cdot)), X(s+\cdot)\big)(\cdot)\end{aligned}$$

holds up to $\mathcal{P}$-indistinguishability. Therefore, the uniqueness of solutions implies that

$$Z_{s+t}^{0,x} = Z_t^{s,Z_s^{0,x}} = Z_t^{s,y}\big|_{y=Z_s^{0,x}}$$

holds $\mathbb{P}$-almost surely for all $\mathbb{P} \in \mathcal{P}$.

Since our stochastic integral is defined pathwise in Proposition 4.53,

$$Z^{0,x}(X_{s+\cdot}) = \Big(x + \mathcal{J}\big(f(Z^{0,x}), X\big)\Big)(X_{s+\cdot}) = x + \mathcal{J}\Big(f\big(Z^{0,x}(X_{s+\cdot})\big), X_{s+\cdot}\Big)$$

entails that $Z^{0,x}(X_{s+\cdot})$ is another solution of the defining equation for $Z^{s,x}$. Note that

$$Z^{0,x}(X_{s+\cdot}) : \Omega \longrightarrow D(\mathbb{R}_+)$$

is Borel-measurable as the composition of the Borel-measurable maps $Z^{0,x} : \Omega \to D(\mathbb{R}_+)$ and $X_{s+\cdot} : \Omega \to \Omega$. Moreover, the $\mathcal{F}_t^{\mathbb{P}}$-measurability of $Z^{0,x}(t)$ implies that $Z^{0,x}(X_{s+\cdot})(t)$ is measurable with respect to $\sigma(X_{s+u} : u \leq t)^{\mathbb{P}} \subset \mathcal{F}_{s+t}^{\mathbb{P}}$. The uniqueness of solutions as in Theorem 4.57 therefore implies that

$$Z^{0,x}(X_{s+\cdot}) = X^{s,x}$$

holds $\mathbb{P}$-almost surely for all $\mathbb{P} \in \mathcal{P}$. This leads to

$$E\Big(\varphi\big(Z_t^{0,x}(X)\big)\Big) = E\Big(\varphi\big(Z_t^{0,x}(X_{s+\cdot})\big)\Big) = E\Big(\varphi\big(Z_t^{s,x}(X)\big)\Big)$$

using the stationarity of the increments of the Lévy process X, which we derived from the Markov property and the spatial homogeneity in Remark 4.38.

According to Lemma 4.59, for every $\check{\mathbb{P}} \in \mathcal{P}$ there exists an $\mathcal{F}_s$-measurable $Y^{\check{\mathbb{P}}} : \Omega \to \mathbb{R}$ such that $Z_s^{0,x} = Y^{\check{\mathbb{P}}}$ holds $\check{\mathbb{P}}$-almost surely, since $Z_s^{0,x}$ is $\mathcal{F}_s^{\check{\mathbb{P}}}$-measurable. In particular, the definition of the completion implies that there exists

$$\mathcal{F}_s \ni \check{\Omega}(\check{\mathbb{P}}) \subset \Big\{\check{\omega} \in \Omega \;\Big|\; Z_s^{0,x}(\check{\omega}) = Y^{\check{\mathbb{P}}}(\check{\omega})\Big\} \in \mathcal{F}_s^{\check{\mathbb{P}}}$$

with $\check{\mathbb{P}}(\check{\Omega}(\check{\mathbb{P}})) = 1$ for all $\check{\mathbb{P}} \in \mathcal{P}$. Note that since $\check{\Omega}(\check{\mathbb{P}}) \in \mathcal{F}_s = \sigma(X_u : u \leq s)$, where

212

$X = (X_t)_{t\geq 0}$ is the coordinate mapping process, [38, Theorem 97, p. 147] implies that $\breve{\omega} \in \check{\Omega}(\check{\mathbb{P}})$ if and only if $(\breve{\omega}_{s\wedge t})_{t\geq 0} \in \check{\Omega}(\check{\mathbb{P}})$. Similarly, $Y^{\check{\mathbb{P}}}(\breve{\omega}) = Y^{\check{\mathbb{P}}}((\breve{\omega}_{s\wedge t})_{t\geq 0})$ for every $\breve{\omega} \in \Omega$ due to the $\mathcal{F}_s$-measurability of $Y^{\check{\mathbb{P}}} : \Omega \to \mathbb{R}$. Altogether,

$$Z_s^{0,x}(\breve{\omega} \otimes_s \hat{\omega}) = Y^{\check{\mathbb{P}}}(\breve{\omega} \otimes_s \hat{\omega}) = Y^{\check{\mathbb{P}}}(\breve{\omega}) = Z_s^{0,x}(\breve{\omega})$$

holds for all $\breve{\omega} \in \check{\Omega}(\check{\mathbb{P}})$ and $\hat{\omega} \in \Omega$. Analogously, one can show that for every $\hat{\mathbb{P}} \in \mathcal{P}$ there exists $\hat{\Omega}(\hat{\mathbb{P}}) \in \sigma(X_u : u \geq s)$ with $\hat{\mathbb{P}}(\hat{\Omega}(\hat{\mathbb{P}})) = 1$ such that

$$Z_t^{s,y}(\breve{\omega} \otimes_s \hat{\omega}) = Z_t^{s,y}(\hat{\omega})$$

holds for all $\hat{\omega} \in \hat{\Omega}(\hat{\mathbb{P}})$ and $\breve{\omega} \in \Omega$. Finally, these two findings imply that

$$E\Big(E_s\big(\varphi(Z_t^{s,y})\big|_{y=Z_s^{0,x}}\big)\Big)$$
$$= \sup_{\check{\mathbb{P}}\in\mathcal{P}} \int_\Omega \sup_{\hat{\mathbb{P}}\in\mathcal{P}} \int_\Omega \varphi\big(Z_t^{s,y}(\breve{\omega}\otimes_s\hat{\omega})\big)\Big|_{y=Z_s^{0,x}(\breve{\omega}\otimes_s\hat{\omega})} \hat{\mathbb{P}}(d\hat{\omega})\,\check{\mathbb{P}}(d\breve{\omega})$$
$$= \sup_{\check{\mathbb{P}}\in\mathcal{P}} \int_\Omega \mathbf{1}_{\check{\Omega}(\check{\mathbb{P}})}(\breve{\omega}) \sup_{\hat{\mathbb{P}}\in\mathcal{P}} \int_\Omega \varphi\big(Z_t^{s,y}(\breve{\omega}\otimes_s\hat{\omega})\big)\Big|_{y=Z_s^{0,x}(\breve{\omega}\otimes_s\hat{\omega})} \mathbf{1}_{\hat{\Omega}(\hat{\mathbb{P}})}(\hat{\omega})\,\hat{\mathbb{P}}(d\hat{\omega})\,\check{\mathbb{P}}(d\breve{\omega})$$
$$= \sup_{\check{\mathbb{P}}\in\mathcal{P}} \int_\Omega \sup_{\hat{\mathbb{P}}\in\mathcal{P}} \int_\Omega \varphi\big(Z_t^{s,y}(\breve{\omega}\otimes_s\hat{\omega})\big)\Big|_{y=Z_s^{0,x}(\breve{\omega}\otimes_s\hat{\omega})} \mathbf{1}_{\hat{\Omega}(\hat{\mathbb{P}})}(\hat{\omega})\mathbf{1}_{\check{\Omega}(\check{\mathbb{P}})}(\breve{\omega})\,\hat{\mathbb{P}}(d\hat{\omega})\,\check{\mathbb{P}}(d\breve{\omega})$$
$$= \sup_{\check{\mathbb{P}}\in\mathcal{P}} \int_\Omega \Big(\sup_{\hat{\mathbb{P}}\in\mathcal{P}} \int_\Omega \varphi\big(Z_t^{s,y}(\hat{\omega})\big)\,\hat{\mathbb{P}}(d\hat{\omega})\Big)\Big|_{y=Z_s^{0,x}(\breve{\omega})} \check{\mathbb{P}}(d\breve{\omega}) = E\Big(E\big(\varphi(Z_t^{s,y})\big)\big|_{y=Z_s^{0,x}}\Big)$$

using the construction of the sublinear conditional expectation from Theorem 4.29. □

Proposition 4.61 (Dependence on Initial Values). *Suppose that $X = (X_t)_{t\geq 0}$ is a d-dimensional stochastic process on a measurable space $(\Omega, \mathcal{A})$ with càdlàg paths, and that $\mathcal{P} \subset \mathfrak{P}_{sem}(\Omega)$ such that the family of probability measures*

$$\mathcal{P} \circ X^{-1} = \{\mathbb{P} \circ X^{-1} \mid \mathbb{P} \in \mathcal{P}\} \subset \mathfrak{P}(D(\mathbb{R}_+; \mathbb{R}^d))$$

is sequentially compact in itself and $(\mathrm{var}(B^\mathbb{P})_t)_{\mathbb{P}\in\mathcal{P}} \subset \mathbb{R}$ is tight for all $t \geq 0$, where $(B^\mathbb{P}, C^\mathbb{P}, \nu^\mathbb{P})$ are the semimartingale characteristics of $X = (X_t)_{t\geq 0}$ with respect to $\mathbb{P} \in \mathcal{P}$ as in Definition 4.13. If $f : \mathbb{R}^k \to \mathbb{R}^{k\times d}$ is globally Lipschitz-continuous and $(Z_t^{0,x})_{t\geq 0}$ the (unique up to $\mathcal{P}$-indistinguishability) solution of

$$Z^{0,x} = x + \mathcal{J}\big(f(Z^{0,x}), X\big)$$

for $x \in \mathbb{R}^k$ from Theorem 4.57, then for bounded and continuous $\phi : D(\mathbb{R}_+; \mathbb{R}^k) \to \mathbb{R}$ (with respect to the Skorokhod topology from Lemma 4.2) the functions

$$\mathbb{R}^k \ni x \longmapsto \sup_{\mathbb{P}\in\mathcal{P}} \mathbb{E}_\mathbb{P}\big(\phi\big(Z^{0,x}\big)\big) \in \mathbb{R}$$

are bounded and continuous.

4. Stochastic Processes

Proof. Suppose that $\phi \in C_b(D(\mathbb{R}_+;\mathbb{R}^k))$ and $(x_n)_{n \in \mathbb{N}} \subset \mathbb{R}^K$ with $\lim_{n \to \infty} x_n = x$. Since

$$\sup_{x \in \mathbb{R}^k} \left| \sup_{\mathbb{P} \in \mathcal{P}} \mathbb{E}_{\mathbb{P}}\left(\phi\left(Z^{0,x}\right)\right) \right| \leq \sup_{x \in \mathbb{R}^k} \sup_{\mathbb{P} \in \mathcal{P}} \mathbb{E}_{\mathbb{P}}\left(\left|\phi\left(Z^{0,x}\right)\right|\right) \leq \sup_{x \in \mathbb{R}^k} |\phi(x)| < \infty,$$

it remains to prove the continuity of the resulting function.

For a fixed probability measure $\mathbb{P} \in \mathcal{P}$, [73, Chapter 9, Theorem 6.9, p. 578] shows

$$\lim_{n \to \infty} \mathbb{P} \circ (Z^{0,x_n})^{-1} = \mathbb{P} \circ (Z^{0,x})^{-1}$$

with respect to the weak convergence of measures (as in Remark 4.16). In particular,

$$\mathbb{E}_{\mathbb{P}}\left(\phi\left(Z^{0,x}\right)\right) = \lim_{n \to \infty} \mathbb{E}_{\mathbb{P}}\left(\phi\left(Z^{0,x_n}\right)\right) \leq \liminf_{n \to \infty} \sup_{\mathbb{P} \in \mathcal{P}} \mathbb{E}_{\mathbb{P}}\left(\phi\left(Z^{0,x_n}\right)\right),$$

which implies the lower semicontinuity of $x \mapsto \sup_{\mathbb{P} \in \mathcal{P}} \mathbb{E}_{\mathbb{P}}(\phi(Z^{0,x}))$.

The definition of the limit superior and the sequential compactness of $\mathcal{P} \circ X^{-1}$ lead to

$$\limsup_{n \to \infty} \sup_{\mathbb{P} \in \mathcal{P}} \mathbb{E}_{\mathbb{P}}\left(\phi\left(Z^{0,x_n}\right)\right) = \lim_{k \to \infty} \mathbb{E}_{\mathbb{P}_{n_k}}\left(\phi\left(Z^{0,x_{n_k}}\right)\right)$$

for some sequence $(\mathbb{P}_{n_k})_{k \in \mathbb{N}} \subset \mathcal{P}$ and $\mathbb{P} \in \mathcal{P}$ such that $\lim_{k \to \infty} \mathbb{P}_{n_k} \circ X^{-1} = \mathbb{P} \circ X^{-1}$ weakly. According to [73, Chapter 6, Theorem 6.21, p. 382], the tightness of $(\text{var}(B^{\mathbb{P}})_t)_{\mathbb{P} \in \mathcal{P}} \subset \mathbb{R}$ for $t \geq 0$ implies that the sequence $(\mathbb{P}_{n_k} \circ X^{-1})_{k \in \mathbb{N}}$ is predictably uniformly tight. This enables us to apply [73, Chapter 9, Theorem 6.9, p. 578], in order to obtain

$$\lim_{k \to \infty} \mathbb{P}_{n_k} \circ (Z^{0,x_{n_k}})^{-1} = \mathbb{P} \circ (Z^{0,x})^{-1}$$

with respect to weak convergence of measures. Finally, this shows that

$$\limsup_{n \to \infty} \sup_{\mathbb{P} \in \mathcal{P}} \mathbb{E}_{\mathbb{P}}\left(\phi\left(Z^{0,x_n}\right)\right) = \lim_{k \to \infty} \mathbb{E}_{\mathbb{P}_{n_k}}\left(\phi\left(Z^{0,x_{n_k}}\right)\right) = \mathbb{E}_{\mathbb{P}}\left(\phi\left(Z^{0,x}\right)\right) \leq \sup_{\mathbb{P} \in \mathcal{P}} \mathbb{E}_{\mathbb{P}}\left(\phi\left(Z^{0,x}\right)\right)$$

and hence the upper semicontinuity $x \mapsto \sup_{\mathbb{P} \in \mathcal{P}} \mathbb{E}_{\mathbb{P}}(\phi(Z^{0,x}))$, as required. $\square$

Remark 4.62 (Lévy-Type Processes). Suppose that $(X_t)_{t \geq 0}$ is the d-dimensional Lévy process for sublinear expectations from Remark 4.38 starting at $X_0 = 0$, which corresponds to the uncertainty coefficients $(b_\alpha, c_\alpha, F_\alpha)_{\alpha \in \mathcal{A}} \subset \mathbb{R}^d \times \mathbb{S}_+^{d \times d} \times \mathfrak{L}(\mathbb{R}^d)$ such that

$$\sup_{\alpha \in \mathcal{A}} \left(|b_\alpha| + |c_\alpha| + \int_{\mathbb{R}^d} \left(|z| \wedge |z|^2\right) F_\alpha(dz) \right) < \infty,$$

together with the sublinear expectation $E(\cdot) := \sup_{\mathbb{P} \in \mathcal{P}} \mathbb{E}_{\mathbb{P}}(\cdot)$ and the uncertainty subset

$$\mathcal{P} := \left\{ \mathbb{P} \in \mathfrak{P}_{\text{sem}}^{\text{ac}}(D_0(\mathbb{R}_+)) \ \middle| \ (b_s^{\mathbb{P}}, c_s^{\mathbb{P}}, F_s^{\mathbb{P}})(\overline{\omega}) \in \bigcup_{\alpha \in \mathcal{A}} (b_\alpha, c_\alpha, F_\alpha) \ \lambda(ds) \times \mathbb{P}(d\overline{\omega})\text{-a.e.} \right\}.$$

If the function $f : \mathbb{R}^k \to \mathbb{R}^{k \times d}$ is globally Lipschitz-continuous and $(Z_t^x)_{t \geq 0}$ is the unique (up to $\mathcal{P}$-indistinguishability) solution of the stochastic differential equation

$$Z^x = x + \mathcal{J}(f(Z^x), X)$$

for $x \in \mathbb{R}^d$ from Theorem 4.57, then $(T_t)_{t \geq 0}$ is a sublinear Markov semigroup on the convex cone of bounded, upper semi-analytic functions due to Proposition 4.60, where

$$T_t \varphi(x) := E(\varphi(Z^x))$$

for $x \in \mathbb{R}^k$ and $\varphi : \mathbb{R}^k \to \mathbb{R}$ upper semi-analytic. Further, if the tightness condition

$$\lim_{r \to 0} \sup_{\alpha \in \mathcal{A}} \int_{|z| < r} (1 \wedge |z|^2) \, F_\alpha(dz) = 0$$

holds, and $\cup_{\alpha \in \mathcal{A}}(b_\alpha, c_\alpha, F_\alpha) \subset \mathbb{R}^d \times \mathbb{S}_+^{d \times d} \times \mathfrak{L}(\mathbb{R}^d)$ is convex and closed, then Theorem 4.41 implies that $\cup_{x \in K} \mathcal{P}_x$ is sequentially compact in itself for all $K \subset \mathbb{R}^d$ compact. Since

$$\text{var}(B^{\mathbb{P}})_t = \int_0^t |b_s^{\mathbb{P}}| \, ds \leq t \sup_{\alpha \in \mathcal{A}} |b_\alpha| < \infty$$

implies the tightness of $(\text{var}(B^{\mathbb{P}})_t)_{\mathbb{P} \in \mathcal{P}}$ for all $t \geq 0$, Proposition 4.61 shows that $(T_t)_{t \geq 0}$ satisfies the C_b-Feller property (i.e. $T_t C_b(\mathbb{R}^d) \subset C_b(\mathbb{R}^d)$ for all $t \geq 0$) under the additional tightness and convexity assumptions. In particular, $(Z_t)_{t \geq 0}$ with

$$Z(x, \omega) := Z^x(\omega)$$

for $(x, \omega) \in \hat{\Omega} := \mathbb{R}^k \times D(\mathbb{R}_+; \mathbb{R}^d)$ is a Markov process for sublinear expectations on $(\hat{\Omega}, \hat{\mathcal{H}}, \hat{E}^x)_{x \in \mathbb{R}^d}$ with the space of integrands $\hat{\mathcal{H}} := \{Y : \hat{\Omega} \to \mathbb{R} \mid Y \text{ is Borel-measurable}\}$,

$$\hat{E}^x(Y) := E(Y(x, \cdot))$$

for $Y \in \hat{\mathcal{H}}$ and $x \in \mathbb{R}^k$, and the family of test functions $\hat{\mathcal{T}} := C_b(\mathbb{R}^k)$. In this thesis, we refer to such processes $(Z_t)_{t \geq 0}$ as *Lévy-type processes for sublinear expectations*.

Fortunately, this construction (in contrast to the one from Remark 4.33, as discussed in Remark 4.43) allows us to apply the general theory from Section 4.1 to relate the spatially inhomogeneous, sublinear Markov semigroup to its sublinear generator equation: Theorem 4.57 shows that Z^x is a k-dimensional $\mathbb{P}$-$\mathcal{F}^{\mathbb{P}}$-semimartingale for every $\mathbb{P} \in \mathcal{P}$ and $x \in \mathbb{R}^k$. Furthermore, Lemma 4.48 implies that

$$\hat{\mathcal{P}}_x \subset \left\{ \hat{\mathbb{P}} \in \mathfrak{P}_{\text{sem}}^{\text{ac}}(D_x(\mathbb{R}_+)) \;\middle|\; (b_s^{\hat{\mathbb{P}}}, c_s^{\hat{\mathbb{P}}}, F_s^{\hat{\mathbb{P}}})(\hat{\omega}) \in \bigcup_{\alpha \in \mathcal{A}} (\hat{b}_\alpha, \hat{c}_\alpha, \hat{F}_\alpha)(\omega_s) \quad \lambda(ds) \times \hat{\mathbb{P}}(d\hat{\omega})\text{-a.e.} \right\}$$

with the push-forward uncertainty subsets $\hat{\mathcal{P}}_x := \mathcal{P} \circ (Z^x)^{-1}$ and uncertainty coefficients

$$\hat{b}_\alpha(x) := f(x) b_\alpha + \int_{\mathbb{R}^d} \left(\hat{h}(f(x)z) - f(x)h(z) \right) F_\alpha(dz)$$

$$\hat{c}_\alpha(x) := f(x) c_\alpha f(x)^T$$

$$\hat{F}_\alpha(x)(B) := \int_{\mathbb{R}^d} \mathbf{1}_B(f(x)z) \, F_\alpha(dz)$$

for $x \in \mathbb{R}^k$ and Borel sets $B \in \mathcal{B}(\mathbb{R}^k)$, where $h : \mathbb{R}^d \to \mathbb{R}^d$ is the truncation function for

4. Stochastic Processes

semimartingale characteristics in $\mathbb{R}^d$ and $\hat{h} : \mathbb{R}^k \to \mathbb{R}^k$ the truncation function in $\mathbb{R}^k$. Note that we tacitly used

$$\int_0^t \Gamma(\hat{\omega}_{s-})\, ds = \int_0^t \Gamma(\hat{\omega}_s)\, ds$$

for all measurable $\Gamma : \mathbb{R}^k \to \mathbb{R}$ and $\hat{\omega} \in D(\mathbb{R}_+; \mathbb{R}^k)$, which holds since càdlàg paths have at most countably many jumps. In particular, if $\mathbb{P}_\alpha \in \mathcal{P}$ such that $X = (X_t)_{t \geq 0}$ is a classical Lévy process (cf. [117, Chapter 2, Corollary 11.6, p. 63]) for $\alpha \in \mathcal{A}$ with

$$(b_s^{\mathbb{P}_\alpha}, c_s^{\mathbb{P}_\alpha}, F_s^{\mathbb{P}_\alpha})(\omega) = (b_\alpha, c_\alpha, F_\alpha)$$

$\lambda(ds) \times \mathbb{P}_\alpha(d\omega)$-almost everywhere, then $\hat{\mathbb{P}}_\alpha := \mathbb{P}_\alpha \circ (Z^x)^{-1} \in \hat{\mathcal{P}}_x$ for $x \in \mathbb{R}^k$ with

$$(b_s^{\hat{\mathbb{P}}_\alpha}, c_s^{\hat{\mathbb{P}}_\alpha}, F_s^{\hat{\mathbb{P}}_\alpha})(\hat{\omega}) = (\hat{b}_\alpha, \hat{c}_\alpha, \hat{F}_\alpha)(\omega_s)$$

$\lambda(ds) \times \hat{\mathbb{P}}(d\hat{\omega})$-almost everywhere. Therefore, if the function $f : \mathbb{R}^k \to \mathbb{R}^{k \times d}$ is not only global Lipschitz-continuous but also bounded, Theorem 4.37 implies that the sublinear generator of $(T_t)_{t \geq 0}$ is of the form

$$\begin{aligned}
\hat{A}(\psi)(x) &= \lim_{\delta \downarrow 0} \frac{T_\delta \psi(x) - T_0 \psi(x)}{\delta} = \lim_{\delta \downarrow 0} \frac{\hat{E}^x(\psi(Z_\delta)) - \hat{E}^x(\psi(Z_0))}{\delta} \\
&= \sup_{\alpha \in \mathcal{A}} \left(\hat{b}_\alpha(x) D\psi(x) + \frac{1}{2} \operatorname{tr}\left(\hat{c}_\alpha(x) D^2 \psi(x)\right) \right. \\
&\quad \left. + \int_{\mathbb{R}^d} \left(\psi(x+z) - \psi(x) - D\psi(x)\hat{h}(z)\right) \hat{F}_\alpha(dz) \right)
\end{aligned}$$

for $\psi \in C_b^\infty(\mathbb{R}^k)$. Note that Theorem 4.37 is applicable, since its proof does not depend on the specific form of the given filtration. For $\varphi \in C_b^{\operatorname{Lip}}(\mathbb{R}^k)$, define $u^\varphi : \mathbb{R}_+ \times \mathbb{R}^k \to \mathbb{R}$ by

$$u^\varphi(t, x) := \hat{E}^x(\varphi(Z_t)) = E(\varphi(Z_t^x)) = T_t \varphi(x)$$

for $(t, x) \in \mathbb{R}_+ \times \mathbb{R}^k$. According to Proposition 4.36, there exists $C > 0$ such that

$$|u^\varphi(t, x) - u^\varphi(s, x)| \leq L \cdot \hat{E}^x(|Z_t - Z_s|) \leq L \cdot C \cdot \left(|t-s|^{1/2} + |t-s|\right)$$

holds for all $t, s \geq 0$ and $x \in \mathbb{R}^k$. Since $(T_t)_{t \geq 0}$ satisfies the C_b-Feller property (cf. the first paragraph of this remark), we also know that

$$x \longmapsto u^\varphi(t, x) = T_t \varphi(x)$$

are continuous for all $t \geq 0$. Combining these two findings shows that $u^\varphi \in C_b(\mathbb{R}_+ \times \mathbb{R}^k)$. Finally, Proposition 4.10 implies that $u^\varphi : \mathbb{R}_+ \times \mathbb{R}^k \to \mathbb{R}$ is a viscosity solution (in terms of Proposition 4.10) in $(0, \infty) \times \mathbb{R}^k$ of the sublinear generator equation

$$\partial_t u^\varphi(t, x) - \hat{A}(u^\varphi(t, \cdot))(x) = 0$$

with $u^\varphi(0, \cdot) = \varphi(\cdot) \in C_b^{\operatorname{Lip}}(\mathbb{R}^k)$.

4.4. Stochastic Differential Equations

As a final point, we will apply our results from Chapter 2, to show that the sublinear generator equations of Lévy-type processes for sublinear expectations have unique viscosity solutions: A simple calculation entails that

$$-\hat{A}(\psi)(x) = G(x, D\psi(x), D^2\psi(x), \psi) := \inf_{\alpha \in \mathcal{A}} G_\alpha(x, D\psi(x), D^2\psi(x), \psi)$$

for $\psi \in C_b^\infty(\mathbb{R}^k)$ and $x \in \mathbb{R}^k$ with the linear operators

$$G_\alpha(x, p, X, \phi) := -\mathcal{L}_\alpha(x, p, X) - \mathcal{I}_\alpha(x, \phi)$$

$$\mathcal{L}_\alpha(x, p, X) := b'_\alpha(x)p + \frac{1}{2} \mathrm{tr}\left(\sigma_\alpha(x)\sigma_\alpha(x)^T X\right)$$

$$\mathcal{I}_\alpha(x, \phi) := \int_{|z| \leq 1} \left(\phi(x + j(x,z)) - \phi(x) - D\phi(x)j(x,z)\right) F_\alpha(dz)$$

$$+ \int_{|z| > 1} \left(\phi(x + j(x,z)) - \phi(x)\right) F_\alpha(dz)$$

for $(x, p, X, \phi) \in \mathbb{R}^k \times \mathbb{R}^k \times \mathbb{S}^{k \times k} \times C_b^2((0, \infty) \times \mathbb{R}^k)$ and $\alpha \in \mathcal{A}$, where $j(x, z) := f(x)z$,

$$b'_\alpha(x) := f(x)b_\alpha + \int_{|z| \leq 1} f(x)(z - h(z)) F_\alpha(dz) - \int_{|z| > 1} f(x)h(z) F_\alpha(dz)$$

and $\sigma_\alpha(x) := f(x)c_\alpha^{1/2}$ with the (unique) square root $c_\alpha^{1/2}$ of positive semi-definite, symmetric matrices as in [89, Section 6.1.4, p. 88]. This shows that the sublinear generator equation is a Hamilton-Jacobi-Bellman equation as in Definition 2.27, which satisfies all assumptions from Remark 2.28. Therefore, Corollary 2.34 implies that

$$\mathbb{R}_+ \times \mathbb{R}^k \ni (t, x) \longmapsto u^\varphi(t, x) = \hat{E}^x(\varphi(Z_t)) = E(\varphi(Z_t^x)) = T_t\varphi(x)$$

is the unique viscosity solution in $(0, \infty) \times \mathbb{R}^k$ of

$$\partial_t u^\varphi(t,x) - \sup_{\alpha \in \mathcal{A}} \left(f(x)b_\alpha D u^\varphi(t,x) + \frac{1}{2} \mathrm{tr}\left(f(x)c_\alpha f(x)^T D^2 u^\varphi(t,x)\right) \right.$$

$$\left. + \int_{\mathbb{R}^d} \left(u^\varphi(t, x + f(x)z) - u^\varphi(t,x) - Du^\varphi(t,x)f(x)h(z)\right) F_\alpha(dz) \right) = 0$$

with $u^\varphi(0, \cdot) = \varphi(\cdot) \in C_b^{\mathrm{Lip}}(\mathbb{R}^k)$. Similar as in Remark 4.38, note that, in order to apply the comparison principle from Chapter 2, we have to use Lemma 2.4, Remark 2.5 and Lemma 2.6, to show that the two different notions of viscosity solutions in Definition 2.1 and Proposition 4.10 coincide.

Since Lévy-type processes for sublinear expectations are constructed in a similar way as classical Lévy-type processes – as solutions of a stochastic differential equation driven by a Lévy processes for sublinear expectations instead of a classical Lévy processes – it seems natural to interpret them as Lévy-type processes under uncertainty in their characteristics. Since classical Lévy-type processes are used frequently in financial modeling (see [31] for an overview), our intrinsic approach could potentially be interesting in many applications to obtain best-case or worst-case bounds for models under uncertainty. For an introduction and fairly complete survey on the recent developments in the theory of classical Lévy-type processes, we refer the interested reader to [21].

A. Appendix

In this chapter, we collect statements of well-known results from convex analysis and functional analysis in the form, in which we require them in other parts of this thesis. Note that the exposition is supposed to be neither introductory nor exhaustive.

A.1. Convex Analysis

Convex analysis is an extension of classical differential calculus to convex functions. For a comprehensive introduction, we refer the interested reader to [111].

In this section, we will recall the strong connection between the convexity of real-valued functions and their differentiability. Moreover, we will present an application of this connection from Robert Jensen in [77], which plays an essential role in the viscosity solution theory of second-order nonlinear equations.

Lemma A.1 (Stability under Supremum). *If $(f_\alpha)_{\alpha \in \mathcal{A}}$ is a family of convex functions*

$$f_\alpha : \Omega \longrightarrow \mathbb{R}$$

on a common, convex domain of definition $\Omega \subset \mathbb{R}^d$, then the pointwise supremum

$$f(x) := \sup_{\alpha \in \mathcal{A}} f_\alpha(x) \in \mathbb{R} \cup \{\infty\}$$

is again a convex function on Ω.

Proof. ↪ [111, Theorem 5.5, p. 35] □

Proposition A.2 (Convexity $\implies$ Continuity). *Every convex function $f : \Omega \to \mathbb{R}$ defined on an open, convex set $\Omega \subset \mathbb{R}^d$ is locally Lipschitz-continuous.*

Proof. ↪ [94, Proposition 3.5.2, p. 121] □

Theorem A.3 (Rademacher). *Every locally Lipschitz-continuous function $f : \Omega \to \mathbb{R}$ defined on an open set $\Omega \subset \mathbb{R}^d$ is differentiable almost everywhere, i.e. the set of points where f is not differentiable has Lebesgue measure zero.*

Proof. ↪ [94, Theorem 3.11.1, p. 151] □

Corollary A.4 (Convexity $\implies$ Differentiability). *Every convex function $f : \Omega \to \mathbb{R}$ defined on an open, convex set $\Omega \subset \mathbb{R}^d$ is differentiable almost everywhere, i.e. the set of points where f is not differentiable has Lebesgue measure zero.*

Proof. This is a straightforward combination of Theorem A.3 and Proposition A.2. □

A. Appendix

Definition A.5 (Semiconvexity). A real-valued function on $\Omega \subset \mathbb{R}^d$
$$f : \Omega \longrightarrow \mathbb{R}$$
is called *semiconvex* if there exists a finite constant $C \geq 0$ for every convex, compactly contained subset $\Omega' \subset \Omega$ such that
$$x \longmapsto f(x) + C|x|^2$$
defines a convex function in Ω'.

Lemma A.6 (Twice Differentiability $\implies$ Semiconvexity). *Every real-valued function $f \in C^2(\Omega)$ defined on an open set $\Omega \subset \mathbb{R}^d$ is semiconvex.*

Proof. Suppose $\Omega' \subset \Omega$ is convex and compactly contained in Ω. Since the second derivative $D^2 f : \Omega \to \mathbb{S}^{d \times d}$ is continuous and Ω' compact, the constant
$$c := \frac{1}{2} \sup_{x \in \Omega'} \|D^2 f(x)\| \geq 0$$
is finite. Moreover, the Cauchy-Schwarz inequality implies
$$\left\langle D^2(f + c | \cdot |^2)(x)\xi, \xi \right\rangle = \left\langle D^2 f(x)\xi, \xi \right\rangle + 2c \left\langle \xi, \xi \right\rangle \geq \left(2c - \|D^2 f(x)\|\right) |\xi|^2 \geq 0$$
for all $\xi \in \mathbb{R}^d$ and $x \in \Omega'$, which means that $D^2(f + c | \cdot |^2)(x) \geq 0$ for every $x \in \Omega'$. In particular, we obtain the convexity of $x \mapsto f(x) + c|x|^2$ in Ω', as required. $\square$

Theorem A.7 (Alexandrov). *Every semiconvex function $f : \Omega \to \mathbb{R}$ defined on an open set $\Omega \subset \mathbb{R}^d$ is twice differentiable almost everywhere, i.e. there exists*
$$\Omega' \subset \{x \in \Omega : f \text{ is differentiable in } x\}$$
such that $\Omega \setminus \Omega'$ is a Lebesgue-null set and for all $x \in \Omega'$ there exists $A \in \mathbb{S}^{d \times d}$ such that
$$Df(y) = Df(x) + A(y - x) + o(|y - x|)$$
holds as $y \to x$.

Proof. $\hookrightarrow$ [94, Theorem 3.11.2, p. 154] $\square$

Lemma A.8 (Jensen). *Suppose that $\Omega \subset \mathbb{R}^d$ is bounded and open, and that $f : \overline{\Omega} \to \mathbb{R}$ is Lipschitz-continuous with an interior maximum that exceeds the boundary values, i.e. there exists $x_0 \in \Omega$ such that*
$$f(x_0) = \max_{x \in \Omega} f(x) > \max_{x \in \partial \Omega} f(x).$$
If $f : \overline{\Omega} \to \mathbb{R}$ is semiconvex in Ω and $\varepsilon > 0$, then there exists $p \in \mathbb{R}^d$ with $|p| \leq \varepsilon$ and an interior point $z \in \Omega$ such that the function
$$x \longmapsto f(x) - \langle p, x \rangle$$
has a global maximum and is twice differentiable at $z \in \Omega$, i.e.
$$f(x) = f(z) + \langle p, x - z \rangle + \tfrac{1}{2} \langle X(x - z), (x - z) \rangle + o(|x - z|^2)$$
holds for some $X \in \mathbb{S}^{d \times d}$ as $x \to z$.

Proof. $\hookrightarrow$ [70, Lemma 5.2] $\square$

A.2. Functional Analysis

Functional analysis can be seen as a generalization of results from linear algebra, and measure and integration theory from finite-dimensional to infinite-dimensional spaces. For comprehensive introductions, we refer the interested reader to [86], [115] and [130].

In this section, we will recall the classical Hahn-Banach extension theorem and the related separation of normed spaces by hyperplanes. Moreover, we will give a brief overview over the Bochner integration theory (which is essentially a generalization of the Lebesgue integration theory for functions with values in a Banach space) and discuss its applicability for differential characteristics in Section 4.2. At last, we will recall the definition of Hausdorff distances as a method to lift metrics to power sets.

Theorem A.9 (Hahn-Banach Extension). *Let $\mathcal{H}$ be a real linear space, $\mathcal{L} \subset \mathcal{H}$ a real linear subspace, $p : \mathcal{H} \to \mathbb{R}$ a sublinear functional and $f : \mathcal{L} \to \mathbb{R}$ a linear functional that is dominated by p (i.e. $f(x) \leq p(x)$ for all $x \in \mathcal{L}$). There exists a linear functional*

$$F : \mathcal{H} \to \mathbb{R}$$

that extends f (i.e. $F|_\mathcal{L} = f$) and is still dominated by p (i.e. $F(x) \leq p(x)$ for all $x \in \mathcal{H}$).

Proof. ↪ [115, Theorem 3.2, p. 57] □

Theorem A.10 (Hyperplane Separation). *Suppose that $(X, \|\cdot\|)$ is a real normed space and $A, B \subset X$ are disjoint, convex sets. If A is closed and B is compact, then there exists a continuous, linear $T : X \to \mathbb{R}$ and constants $c_1, c_2 \in R$ such that*

$$T(a) \leq c_1 < c_2 \leq T(b)$$

for all $a \in A$ and $b \in B$.

Proof. ↪ [115, Theorem 3.4, p. 59] □

Remark A.11 (Bochner Integration). Suppose that $(X, \|\cdot\|)$ is a real Banach space and $(\Omega, \mathcal{A}, \mu)$ a finite measure space. It is possible to show (cf. [44, Chapter 2, p. 41]) that there exists a generalization of the Lebesgue integral to X-valued integrands

$$L^1(\mu; X) := \{f : \Omega \to X \mid f \text{ is strongly } \mu\text{-measurable and } \|f\| \in L^1(\mu)\},$$

which is typically referred to as Bochner integral or Dunford integral, such that

$$\int_\Omega T(f(x)) \, \mu(dx) = T\left(\int_\Omega f(x) \, \mu(dx)\right)$$

holds for all $f \in L^1(\mu; X)$ and continuous, linear operators $T : X \to \mathbb{R}$. A function $f : \Omega \to X$ is called strongly μ-measurable if there exists a sequence $(f_n)_{n \in \mathbb{N}}$ of Borel-measurable functions $f_n : \Omega \to X$ such that $f_n(\Omega) \subset X$ is countable and

$$\lim_{n \to \infty} f_n = f$$

μ-almost surely. In particular, if $(X, \|\cdot\|)$ is separable, then Petti's measurability test

A. Appendix

(cf. [44, Chapter 2, Section 1, Theorem 2, p. 42] and [119, Theorem 8.8, p. 61]) shows that every Borel-measurable $f : \Omega \to X$ is strongly μ-measurable. It is possible to show that most of the properties of the Lebesgue integral, such as the triangle inequality

$$\left\| \int_\Omega f(x)\,\mu(dx) \right\| \leq \int_\Omega \|f(x)\|\,\mu(dx)$$

for $f \in L^1(\mu; X)$ and dominated convergence, carry over to the Bochner integral.

In order to talk work with differential characteristics, as defined in Remark 4.19, we would like to integrate functions taking values in $\mathfrak{L}(\mathbb{R}^d)$, whose topology is given by an isometry into $\mathfrak{M}_b(\mathbb{R}^d)$ with respect to weak convergence of measures as in Remark 4.17. Many publications (cf. e.g. [73, Chapter 2, Proposition 2.9, p. 77] and [21, Definition 2.43, p. 64]) work with a notion, where a function $f : [0,T] \to \mathfrak{M}_b(\mathbb{R}^d)$ with $T > 0$ is called integrable with integral $F \in \mathfrak{M}_b(\mathbb{R}^d)$ if the equality

$$\int_0^T \int_{\mathbb{R}^d} \varphi(x)\,f(t)(dx)\,dt = \int_{\mathbb{R}^d} \varphi(x)\,F(dx)$$

holds for all $\varphi \in C_b(\mathbb{R}^d)$. This corresponds to the notion of the Pettis integral as discussed in [44, Chapter 2, Section 3, p. 52] or [124, Chapter 4, p. 45], which only requires the weaker measurability for $f : \mathbb{R}_+ \to \mathfrak{M}_b(\mathbb{R}^d)$ in that the mappings

$$t \longmapsto \int_{\mathbb{R}^d} \varphi(x)\,f(t)(dx)$$

are Lebesgue-measurable for all $\varphi \in C_b(\mathbb{R}^d)$. According to [20, Section 8.3, p. 191] the linear space $\mathfrak{M}_b(\mathbb{R}^d)$ with the Kantorovich-Rubinstein norm

$$\|\mu\|_{\mathfrak{M}_b} := \sup \left\{ \int_{\mathbb{R}^d} \varphi(x)\,\mu(dx) : \sup_{x \in \mathbb{R}^d} |\varphi(x)| \leq 1, \sup_{x \neq y} \frac{|\varphi(x) - \varphi(y)|}{|x - y|} \leq 1 \right\}$$

is a normed space. Hence, the (topological) completion $X := \overline{\mathfrak{M}_b(\mathbb{R}^d)}$ with respect to $\|\cdot\|_{\mathfrak{M}_b}$ is a Banach space. Moreover, as we have already noted in Remark 4.16, on the convex and closed subset of non-negative, bounded Borel measures

$$\mathfrak{M}_b^+(\mathbb{R}^d) \subset X,$$

the Kantorovich-Rubinstein norm generates the topology of weak convergence of measures, which is separable due to the separability of the Euclidean space $\mathbb{R}^d$. Therefore, Pettis's measurability test (cf. [44, Chapter 2, Section 1, Theorem 2, p. 42]) shows that if the function $f : [0,T] \to \mathfrak{M}_b^+(\mathbb{R}^d)$ satisfies the boundedness condition

$$\int_0^T \|f(t)\|_{\mathfrak{M}_b}\,dt = \int_0^T f(t)(\mathbb{R}^d)\,dt < \infty,$$

then $f : [0,T] \to \mathfrak{M}_b^+(\mathbb{R}^d)$ is Bochner integrable if and only if $t \mapsto \int_{\mathbb{R}^d} \varphi(x)\,f(t)(dx)$ is Lebesgue integrable for all $\varphi \in C_b(\mathbb{R}^d)$. In particular, all integrals in Chapter 4 are not only Pettis integrals but even Bochner integrals.

A.2. Functional Analysis

Lemma A.12 (Range of Integrals). *If $(X, \mathcal{A}, \mu)$ is a measure space and $(Y, \|\cdot\|)$ is a Banach space, then for all $A \in \mathcal{A}$ with $0 < \mu(A) < \infty$ and every $f \in L^1(X;Y)$*

$$\frac{1}{\mu(A)} \int_A f(x)\, \mu(dx) \in \overline{\operatorname{conv}}(f(A)),$$

where $\overline{\operatorname{conv}}(f(A))$ is the closure of the convex hull of $f(A)$.

Proof. Suppose that $y := (\mu(A))^{-1} \int_A f(x)\, \mu(dx) \notin \overline{\operatorname{conv}}(f(A))$. The hyperplane separation theorem (cf. Theorem A.10) implies the existence of a continuous, linear $T: Y \to \mathbb{R}$ and a constant $c \in \mathbb{R}$ such that

$$T(z) \leq c < T(y) = T\left(\frac{1}{\mu(A)} \int_A f(x)\, \mu(dx)\right)$$

holds for all $z \in \overline{\operatorname{conv}}(f(A))$. In particular, the properties from Remark A.11 imply

$$c < T\left(\frac{1}{\mu(A)} \int_A f(x)\, \mu(dx)\right) = \frac{1}{\mu(A)} \int_A T(f(x))\, \mu(dx) \leq \frac{1}{\mu(A)} \int_A c\, \mu(dx) = c,$$

contradicting our assumption that $y = \frac{1}{\mu(A)} \int_A f(x)\, \mu(dx) \notin \overline{\operatorname{conv}}(f(A))$. □

Remark A.13 (Absolute Continuity). A function $F: \mathbb{R}_+ \to X$ mapping into a Banach space $(X, \|\cdot\|)$ is called absolutely continuous if there exists

$$f \in L^1_{\operatorname{loc}}(\mathbb{R}_+; X) := \{f: \mathbb{R}_+ \to X \mid \forall T > 0 : f|_{[0,T]} \in L^1([0,T];X)\}$$

such that $F(t) = \int_0^t f(s)\, ds$ for all $t \geq 0$. In other words, $F: \mathbb{R}_+ \to X$ is absolutely continuous, if $F \in \Upsilon(L^1_{\operatorname{loc}}(\mathbb{R}_+; X))$ with

$$\Upsilon: L^1_{\operatorname{loc}}(\mathbb{R}_+; X) \longrightarrow C(\mathbb{R}_+; X)$$

$$\theta \longmapsto \Upsilon(\theta) := \int_0^{\cdot} \theta_s\, ds,$$

which we will refer to as the (Bochner) integration operator. A standard argument shows that if a function $F: \mathbb{R}_+ \to X$ is absolutely continuous, then for every $T > 0$ and $\varepsilon > 0$ there exists $\delta > 0$ such that

$$\sum_{j=1}^n |F(x_j) - F(y_j)| \leq \varepsilon$$

for every finite sequence of pairwise disjoint subintervals $((x_j, y_j))_{j \leq n} \subset [0,T]$ with $\sum_{j=1}^n |x_j - y_j| \leq \delta$ and $n \in \mathbb{N}$. (Note that [44, Chapter 1, Section 2, Theorem 1, p. 10] implies that the previous condition is equivalent to $\int_B f(s)\, ds = 0$ for all Lebesgue-null sets $B \subset \mathbb{R}_+$, which obviously holds due to the properties of the Bochner integral, cf. [44, Chapter 2, Section 2, Corollary 5, p. 47].) Unfortunately, [44, Chapter 2, Section 2, Example 10, p. 50] shows that the converse does not hold for arbitrary Banach spaces $(X, \|\cdot\|)$. Banach spaces $(X, \|\cdot\|)$ for which the converse holds are said to have the Radon–Nikodym property (with respect to the Lebesgue measure), and include all reflexive Banach spaces and separable dual spaces, cf. [44, Chapter 3, Section 2, Corollary 13, p. 76] and [44, Chapter 3, Section 3, Theorem 1, p. 79].

A. Appendix

The previous remark shows that we have to be careful, when we want to conclude absolute continuity from Lipschitz-continuity for Bochner integration. Throughout this thesis, we therefore applied this argument only for finite dimensional Banach spaces, which are reflexive and hence have the Radon-Nikodym property. Note in particular that even though weakly compact, convex sets have the Radon-Nikodym property according to [15, Theorem 5.11, p. 106], we equipped the non-negative, bounded Borel measures $\mathfrak{M}_b^+(\mathbb{R}^d)$ with the $C_b(\mathbb{R}^d)$-weak convergence of measures in Remark 4.16, which is actually the weak-* and not the weak convergence in the terms of functional analysis.

Lemma A.14 (Lipschitz Constant Map). *The map* $\Lambda : D(\mathbb{R}_+; X) \to [0, \infty]$ *for a metric space* (X, d) *that assigns the optimal Lipschitz-continuity constant*

$$\Lambda(\omega) := \sup \left\{ \frac{d(\omega(t), \omega(s))}{|t - s|} : t > s \geq 0 \right\}$$

to a càdlàg path $\omega \in D(\mathbb{R}_+)$ *is lower semicontinuous with respect to the Skorokhod topology as introduced in Lemma 4.2.*

Proof. Suppose that $(\omega^n)_{n \in \mathbb{N}} \subset D(\mathbb{R}_+; X)$ with $\lim_{n \to \infty} \omega^n = \omega \in D(\mathbb{R}_+; X)$ with respect to the Skorokhod topology, i.e. there exists a sequence $(\lambda^n)_{n \in \mathbb{N}}$ of strictly increasing functions $\lambda^n : \mathbb{R}_+ \to \mathbb{R}_+$ with $\lim_{n \to \infty} \sup_{t \geq 0} |\lambda^n(t) - t| = 0$ and

$$\lim_{n \to \infty} \sup_{t \in [0,T]} d(\omega^n(\lambda^n(t)), \omega(t)) = 0$$

for all $T > 0$. Hence, for $t > s \geq 0$ the convergence implies

$$\frac{d(\omega(t), \omega(s))}{|t - s|} = \lim_{n \to \infty} \frac{d(\omega^n(\lambda^n(t)), \omega^n(\lambda^n(s)))}{|\lambda^n(t) - \lambda^n(s)|} \leq \liminf_{n \to \infty} \Lambda(\omega^n)$$

and hence $\Lambda(\omega) \leq \liminf_{n \to \infty} \Lambda(\omega^n)$, as required. □

Remark A.15 (Hausdorff Distances). The *Hausdorff distance* $d_H^X : 2^X \times 2^X \to [0, \infty]$ is the metric on the power set 2^X of a metric space (X, d) defined by

$$d_H^X(A, B) := \max \left\{ \sup_{a \in A} \inf_{b \in B} d(a, b), \sup_{b \in B} \inf_{a \in A} d(a, b) \right\} = \inf \{ \varepsilon > 0 \mid A \subset B_\varepsilon, B \subset A_\varepsilon \}$$

for $A, B \subset X$, where $M_\varepsilon := \{ x \in X \mid d(x, M) \leq \varepsilon \}$ for subsets $M \subset X$ and $\varepsilon \geq 0$, cf. e.g. [50, Problem 4.5.23, p. 298] and [112, Chapter 4, Section C, p. 117]. It is easy to check that the following triangle inequality

$$d(x, A) \leq d(x, B) + d_H^X(A, B)$$

holds for all $x \in X$ and $A, B \subset X$.

Outlook

Based on the novel results from this thesis, there are a variety of possible directions for subsequent research. In this short outlook, we collect some of the potentially promising ideas, we stumbled upon during the creation of this thesis:

In the field of viscosity solution theory for fully nonlinear partial integro-differential equations, it would be interesting to study comparison principles and regularity results for general elliptic problems and problems with more general boundary conditions (in the spirit of [74] and [11]). In particular, our general maximum principle from Section 2.2 seems to provide the correct tool to extend some of the existing results to our generalized set-up in an easy manner. On top of that, it would be important for applications of Lévy-type processes for sublinear expectations, to extend some of the recent results (such as [27]) on the numerical analysis of nonlocal equations to our generalized set-up.

In the field of nonlinear expectations, one could try to generalize our construction of sublinear Markov semigroups to Markov-type semigroups via backwards stochastic differential equations (based on [106]). Compared to Remark 4.62, this generalization would correspond to path-dependent uncertainty coefficients and resulting path-dependent nonlocal equations. Another approach could be to extend our construction of Lévy-type processes for sublinear expectations to unbounded uncertainty coefficients or processes with finite lifetime (similar to the classical Markov process theory in [21]). Moreover, one could generalize the strong connection between Lévy-type processes and elliptical Hamilton-Jacobi-Bellman equation to sublinear expectations using an optimal stopping argument – imitating the relationship between classical Brownian motion and the Laplace equation. Alternatively, one could focus on applications (such as stochastic optimal control, stochastic game theory or financial modeling) and employ our intrinsic calculus for stochastic processes on sublinear expectation spaces.

Bibliography

[1] ALVAREZ, O., TOURIN, A. Viscosity solutions of nonlinear integro-differential equations. *Annales de l'Institut Henri Poincare (C) Non Linear Analysis 13*, 3 (1996), 293–317. (p. 15, 58, 60, 62, 65, 69, 72, 104, 111, 116)

[2] APOSTOL, T. *Calculus*, 2nd ed., vol. 1. Wiley, 1967. (p. 121, 128)

[3] ARISAWA, M. A new definition of viscosity solutions for a class of second-order degenerate elliptic integro-differential equations. *Annales de l'Institut Henri Poincare (C) Non Linear Analysis 23*, 5 (2006), 695–711. (p. 65)

[4] ARISAWA, M. Corrigendum for the comparison theorems in: "A new definition of viscosity solutions for a class of second-order degenerate elliptic integro-differential equations". *Annales de l'Institut Henri Poincare (C) Non Linear Analysis 24*, 1 (2007), 167–169. (p. 65)

[5] ARTZNER, P., DELBAEN, F., EBER, J., HEATH, D. Coherent measures of risk. *Mathematical Finance 9*, 3 (1999), 203–228. (p. 135, 138)

[6] BAIRE, R. *Leçons Sur Les Fonctions Discontinues*. Gauthier-Villars, 1905. (p. 38)

[7] BARDI, M., CAPUZZO-DOLCETTA, I. *Optimal Control and Viscosity Solutions of Hamilton-Jacobi-Bellman Equations*. Birkhäuser, 1997. (p. 113)

[8] BARDI, M., RAGHAVAN, T., PARTHASARATHY, T., Eds. *Stochastic and Differential Games: Theory and Numerical Methods*. Springer, 2012. (p. 116)

[9] BARLES, G., BUCKDAHN, R., PARDOUX, E. Backward stochastic differential equations and integral-partial differential equations. *Stochastics: An International Journal of Probability and Stochastic Processes 60* (1997), 57–83. (p. 58, 68, 132)

[10] BARLES, G., CHASSEIGNE, E., CIOMAGA, A., IMBERT, C. Lipschitz regularity of solutions for mixed integro-differential equations. *Journal of Differential Equations 252*, 11 (2012), 6012–6060. (p. 72)

[11] BARLES, G., CHASSEIGNE, E., IMBERT, C. On the Dirichlet problem for second-order elliptic integro-differential equations. *Indiana University Mathematics Journal 57* (2008), 213–246. (p. 70, 225)

[12] BARLES, G., IMBERT, C. Second-order elliptic integro-differential equations: Viscosity solutions' theory revisited. *Annales de l'Institut Henri Poincare (C) Non Linear Analysis 25*, 3 (2008), 567–585. (p. 15, 26, 58, 60, 62, 65, 72, 104)

[13] BAYRAKTAR, E., MUNK, A. An α-stable limit theorem under sublinear expectation. arXiv:1409.7960v3, 2015. (p. 145)

[14] BELLMAN, R. *Dynamic Programming*. Princeton University Press, 1957. (p. 113)

[15] BENYAMINI, Y., LINDENSTRAUSS, J. *Geometric Nonlinear Functional Analysis*, vol. 1. American Mathematical Socitety, 2000. (p. 224)

[16] BERTSEKAS, D., SHREVE, S. *Stochastic Optimal Control: The Discrete-Time Case*. Academic Press, 1996. (p. 167, 170, 211)

[17] BICHTELER, K. Stochastic integration and L^p-theory of semimartingales. *Annals of Probability 9*, 1 (1981), 49–89. (p. 208)

[18] BILLINGSLEY, P. *Convergence of Probability Measures*. Wiley, 1968. (p. 148, 205)

[19] BILLINGSLEY, P. *Probability and Measure*, 3rd ed. Wiley, 1995. (p. 163)

[20] BOGACHEV, V. *Measure Theory*, vol. 2. Springer, 2007. (p. 158, 159, 222)

[21] BÖTTCHER, B., SCHILLING, R., WANG, J. *Lévy-Type Processes: Construction, Approximation and Sample Path Properties*, vol. 2099, Lévy Matters III of *Lecture Notes in Mathematics*. Springer, 2014. (p. 114, 117, 150, 151, 152, 154, 160, 181, 217, 222, 225)

[22] BUCKDAHN, R., LI, J. Stochastic differential games and viscosity solutions of Hamilton–Jacobi–Bellman–Isaacs equations. *SIAM Journal on Control and Optimization 47*, 1 (2008), 444–475. (p. 116)

[23] CAMERER, C., WEBER, M. Recent developments in modeling preferences: Uncertainty and ambiguity. *Journal of Risk and Uncertainty 5* (1992), 325–370. (p. 10)

[24] CARATHÉODORY, C. *Variationsrechnung und partielle Differentialgleichungen erster Ordnung*. Teubner, 1935. (p. 113)

[25] CHEN, Z. Strong laws of large numbers for sub-linear expectations. *Science China Mathematics 59*, 5 (2016), 945–954. (p. 145)

[26] CHOQUET, G. Theory of capacities. *Annales de l'Institut Fourier 5* (1954), 131–295. (p. 140)

[27] COCLITE, G., REICHMANN, O., RISEBRO, N. A convergent difference scheme for a class of partial integro-differential equations modeling pricing under uncertainty. *SIAM Journal on Numerical Analysis 54*, 2 (2016), 588–605. (p. 225)

[28] COHEN, S. Quasi-sure analysis, aggregation and dual representations of sublinear expectations in general spaces. *Electronic Journal of Probability 17*, 62 (2012), 1–15. (p. 162)

[29] COHEN, S., ELLIOTT, R. *Stochastic Calculus and Applications*, 2nd ed. Birkhäuser, 2015. (p. 202)

[30] COHN, D. *Measure Theory*, 2nd ed. Birkhäuser, 2013. (p. 164, 170, 210)

[31] CONT, R., TANKOV, P. *Financial Modelling with Jump Processes*. CRC Press, 2004. (p. 13, 147, 217)

[32] COURRÈGE, P. Sur la forme intégro-différentielle des opérateurs de C_k^∞ dans C satisfaisant au principe du maximum. *Séminaire Brelot-Choquet-Deny. Théorie du Potentiel 10*, 1 (1965), 1–38. (p. 114)

[33] CRANDALL, M. Semidifferentials, quadratic forms and fully nonlinear elliptic equations of second order. *Annales de l'Institut Henri Poincare (C) Non Linear Analysis 6*, 6 (1989), 419–435. (p. 81, 90, 93)

[34] CRANDALL, M., ISHII, H. The maximum principle for semicontinuous functions. *Differential Integral Equations 3*, 6 (1990), 1001–1014. (p. 96)

[35] CRANDALL, M., ISHII, H., LIONS, P. User's guide to viscosity solutions of second order partial differential equations. *Bulletin of the American Mathematical Society 27*, 1 (1992), 1–67. (p. 26, 48, 58, 69, 70, 93, 104)

[36] CRANDALL, M., LIONS, P. Viscosity solutions of Hamilton-Jacobi equations. *Transactions of the American Mathematical Society 277*, 1 (1983), 1–42. (p. 8, 13, 58, 72, 113)

[37] DANIELL, P. A general form of integral. *Annals of Mathematics 19*, 4 (1918), 279–294. (p. 139)

[38] DELLACHERIE, C., MEYER, P. *Probabilities and Potential*, vol. A. Elsevier, 1979. (p. 163, 184, 197, 209, 213)

[39] DELLACHERIE, C., MEYER, P. *Probabilities and Potential*, vol. C. Elsevier, 1988. (p. 201)

[40] DENIS, L., HU, M., PENG, S. Function spaces and capacity related to a sublinear expectation: Application to G-Brownian motion paths. *Potential Analysis 34*, 2 (2011), 139–161. (p. 142, 155)

[41] DENKOWSKI, Z., MIGORSKI, S., PAPAGEORGIOU, N. *An Introduction to Nonlinear Analysis: Theory*. Springer, 2003. (p. 194)

[42] DENNEBERG, D. *Non-Additive Measure and Integral*. Springer, 1994. (p. 140)

[43] DENNEBERG, D. Non-additive measure and integral, basic concepts and their role for applications. In *Fuzzy Measures and Integrals - Theory and Applications* (2000), M. Grabisch, T. Murofushi, and M. Sugeno, Eds., Studies in Fuzziness and Soft Computing, Springer. (p. 10, 135)

[44] DIESTEL, J., UHL, J. *Vector Measures*. American Mathematical Socitety, 1977. (p. 221, 222, 223)

[45] DINI, U. *Fondamenti Per La Teorica Delle Funzioni Di Variabili Reali*. Pisa, T. Nistri, 1878. (p. 54)

[46] DUDLEY, R. *Real Analysis and Probability*. Cambridge University Press, 2004. (p. 139, 203)

[47] DUNFORD, N., SCHWARTZ, J. *Linear Operators*, vol. 1: General Theory. Wiley Interscience, 1957. (p. 54)

[48] EINSTEIN, A. Über die von der molekularkinetischen Theorie der Wärme geforderte Bewegung von in ruhenden Flüssigkeiten suspendierten Teilchen. *Annalen der Physik 322*, 8 (1905), 549–560. (p. 117, 206)

[49] ELLSBERG, D. Risk, ambiguity, and the savage axioms. *The Quarterly Journal of Economics 75*, 4 (1961), 643–669. (p. 9)

[50] ENGELKING, R. *General Topology*. Heldermann, 1989. (p. 187, 224)

[51] EVANS, L. On solving certain nonlinear partial differential equations by accretive operator methods. *Israel Journal of Mathematics 36* (1980), 225–247. (p. 58)

[52] EVANS, L. *Partial Differential Equations*. American Mathematical Socitety, 1998. (p. 58)

[53] FENCHEL, W. On conjugate convex functions. *Canadian Journal of Mathematics 1* (1949), 73–77. (p. 25)

[54] FLEMING, W., SONER, H. *Controlled Markov Processes and Viscosity Solutions*, 2nd ed. Springer, 2006. (p. 149, 154)

[55] FÖLLMER, H., SCHIED, A. *Stochastic Finance: An Introduction in Discrete Time*. Walter de Gruyter, 2002. (p. 135, 140)

[56] FOLLAND, G. *Real Analysis*, 2nd ed. Wiley, 1999. (p. 20, 186)

[57] FRIEDRICHS, K. The identity of weak and strong extensions of differential operators. *Transactions of the American Mathematical Society 55*, 1 (1944), 132–151. (p. 19)

[58] GAUSS, C. Allgemeine Theorie des Erdmagnetismus. In *Resultate aus den Beobachtungen des magnetischen Vereins im Jahre 1838* (1839), 1–57. (p. 71)

[59] GRECO, G. Sulla rappresentazione di funzionali mediante integrali. *Rendiconti del Seminario Matematico della Università di Padova 66* (1982), 21–42. (p. 140)

[60] HAMILTON, W. On the application to dynamics of a general mathematical method previously applied to optics. In *Report of the Fourth Meeting of the British Association for the Advancement of Science* (1834), 513–518. (p. 113)

[61] HAUSDORFF, F. Über halbstetige Funktionen und deren Verallgemeinerung. *Mathematische Zeitschrift 5*, 3 (1919), 292–309. (p. 16, 38, 52)

[62] HOCKING, J., YOUNG, G. *Topology*. Dover, 1988. (p. 159)

[63] HOH, W., JACOB, N. On the Dirichlet problem for pseudodifferential operators generating Feller semigroups. *Journal of Functional Analysis 137*, 1 (1996), 19–48. (p. 65)

[64] HOPF, E. Elementare Bemerkungen über die Lösungen partieller Differentialgleichungen zweiter Ordnung vom elliptischen Typus. *Sitzungsberichte der Berliner Akademie der Wissenschaften 19* (1927), 147–152. (p. 71)

[65] HOPF, E. Generalized solutions of non-linear equations of first order. *Journal of Applied Mathematics and Mechanics 14* (1965), 951–973. (p. 25)

[66] HÖRMANDER, L. *The Analysis of Linear Partial Differential Operators*, 2nd ed., vol. 1. Springer, 1990. (p. 19, 20)

[67] HU, M., PENG, S. G-Lévy processes under sublinear expectations. arXiv:0911.3533v1, 2009. (p. 7, 8, 11, 12, 17, 58, 62, 72, 81, 110, 111, 118, 147, 155, 178, 179)

[68] ISAACS, R. *Differential Games*. Wiley, 1965. (p. 113)

[69] ISHII, H. Hamilton-Jacobi equations with discontinuous Hamiltonians on arbitrary open sets. *Bulletin of the Faculty of Science and Engineering, Chuo University 28* (1985), 33–77. (p. 116)

[70] ISHII, H. On uniqueness and existence of viscosity solutions of fully nonlinear second order elliptic PDE's. *Communications on Pure and Applied Mathematics 42* (1989), 15–45. (p. 220)

[71] ITÔ, K. Stochastic integral. *Proceedings of the Imperial Academy 20*, 8 (1944), 519–524. (p. 196, 206)

[72] JACOBI, C. Zur Theorie der Variations-Rechnung und der Differential-Gleichungen. *Journal für die reine und angewandte Mathematik 17* (1837), 68–82. (p. 113)

[73] JACOD, J., SHIRYAEV, A. *Limit Theorems for Stochastic Processes*, 2nd ed. Springer, 2003. (p. 148, 156, 157, 158, 160, 161, 172, 174, 176, 180, 187, 189, 190, 195, 197, 198, 199, 200, 214, 222)

[74] JAKOBSEN, E., KARLSEN, K. Continuous dependence estimates for viscosity solutions of integro-PDEs. *Journal of Differential Equations 212*, 2 (2005), 278–318. (p. 72, 96, 225)

Bibliography

[75] JAKOBSEN, E., KARLSEN, K. A "maximum principle for semicontinuous functions" applicable to integro-partial differential equations. *Nonlinear Differential Equations and Applications 13*, 2 (2006), 137–165. (p. 8, 14, 15, 16, 26, 50, 58, 60, 62, 65, 72, 74, 90, 93, 104)

[76] JARROW, R., PROTTER, P. A short history of stochastic integration and mathematical finance: The early years, 1880-1970. *Institute of Mathematical Statistics Lecture Notes - Monograph Series 45* (2004), 75–91. (p. 196, 206)

[77] JENSEN, R. The maximum principle for viscosity solutions of fully nonlinear second order partial differential equations. *Archive for Rational Mechanics and Analysis 101*, 1 (1988), 1–27. (p. 50, 58, 72, 75, 77, 219)

[78] JENSEN, R., LIONS, P., SOUGANIDIS, P. A uniqueness result for viscosity solutions of second order fully nonlinear partial differential equations. *Proceedings of the American Mathematical Society 102* (1988), 975–978. (p. 72)

[79] JOHNSON, W., LINDENSTRAUSS, J., Eds. *Handbook of the Geometry of Banach Spaces*, vol. 2. Elsevier, 2003. (p. 26)

[80] KARANDIKAR, R. On pathwise stochastic integration. *Stochastic Processes and their Applications 57* (1995), 11–18. (p. 202, 205)

[81] KEYNES, J. *A Treatise On Probability*. Macmillan, 1921. (p. 9)

[82] KNIGHT, F. *Risk, Uncertainty and Profit*. Sentry Press, 1921. (p. 9)

[83] KOLOKOLTSOV, V. *Markov Processes, Semigroups and Generators*. De Gruyter, 2011. (p. 117, 149)

[84] LASRY, J., LIONS, P. A remark on regularization in Hilbert spaces. *Israel Journal of Mathematics 55*, 3 (1986), 257–266. (p. 25)

[85] LAX, P. Hyperbolic systems of conservation laws II. *Communications on Pure and Applied Mathematics 10*, 4 (1957), 537–566. (p. 25)

[86] LAX, P. *Functional Analysis*. Wiley, 2002. (p. 221)

[87] LIONS, P. Fully nonlinear elliptic equations and applications. *Nonlinear Analysis, Function Spaces and Applications 2* (1982), 126–149. (p. 58)

[88] LIONS, P. Optimal control of diffusion processes and Hamilton-Jacobi-Bellman equations. *Communications in Partial Differential Equations 8*, 11 (1983), 1229–1276. (p. 72)

[89] LÜTKEPOHL, H. *Handbook of Matrices*. Wiley, 1996. (p. 119, 120, 127, 181, 217)

[90] MARKOV, A. Ausdehnung der Sätze über die Grenzwerte in der Wahrscheinlichkeitsrechnung auf eine Summe verketteter Größen. In *Wahrscheinlichkeitsrechnung* (1912), H. Liebmann, Ed., Teubner, 272–298. (p. 149)

[91] MOREAU, J. Proximité et dualité dans un espace Hilbertien. *Bulletin de la Société Mathématique de France 93* (1965), 273–299. (p. 25)

[92] NEUFELD, A., NUTZ, M. Measurability of semimartingale characteristics with respect to the probability law. *Stochastic Processes and their Applications 124*, 11 (2014), 3819–3845. (p. 156, 158, 159, 160, 161)

[93] NEUFELD, A., NUTZ, M. Nonlinear Lévy processes and their characteristics. arXiv:1401.7253v2, 2015. (p. 7, 8, 11, 12, 15, 17, 62, 118, 147, 156, 168, 169, 178)

[94] NICULESCU, C., PERSSON, L. *Convex Functions and Their Applications.* Springer, 2006. (p. 219, 220)

[95] NISIO, M. On a non-linear semi-group attached to stochastic optimal control. *Publications of the Research Institute for Mathematical Sciences 13* (1976), 513–537. (p. 12, 149)

[96] NORRIS, J. *Markov Chains.* Cambridge University Press, 1997. (p. 149)

[97] NUTZ, M. Pathwise construction of stochastic integrals. *Electronic Communications in Probability 17*, 24 (2012), 1–7. (p. 17, 196, 200, 201, 209)

[98] NUTZ, M., VAN HANDEL, R. Constructing sublinear expectations on path space. *Stochastic Processes and their Applications 123*, 8 (2013), 3100–3121. (p. 136, 142, 150, 156, 162, 163, 166, 168, 169)

[99] OTHER. Non-Local Equations Wiki. http://www.ma.utexas.edu/mediawiki, 2013. Accessed on 05.04.2016. (p. 57)

[100] OTHER. AMS MathSciNet. http://www.ams.org/mathscinet/citations.html, 2016. Accessed on 06.04.2016. (p. 58)

[101] PACZKA, K. Itô calculus and jump diffusions for G-Lévy processes. arXiv:1211.2973v1, 2012. (p. 196)

[102] PENG, S. *G*-expectation, *G*-Brownian motion and related stochastic calculus of Itô type. In *Stochastic Analysis and Applications - The Abel Symposium 2005* (2007), vol. 2, Springer, 541–567. (p. 7, 8, 11, 12, 15, 17, 118, 135, 141, 144, 146, 147, 155, 162, 179, 196)

[103] PENG, S. Law of large numbers and central limit theorem under nonlinear expectations. arXiv:math/0702358v1, 2007. (p. 141)

Bibliography

[104] PENG, S. Survey on normal distributions, central limit theorem, Brownian motion and the related stochastic calculus under sublinear expectations. *Science in China Series A: Mathematics 52*, 7 (2009), 1391–1411. (p. 135, 143, 145)

[105] PENG, S. Nonlinear expectations and stochastic calculus under uncertainty. arXiv:1002.4546v1, 2010. (p. 11, 118, 136, 142, 146)

[106] PENG, S., WANG, F. BSDE, path-dependent PDE and nonlinear Feynman-Kac formula. *Science China Mathematics 59*, 1 (2016), 19–36. (p. 225)

[107] PERRON, O. Eine neue Behandlung der ersten Randwertaufgabe für $\Delta u = 0$. *Mathematische Zeitschrift 18*, 1 (1923), 42–54. (p. 111)

[108] PHAM, H. Optimal stopping of controlled jump diffusion processes: A viscosity solution approach. *Journal of Mathematical Systems, Estimation, and Control 8*, 1 (1998), 1–27. (p. 58, 62, 72, 104, 110, 132)

[109] PRIGENT, J. *Portfolio Optimization and Performance Analysis*. CRC Press, 2007. (p. 140)

[110] PROTTER, P. *Stochastic Integration and Differential Equations*, 2nd ed. Springer, 2004. (p. 173, 176, 197, 207, 208)

[111] ROCKAFELLAR, R. *Convex Analysis*. Princeton University Press, 1970. (p. 25, 219)

[112] ROCKAFELLAR, R., WETS, R. *Variational Analysis*. Springer, 2009. (p. 27, 44, 224)

[113] RUDIN, W. *Principles of Mathematical Analysis*, 3rd ed. McGraw-Hill, 1976. (p. 54)

[114] RUDIN, W. *Real and Complex Analysis*, 3rd ed. McGraw-Hill, 1987. (p. 185)

[115] RUDIN, W. *Functional Analysis*, 2nd ed. McGraw-Hill, 1991. (p. 21, 24, 221)

[116] RUFFINO, D. Some implications of Knightian uncertainty for finance and regulation. In *FEDS Notes* (2014). (p. 11)

[117] SATO, K. *Lévy Processes and Infinitely Divisible Distributions*. Cambridge University Press, 1999. (p. 117, 180, 216)

[118] SAYAH, A. Equations d'Hamilton-Jacobi du premier ordre avec termes intégro-différentiels. *Communications in Partial Differential Equations 16* (1991), 1057–1074. (p. 58, 68, 72)

[119] SCHILLING, R. *Measures, Integrals and Martingales*. Cambridge University Press, 2005. (p. 185, 222)

[120] SONER, H. Optimal control with state-space constraint. *SIAM Journal on Control and Optimization* 24, 6 (1985), 1110–1122. (p. 58)

[121] STONE, M. Applications of the theory of Boolean rings to general topology. *Transactions of the American Mathematical Society* 41, 3 (1937), 375–481. (p. 54)

[122] STONE, M. Notes on integration, II. *Proceedings of the National Academy of Sciences of the United States of America* 34, 9 (1948), 447–455. (p. 139)

[123] STROOCK, D., VARADHAN, S. *Multidimensional Diffusion Processes*. Springer, 1997. (p. 163)

[124] TALAGRAND, M. *Pettis Integral and Measure Theory*. American Mathematical Socitety, 1984. (p. 222)

[125] TIETZE, H. Über Funktionen, die auf einer abgeschlossenen Menge stetig sind. *Journal für die reine und angewandte Mathematik* 145 (1915), 9–14. (p. 52)

[126] URYSOHN, P. Über die Mächtigkeit der zusammenhängenden Mengen. *Mathematische Annalen* 94, 1 (1925), 262–295. (p. 52)

[127] VAN ROOIJ, A., SCHIKHOF, W. *A Second Course on Real Functions*. Cambridge University Press, 1982. (p. 164)

[128] WANG, Z., KLIR, G. *Generalized Measure Theory*. Springer, 2009. (p. 135)

[129] WEIERSTRASS, K. Über die analytische Darstellbarkeit sogenannter willkürlicher Functionen einer reellen Veränderlichen. *Sitzungsberichte der Königlich Preußischen Akademie der Wissenschaften zu Berlin* 2 (1885), 633–639. (p. 54)

[130] YOSIDA, K. *Functional Analysis*, 6th ed. Springer, 1980. (p. 221)

List of Symbols

Assumptions and Equations

(A1) – (A5)	assumptions for general nonlinear nonlocal equations (p. 60)
(B1) – (B5)	assumptions for Hamilton-Jacobi-Bellman equations (p. 115)
(C1) – (C3)	assumptions for sublinear conditional expectations (p. 165)
(D1) – (D3)	assumptions for uncertainty coefficients (p. 182)
(E1)	general elliptic, nonlinear nonlocal equation (p. 59)
(E2)	general parabolic, nonlinear nonlocal equation (p. 96)
(E3)	Hamilton-Jacobi-Bellman equation (p. 114)

Abbreviations

a.e.	almost everywhere
a.s.	almost surely
càdlàg	continue à droite limite à gauche, right continuous with left limits
càglàd	continue à gauche limite à droite, left continuous with right limits
w.l.o.g.	without loss of generality

Number Ranges

$\varnothing = \{\,\}$	empty set, void set
$\mathbb{N} = \{1, 2, 3, \ldots\}$	natural numbers, positive integers
$\mathbb{N}_0 = \mathbb{N} \cup \{0\}$	non-negative integers
$\mathbb{Q}$	rational numbers
$\mathbb{Q}_+ = \mathbb{Q} \cap [0, \infty)$	non-negative rational numbers
$\mathbb{R}$	real numbers
$\mathbb{R}_+$	non-negative real numbers
$\overline{\mathbb{R}} = \mathbb{R} \cup \{\pm\infty\}$	two-point compactification of the real numbers (p. 162)
$\mathbb{R}^d$	d-dimensional Euclidean space
$\mathbb{S}^{d \times d}$	d-dimensional symmetric matrices
$\mathbb{S}^{d \times d}_+$	d-dimensional positive semi-definite matrices
$B(x, r)$, $B[x, r]$	open, closed balls of radius r with center x
(a, b), $[a, b)$, $[a, b]$	open, half-open and closed intervals with endpoints a and b

List of Symbols

Elementary Operations

$a \vee b = \max\{a, b\}$	maximum of a and b
$a \wedge b = \min\{a, b\}$	minimum of a and b
$a^+ = a \vee 0$	positive part of a
$a^- = -(a \wedge 0)$	negative part of a
$|a| = a^+ + a^-$	modulus of a, absolute value of a
$\lim_{n \to \infty} x_n$	limit of sequence $(x_n)_{n \in \mathbb{N}}$
$\liminf_{n \to \infty} x_n$	limit inferior of sequence $(x_n)_{n \in \mathbb{N}}$
$\limsup_{n \to \infty} x_n$	limit superior of sequence $(x_n)_{n \in \mathbb{N}}$
$x \to a-,\ x \to a+$	x converges to a with $x < a$ (or $x > a$)
$f_n \uparrow f,\ f_n \downarrow f$	f_n converges monotonically to f from below (or from above)
2^Ω	power set of Ω, containing all subsets of Ω
$A^C = \Omega \setminus A$	complement of set $A \subset \Omega$
$\overline{A}$	closure of set A
$\mathring{A} = \mathrm{int}(A)$	interior of set A
$\overline{\mathrm{conv}}(A)$	closure of convex hull of set A
$\mathbf{1}_A(\cdot)$	indicator function of set A
$\langle x, y \rangle,\ x^T y,\ x \cdot y$	scalar product of x and y
$\mathrm{id} = \mathrm{diag}(1, \ldots, 1)$	identity matrix
$X^T = (x_{ij})_{ji}$	transpose of matrix $X = (x_{ij})_{ij}$
$\mathrm{tr}(X) = \sum_i x_{ii}$	trace of matrix $X = (x_{ij})_{ij}$
$\mathrm{diag}(X_1, \ldots, X_k)$	diagonal block matrix with main diagonal $X_1, \ldots, X_k$
$\partial_t u(t, x)$	partial derivative with respect to time t
$Du(t, x)$	gradient operator, first derivative with respect to space x
$D^2 u(t, x)$	Hessian matrix, second derivative with respect to space x
$\mathrm{supp}(f)$	support of function f
$\mathrm{var}(f)$	total variation of function f
$\arg\max(f)$	maximum points of function f

Metrics and Norms

$d_{\mathfrak{M}_b^+(X)}$	Kantorovich-Rubinstein metric on $\mathfrak{M}_b^+(X)$ (p. 158)
$d_{\mathfrak{L}(\mathbb{R}^d)},\ d_{\mathfrak{L}(\mathbb{R}_+ \times \mathbb{R}^d)}$	metric on Lévy measures $\mathfrak{L}(\mathbb{R}^d)$ and $\mathfrak{L}(\mathbb{R}_+ \times \mathbb{R}^d)$ (p. 159)
d_H	Hausdorff distance (p. 224)
$|\cdot|$	Euclidean norm
$\|\cdot\|_\infty$	supremum norm
$\|\cdot\|_p$	optimal constant for p-polynomial growth (p. 33)

List of Symbols

Probability and Measure Theory

$(\Omega, \mathcal{A}, \mathbb{P})$	(classical) probability space
$(\Omega, \mathcal{H}, E)$	sublinear expectation space (p. 136)
$\sigma(\mathcal{A})$	smallest σ-algebra containing all elements of $\mathcal{A} \subset 2^\Omega$
$\mathcal{B}(\mathbb{R}^d)$	Borel σ-algebra of Euclidean space $\mathbb{R}^d$
$\mathfrak{A}(\mathbb{R}^d)$	analytic subsets of Euclidean space $\mathbb{R}^d$ (p. 163)
$\lambda^d = dx$	Lebesgue measure of Euclidean space $\mathbb{R}^d$
$\mathbb{E}(X) = \mathbb{E}_\mathbb{P}(X)$	(classical) expectation of random variable X for probability $\mathbb{P}$
$E_X(\cdot)$	distribution of random variable X (p. 141)
$X \sim Y$	equivalence in distribution of random variables X and Y (p. 142)
$X \perp\!\!\!\perp Y$	independence of random variable X from Y (p. 142)
$[X,Y]_t, [X]_t$	quadratic (co-)variation of semimartingales (p. 172)
$\langle X,Y \rangle_t, \langle X \rangle_t$	predictable quadratic (co-)variation of martingales (p. 172)
$\mathcal{A}^\mathbb{P}$	completion of σ-algebra $\mathcal{A}$ under probability $\mathbb{P}$ (p. 164)
$\mathcal{A}^* = \cap_{\mathbb{P} \in \mathfrak{P}(\Omega)} \mathcal{A}^\mathbb{P}$	universal completion of σ-algebra $\mathcal{A}$ (p. 164)
$\mathcal{A}^\mathcal{P} = \cap_{\mathbb{P} \in \mathcal{P}} \mathcal{A}^\mathbb{P}$	$\mathcal{P}$-universal completion of σ-algebra $\mathcal{A}$ for $\mathcal{P} \subset \mathfrak{P}(\Omega)$ (p. 201)
$\mathcal{F}^\mathbb{P} = (\mathcal{F}_t^\mathbb{P})_{t \geq 0}$	completion of filtration $\mathcal{F}$ under probability $\mathbb{P}$ (p. 197)
$\mathcal{F}_+^\mathbb{P} = (\mathcal{F}_{t+}^\mathbb{P})_{t \geq 0}$	augmentation of filtration $\mathcal{F}$ under probability $\mathbb{P}$ (p. 197)

Special Operations

$\eta_\delta(\cdot)$	standard mollifier, approximate identity (p. 19)
$f * g$	(classical) convolution of function $f(\cdot)$ and $g(\cdot)$ (p. 20)
$\Delta_\varphi^\varepsilon[f], \nabla_\varphi^\varepsilon[f]$	supremal and infimal convolution of function $f(\cdot)$ (p. 25)
$\Delta_\varphi^\varepsilon[F], \nabla_\varphi^\varepsilon[F]$	smudging of operator $F(\cdot)$ in space (p. 73)
$g \circ F$	integration of $g(\cdot)$ under Lévy measure $F \in \mathfrak{L}(\mathbb{R}^d)$ (p. 182)
$g * \nu$	integration of $g(\cdot)$ under Lévy measure $\nu \in \mathfrak{L}(\mathbb{R}_+ \times \mathbb{R}^d)$ (p. 156)
$\mathcal{J}(H, X)$	pathwise stochastic integral of $H_- = (H_{t-})_{t \geq 0}$ (p. 204)
$\mathcal{J}_\mathbb{P}(H, X)$	(classical) stochastic integral for probability $\mathbb{P}$ (p. 197)
$\mathcal{J}_\mathcal{P}(H, X)$	stochastic integral under $\mathcal{P} \subset \mathfrak{P}_{\text{sem}}^{\text{ac}}(\Omega)$ (p. 202)
$\mathcal{J}^m(H, X)$	approximation of pathwise stochastic integral (p. 204)
$\omega \otimes_t \bar{\omega}$	concatenation of càdlàg paths at time t (p. 162)
$\lim \text{med}_{n \to \infty} X_n$	medial limit of random variables (p. 201)
$\Upsilon(\cdot)$	Bochner integration operator (p. 223)
$\Lambda(\cdot)$	optimal Lipschitz constant of càdlàg paths (p. 224)

List of Symbols

Spaces of Functions, Measures and Processes

$C(\mathbb{R}^d)$	continuous functions
$C_b^{\text{Lip}}(\mathbb{R}^d)$	bounded Lipschitz-continuous functions
$C^k(\mathbb{R}^d)$	k-times continuously differentiable functions
$C_c^k(\mathbb{R}^d)$	continuously differentiable functions with compact support
$C_b^k(\mathbb{R}^d)$	continuously differentiable functions with bounded derivatives
$\text{LSC}(\mathbb{R}^d)$	lower semicontinuous functions (p. 27)
$\text{USC}(\mathbb{R}^d)$	upper semicontinuous functions (p. 27)
$\text{SC}(\mathbb{R}^d)$	lower and upper semicontinuous functions (p. 27)
$\mathcal{F}_p(\mathbb{R}^d)$	functions in $\mathcal{F}(\mathbb{R}^d)$ with bounded p-polynomial growth (p. 33)
$D(\mathbb{R}_+)$, $D(\mathbb{R}_+;\mathbb{R}^d)$	càdlàg functions, d-dimensional Skorokhod space (p. 148)
$L(\mathbb{R}_+)$, $L(\mathbb{R}_+;\mathbb{R}^d)$	d-dimensional integrable functions under the Lebesgue measure
$L_{\text{loc}}(\mathbb{R}_+)$	locally integrable functions under the Lebesgue measure
$\mathfrak{M}(X)$	Borel measures on metric space (X,d) (p. 158)
$\mathfrak{M}_b(X)$	bounded Borel measures on metric space (X,d) (p. 158)
$\mathfrak{M}_b^+(X)$	bounded, non-negative Borel measures on metric space (X,d)
$\mathfrak{L}(\mathbb{R}^d)$, $\mathfrak{L}(\mathbb{R}_+\times\mathbb{R}^d)$	Lévy measures on $\mathbb{R}^d$ and $\mathbb{R}_+\times\mathbb{R}^d$ (p. 159)
$\mathfrak{P}(\Omega)$	probability measures on measurable space $(\Omega,\mathcal{A})$ (p. 158)
$\mathfrak{P}_{\text{sem}}(\Omega)$	probability measures such that $(X_t)_{t\geq 0}$ is semimartingale (p. 160)
$\mathfrak{P}_{\text{sem}}^{\text{ac}}(\Omega)$	probability measures such that $(X_t)_{t\geq 0}$ is semimartingale with semimartingale characteristics $(B,C,\nu)\ll dt$ (p. 161)
$L_\mathbb{P}(X)$, $L_\mathbb{P}(X;\mathbb{R}^d)$	integrable processes for semimartingale X under $\mathbb{P}$ (p. 197)
$L_\mathcal{P}(X)$, $L_\mathcal{P}(X;\mathbb{R}^d)$	integrable processes under $\mathcal{P}\subset\mathfrak{P}_{\text{sem}}^{\text{ac}}(\Omega)$ (p. 202)
$\mathfrak{D}(\mathbb{R}^d)$, $\mathfrak{D}(\mathbb{R}^d;\mathcal{F})$	d-dimensional $\mathcal{F}$-adapted processes with càdlàg paths (p. 207)

List of Statements

1.1. Lemma (Standard Mollifier) 19
1.2. Theorem (Mollifications) 20
1.3. Proposition (Partitions of Unity) 21
1.4. Theorem (Uniform Approximations) 21
1.5. Remark (Derivatives of Approximations) 22
1.6. Lemma (Cutoff Functions with Compact Support) 22
1.7. Proposition (Cutoff Functions) 23
1.8. Definition (Supremal & Infimal Convolutions) 25
1.9. Definition (Quasidistances) 26
1.10. Lemma (Stability of Quasidistances) 27
1.11. Definition (Semicontinuous Functions) 27
1.12. Lemma (Approximation by Supremal Convolutions) 28
1.13. Definition (Polynomial Growth) 33
1.14. Lemma (Quasidistances with Polynomial Growth) 33
1.15. Lemma (Rate of Convergence) 36
1.16. Lemma (Approximation of Semicontinuous Functions) 38
1.17. Lemma (Smooth Functions with Polynomial Growth) 40
1.18. Lemma (Tamed Scanning Functions) 41
1.19. Lemma (Approximation of Positive Part) 42
1.20. Theorem (Approximation by Scanning Functions) 43
1.21. Lemma (Smoothing of Monotone Functions) 44
1.22. Proposition (Approximation of Monotone Functions) 46
1.23. Theorem (Existence of Scanning Functions) 48
1.24. Corollary (Adapted Scanning Functions) 49
1.25. Lemma (Approximate Maximum Points) 50
1.26. Theorem (Tietze Extension) 52
1.27. Theorem (Dini) 54
1.28. Theorem (Weierstrass) 54
1.29. Corollary (Countable Families of Approximations) 55

2.1. Definition (Viscosity Solutions) 59
2.2. Remark (Notations for Viscosity Solutions) 60
2.3. Remark (Assumptions on F) 60
2.4. Lemma (Choice of Scanning Functions) 62
2.5. Remark (Bounded Derivatives of Scanning Functions) 65
2.6. Lemma (Domain of Solutions) 66

List of Statements

2.7. Lemma (Characterization of Solutions) 68
2.8. Lemma (Stability under Supremal Convolutions) 73
2.9. Corollary (Stability under Infimal Convolutions) 75
2.10. Lemma (Continuity of Smudged Operators) 75
2.11. Proposition (Maximum Principle for Smudged Operators) 78
2.12. Lemma (Matrix Transformation) 82
2.13. Corollary (Matrix Compactification) 83
2.14. Theorem (Maximum Principle) 84
2.15. Lemma (Maximum Principle for Generalized Operators) 91
2.16. Lemma (Perturbed Maximum Points) 93
2.17. Corollary (Variable Augmentation) 95
2.18. Lemma (Parabolic Maximum Principle) 96
2.19. Corollary (Parabolic Maximum Principle for Generalized Operators) ... 101
2.20. Remark (Stability under Parabolic Supremal Convolutions) 102
2.21. Definition (Regularity Condition) 103
2.22. Theorem (Domination Principle) 104
2.23. Corollary (Comparison Principle) 110
2.24. Remark (Perron's Method) 111
2.25. Corollary (Subadditivity of Solutions) 112
2.26. Corollary (Convexity of Solutions) 112
2.27. Definition (Hamilton–Jacobi–Bellman Equations) 114
2.28. Remark (Assumptions on Coefficients) 115
2.29. Remark (Isaacs Equations) 116
2.30. Remark (Connection to Markov Processes) 116
2.31. Remark (Universal Constant $C \geq 0$) 118
2.32. Lemma (Admissibility of Equations) 119
2.33. Proposition (Regularity Condition) 124
2.34. Corollary (Comparison Principle) 132
2.35. Corollary (Subadditivity & Convexity of Solutions) 133

3.1. Definition (Space of Integrands) 136
3.2. Remark ($\mathcal{H}$ are Vector Lattices) 136
3.3. Definition (Expectation Spaces) 136
3.4. Lemma (Properties of Sublinear Expectations) 137
3.5. Theorem (Representation of Sublinear Functionals) 138
3.6. Remark (Properties of Parameter Sets Θ) 139
3.7. Theorem (Daniell-Stone) 139
3.8. Corollary (Representation of Sublinear Expectations) 139
3.9. Remark (Dominated Convergence) 140
3.10. Definition (Random Variables & Test Functions) 141
3.11. Definition (Distributions) 141
3.12. Remark (Equivalence in Distribution) 142
3.13. Definition (Independence) 142
3.14. Remark (Symmetry of Independence) 142

3.15. Definition (Stable Distributions) 143
3.16. Remark (Characterization of Stable Distributions) 143
3.17. Definition (Convergence in Distribution) 144
3.18. Theorem (Central Limit Theorem & Law of Large Numbers) 145

4.1. Definition (Skorokhod Space) 148
4.2. Lemma (Skorokhod Topology) 148
4.3. Definition (Stochastic Processes) 148
4.4. Definition (Markov Semigroups) 149
4.5. Lemma (Properties of Markov Semigroups) 149
4.6. Definition (Markov Processes) 150
4.7. Remark (Notion of Markov Processes) 151
4.8. Corollary (Characterization of Markov Processes) 151
4.9. Definition (Generators) 152
4.10. Proposition (Generator Equations) 152
4.11. Definition (Semimartingales) 156
4.12. Remark (Integral Processes) 156
4.13. Definition (Semimartingale Characteristics) 157
4.14. Remark (Construction of Characteristics) 157
4.15. Lemma (Filtration of Semimartingales) 158
4.16. Remark (Topology on Measures) 158
4.17. Remark (Lévy Measures) 159
4.18. Theorem (Measurability of Characteristics) 160
4.19. Remark (Differential Characteristics) 160
4.20. Theorem (Measurability of Differential Characteristics) 161
4.21. Remark (Non-Integrable Functions) 162
4.22. Definition (Operations on Skorokhod Space) 162
4.23. Remark (Regular Conditional Distributions) 163
4.24. Definition (Analytic Sets) 163
4.25. Remark (Properties of Analytics Sets) 164
4.26. Definition (Semi-Analytic Functions) 164
4.27. Remark (Stability of Semi-Analytic Functions) 164
4.28. Remark (Assumptions on $\mathcal{P}$) 165
4.29. Theorem (Sublinear Conditional Expectation) 166
4.30. Lemma (Regularity of $x \mapsto E^x(\xi)$) 167
4.31. Proposition (Admissibility of $\mathcal{P}^\Theta$) 168
4.32. Lemma (Markov Property) 170
4.33. Remark (Construction of Markov Semigroups) 172
4.34. Remark (Quadratic Variations) 172
4.35. Lemma (Burkholder-Davis-Gundy Inequality for Increments) 172
4.36. Proposition (Continuity in Time) 173
4.37. Theorem (Form of Generators) 175
4.38. Remark (Lévy Processes) 178
4.39. Remark (Assumptions on $(b_\alpha, c_\alpha, F_\alpha)$) 182

List of Statements

4.40. Lemma (Properties of Uncertainty Coefficients) 182
4.41. Theorem (Compactness of Uncertainty Subsets) 186
4.42. Lemma (Upper Semicontinuity in Space) 193
4.43. Remark (Lower Semicontinuity in Space) 195
4.44. Remark (Usual Conditions) 197
4.45. Remark (Classical Stochastic Integration) 197
4.46. Lemma (Characterization of $L_\mathbb{P}(X)$) 199
4.47. Remark (Scalar & Matrix-Valued Integrands) 200
4.48. Lemma (Characteristics of Stochastic Integrals) 200
4.49. Theorem (Medial Limits) 201
4.50. Theorem (Pathwise Stochastic Integration) 201
4.51. Definition (Stochastic Integrals) 202
4.52. Lemma (Measurability of Limits) 203
4.53. Proposition (Measurability of Stochastic Integrals) 204
4.54. Definition (Functional Lipschitz) 207
4.55. Theorem (Classical Stochastic Differential Equations) 207
4.56. Proposition (Picard Iteration Method) 207
4.57. Theorem (Equations with Uniform Lipschitz Coefficients) 208
4.58. Remark (Equations with Functional Lipschitz Coefficients) 209
4.59. Lemma (Measurable Modifications) 210
4.60. Proposition (Markov Property) 210
4.61. Proposition (Dependence on Initial Values) 213
4.62. Remark (Lévy-Type Processes) 214

A.1. Lemma (Stability under Supremum) 219
A.2. Proposition (Convexity $\implies$ Continuity) 219
A.3. Theorem (Rademacher) 219
A.4. Corollary (Convexity $\implies$ Differentiability) 219
A.5. Definition (Semiconvexity) 220
A.6. Lemma (Twice Differentiability $\implies$ Semiconvexity) 220
A.7. Theorem (Alexandrov) 220
A.8. Lemma (Jensen) 220
A.9. Theorem (Hahn-Banach Extension) 221
A.10. Theorem (Hyperplane Separation) 221
A.11. Remark (Bochner Integration) 221
A.12. Lemma (Range of Integrals) 223
A.13. Remark (Absolute Continuity) 223
A.14. Lemma (Lipschitz Constant Map) 224
A.15. Remark (Hausdorff Distances) 224

Index of Subjects

absolute continuity, 223
Alexandrov's theorem, 220
analytic sets, 163
    properties, 164
    semi-analytic functions, 164
approximation of functions
    derivatives, 22
    globally uniform, 21
    in L^p, 20
    locally uniform, 20
    monotonically, 28
    polynomial growth, 41
approximation theory, 19

Bochner integrals, 221
bump functions, 22, 23
Burkholder-Davis-Gundy inequality, 172

central limit theorem, 145
Choquet integrals, 140
classical maximum principle, 77
coherent risk measures, 135
comparison principle, 110
    motivation, 70
completion of σ-algebras, 197
coordinate-mapping process, 161
cutoff functions, 22, 23

Daniell integrals, 139
Daniell-Stone theorem, 139
degenerate ellipticity, 61
distributions, 141, 144
dominated convergence theorem, 140
domination principle, 104

ellipticity, 61
exteme value theorem, 27

Friedrichs mollification, 20

functional Lipschitz coefficients, 207
G-Brownian motions, 155
G-heat equations, 155
G-Lévy processes, 178
G-Lévy-type processes, 214
generalized operators, 60
generator equations, 152
Greco's representation theorem, 140

Hahn-Banach theorem, 221
Hamilton-Jacobi-Bellman equations, 114
    admissability, 119
    assumptions on coefficients, 115
    comparison principle, 132
    connection to Markov processes, 116
    convexity of solutions, 133
    elliptic equations, 116
    examples, 116
    regularity condition, 124
    subadditivity of solutions, 133
Hausdorff distances, 224
hyperplane separation theorem, 221

independence, 142
infimal convolutions, 25
    approximation by, 28
    rate of convergence, 36
integral processes, 156
integro-differential equations, 57
Isaacs equations, 116

Jensen's lemma, 50, 220

Kantorovich-Rubinstein metric, 158

law of large numbers, 145
Legendre-Fenchel transformations, 25
Lévy measures, 159
Lévy processes, 178

Index of Subjects

Lévy-type processes, 214
Lévy–Prokhorov metric, 158
little o-notation, 44
local equations, 58

Markov generators, 152, 175
Markov processes, 150
    characterization, 151
    notion, 151
Markov semigroups, 149
    construction, 172, 210
    continuity, 193, 195
    generator equations, 152
    generators, 152
    properties, 149
    representation of generators, 175
martingale problems, 195
matrix compactification, 83
maximum principle, 84
    classical result, 77
    for generalized operators, 91
    for smudged operators, 78
    motivation, 77
    parabolic version, 96, 101
medial limits, 201
Mokobodzki's medial limit, 201
mollifications, 20
mollifier, 19
monotone functions
    approximation of, 46
    smoothing of, 44
Moreau–Yosida approximations, 25

nonlocal equations, 57
    continuity, 61
    degenerate ellipticity, 61
    general assumptions, 60
    generalized operators, 60
    maximum principle, 84
    parabolic equations, 95
    regularity condition, 103
    supremal convolution, 73

o-notation, 44

parabolic maximum principle, 96
    for generalized operators, 101
parabolic nonlocal equations, 95
    comparison principle, 110
    convexity of solutions, 112
    domination principle, 104
    existence of solutions, 111
    regularity condition, 103
    subadditivity of solutions, 112
    supremal convolution, 102
partitions of unity, 21
Perron's method, 111
Pettis integrals, 222
Picard iteration method, 207
polynomial growth, 33, 40
positive part, 42
previsible quadratic variations, 172

quadratic variations, 172
quasidistances, 26
    polynomial growth, 33
    smooth versions, 33
    stability, 27

Rademacher's theorem, 219
random variables, 141
    convergence in distribution, 144
    distributions, 141
    independence, 142
regular conditional distributions, 163
regularity condition, 103
risk measures, 135

scanning functions, 40
    adaption, 49
    approximation by, 43
    bounded derivatives, 65
    existence, 48
    polynomial growth, 41
    restriction to subfamilies, 62
    taming, 41
second-order superjets, 48
semi-analytic functions, 164
semicontinuity, 27
    approximation, 28, 38

Index of Subjects

semiconvexity, 220
    differentiability, 220
semimartingales, 156
    characteristics, 157
    construction of characteristics, 157
    differential characteristics, 160
    filtrations, 158
    measurability of characteristics, 161
Skorokhod space, 148
    operations, 162
Skorokhod topology, 148
small o-notation, 44
smudged operators, 73
    continuity, 75
stable distributions, 143
standard mollifier, 19
stochastic differential equations, 208
    dependence on initial values, 213
    functional Lipschitz coefficients, 209
    Markov property, 210
    Picard iterations, 207
    uniform Lipschitz coefficients, 208
    uniqueness of solutions, 207
stochastic integrals, 202
    characterization of integrands, 199
    construction, 201, 204
    matrix-valued integrands, 200
    measurability, 204
    properties, 197
    scalar integrands, 200
stochastic processes, 148
sublinear conditional expectations, 166
sublinear expectations, 136
    construction, 166
    convergence in distribution, 144
    distributions, 141
    dominated convergence, 140
    independence, 142
    integrands, 136
    monotone convergence, 139
    properties, 137
    random variables, 141
    representation, 139
    stochastic processes, 148

test functions, 141
    uncertainty subsets, 139
sublinear functionals, 138
subsolutions, 59
superjets, 48
supersolutions, 59
supremal convolutions, 25
    approximation by, 28
    rate of convergence, 36
    stability for viscosity solutions, 73

test functions, 141

uncertainty coefficients, 168
    assumptions, 182
    properties, 182
uncertainty subsets, 139
    assumptions, 165
    compactness, 186
    construction, 168
    Markov property, 170
universal completion of σ-algebras, 201
universal constants, 118
usual conditions, 197
usual hypotheses, 197

variable augmentation, 95
    motivation, 94
vector lattices, 136
viscosity solutions, 59
    additional notation, 60
    alternative characterizations, 68
    comparison principle, 110
    convexity, 112
    domain of functions, 66
    domination principle, 104
    existence, 111
    scanning functions, 62
    smudged equations, 73
    sub- and supersolutions, 59
    subadditivity, 112
    supremal convolution, 73
    uniqueness, 110

weak convergence of measures, 158

Good Scientific Practice

Affirmation

(a) Hereby I affirm, that I wrote the present thesis without any inadmissible help by a third party and without using any other means than indicated. Thoughts that were taken directly or indirectly from other sources are indicated as such. This thesis has not been presented to any other examination board in this or a similar form, neither in this nor in any other country.

(b) The present thesis has been produced since January 2013 at the Institute of Mathematical Stochastics, Departement of Mathematics, Faculty of Science, Technische Universität Dresden under the supervision of Prof. Dr. René L. Schilling.

(c) There have been no prior attempts to obtain a PhD degree at any university.

(d) I accept the requirements for obtaining a PhD (Promotionsordnung) of the Faculty of Science of the Technische Universität Dresden, issued 23rd February 2011.

Dresden, on 13th July 2016

Versicherung

(a) Hiermit versichere ich, dass ich die vorliegende Arbeit ohne unzulässige Hilfe Dritter und ohne Benutzung anderer als der angegebenen Hilfsmittel angefertigt habe. Die aus fremden Quellen direkt oder indirekt übernommenen Gedanken sind als solche kenntlich gemacht. Die Arbeit wurde bisher weder im In- noch im Ausland in gleicher oder ähnlicher Form einer anderen Prüfungsbehörde vorgelegt.

(b) Die vorliegende Dissertation wurde seit Januar 2013 am Institut für Mathematische Stochastik, Fachrichtung Mathematik, Fakultät Mathematik und Naturwissenschaften, Technische Universität Dresden unter der Betreuung von Prof. Dr. René L. Schilling angefertigt.

(c) Es wurden zuvor keine Promotionsvorhaben an Universitäten unternommen.

(d) Ich erkenne die Promotionsordnung der Fakultät Mathematik und Naturwissenschaften der Technischen Universität Dresden vom 23. Februar 2011 an.

Dresden, den 13. Juli 2016

www.ingramcontent.com/pod-product-compliance
Lightning Source LLC
Chambersburg PA
CBHW070314190526
45169CB00005B/1622